MATH/STAT
LIBRARY
Y0-AAA-425
$39.95

MATH/STAT
LIBRARY

Probability, Statistics, and Mathematics

Papers in Honor of Samuel Karlin

Samuel Karlin

Probability, Statistics, and Mathematics

Papers in Honor of Samuel Karlin

Edited by

T. W. Anderson

Departments of Statistics and Economics
Stanford University
Stanford, California

Krishna B. Athreya

Departments of Mathematics and Statistics
Iowa State University
Ames, Iowa

Donald L. Iglehart

Department of Operations Research
Stanford University
Stanford, California

ACADEMIC PRESS, INC.
Harcourt Brace Jovanovich, Publishers
Boston San Diego New York
Berkeley London Sydney
Tokyo Toronto

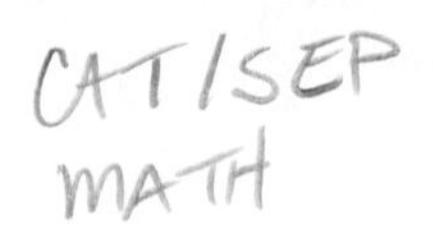

Copyright © 1989 by Academic Press, Inc.
All rights reserved.
No part of this publication may be reproduced or transmitted in any form or by any means, electronic or mechanical, including photocopy, recording, or any information storage and retrieval system, without permission in writing from the publisher.

ACADEMIC PRESS, INC.
1250 Sixth Avenue, San Diego, CA 92101

United Kingdom Edition published by
ACADEMIC PRESS INC. (LONDON) LTD.
24-28 Oval Road, London NW1 7DX

Frontispiece photo courtesy of News and Publications Service, Stanford University.

Library of Congress Cataloging-in-Publication Data

Probability, statistics, and mathematics: papers in honor of Samuel Karlin / edited by T.W. Anderson, Krishna B. Athreya, Donald L. Iglehart.
p. cm.
Includes bibliographical references.
ISBN 0-12-058470-0 (alk. paper)
1. Probabilities. 2. Mathematical statistics. 3. Karlin, Samuel, Date- . I. Karlin, Samuel, Date- . II. Anderson, T.W. (Theodore Wilbur), Date- . III. Athreya, Krishna B., Date- . IV. Iglehart, Donald L., Date- .
QA273.18.P755 1989
519.2 — dc20 89-17552
CIP

Printed in the United States of America
89 90 91 92 9 8 7 6 5 4 3 2 1

QA273
.18
P755
1989
MATH

CONTENTS

LIST OF CONTRIBUTORS

Theodore W. Anderson, Departments of Statistics and Economics, Stanford University, Stanford, California 94305

Kenneth J. Arrow, Department of Economics, Stanford University, Stanford, California 94305

Richard A. Askey, Department of Mathematics, University of Wisconsin, Madison, Wisconsin, 53706

Krishna B. Athreya, Departments of Mathematics and Statistics, Iowa State University, Ames, Iowa 50011

Pierre Baldi, Department of Mathematics, University of California at San Diego, La Jolla, California 92093

David H. Blackwell, Department of Statistics, University of California at Berkeley, Berkeley, California 94720

Peter W. Glynn, Department of Operations Research, Stanford University, Stanford, California 94305

Donald L. Iglehart, Department of Operations Research, Stanford University, Stanford, California 94305

Harry Kesten, Department of Mathematics, Cornell University, Ithaca, New York 14853

John F. C. Kingman, University of Bristol, Bristol, BS8 1TW, United Kingdom

Thomas M. Liggett, Department of Mathematics, University of California at Los Angeles, Los Angeles, California 90024

Charles A. Micchelli, IBM, P.O. Box 128, Yorktown Heights, New York 10598

Marcel F. Neuts, Department of Systems and Industrial Engineering, University of Arizona, Tucson, Arizona 85721

Peter E. Ney, Department of Mathematics, University of Wisconsin, Van Vleck Hall, Madison, Wisconsin 53706

Allan M. Pinkus, Israel Institute of Technology, Technion, Haifa, Israel

John W. Pratt, Graduate School of Business Administration, Harvard University, Boston, Massachusetts 02163

William E. Pruitt, Department of Mathematics, University of Minnesota, Minneapolis, Minnesota 55455

C. R. Rao, Department of Statistics, Pennsylvania State University, University Park, Pennsylvania 16802

Donald Richards, Department of Mathematics, University of Virginia, Charlottesville, Virginia 22903

Yosef Rinott, Department of Statistics, Hebrew University, Jerusalem, Israel

Murray Rosenblatt, Department of Mathematics, University of California at San Diego, La Jolla, California 92093

Gennady Samorodnitsky, Department of Mathematics, Boston University, Boston, Massachusetts 02215

Stephen M. Samuels, Department of Statistics, Purdue University, West Lafayette, Indiana 47907

Paul A. Samuelson, Department of Economics, Massachusetts Institute of Technology, Cambridge, Massachusetts 02139

Herbert Scarf, Department of Economics, Yale University, New Haven, Connecticut 06520

D. N. Shanbhag, Department of Probability and Statistics, The University, Sheffield, S3 7RH, United Kingdom

Burton H. Singer, Department of Epidemiology and Public Health, Yale University, New Haven, Connecticut 06520

Charles M. Stein, Department of Statistics, Stanford University, Stanford, California 94305

Charles J. Stone, Department of Statistics, University of California at Berkeley, Berkeley, California 94720

William J. Studden, Department of Statistics, Purdue University, West Lafayette, Indiana 47907

Murad S. Taqqu, Department of Mathematics, Boston University, Boston, Massachusetts 02215

A. P. N. Weerasinghe, Departments of Mathematics and Statistics, Iowa State University, Ames, Iowa, 50011

Roy E. Welsch, Statistical Center, Sloan School of Management, Massachusetts Institute of Technology, Cambridge, Massachusetts 02139

FOREWORD

This volume consists of 24 papers written by mathematicians, probabalists, statisticians, and economists in honor of Samuel Karlin on the occasion of his sixty–fifth birthday. These contributors are only a small subset of Sam's many friends, colleagues, collaborators, and students who wish him well as he reaches this milestone. Sam's enormous influence and activity is indicated by the fact that he has had almost 100 co–authors on his books and papers and has directed the Ph.D. dissertations of more than 50 students. It was not possible to invite all of these collaborators and students to contribute to this volume.

We dedicate this publication with affection, admiration, and appreciation to Sam, who has impacted our lives, careers, and research in many ways. We know he will continue to be active and energetic in pursuing truth and knowledge in many directions.

Samuel Karlin was born June 8, 1924, in Yonova, Poland. He received his B.S. at the Illinois Institute of Technology, Chicago, in 1944 and his Ph.D. at Princeton University in 1947. He spent the next nine years at the California Institute of Technology, Pasadena, attaining the rank of professor in 1955. In 1956 he moved to Stanford University as Professor of Mathematics and Statistics; in 1978 he became the Robert Grimmet Professor of Mathematics.

Sam has received numerous honors, including an honorary Doctor of Science from Technion–Israel Institute of Technology, membership in the National Academy of Sciences, and fellowship in the American Academy of Arts and Sciences. He has received the National Academy of Sciences Award in Applied Mathematics and Numerical Analysis, the Lester R. Ford Award of the Mathematical Association of America, and the John von Neumann Theory Prize of the Operations Research Society of America. Among the many named and invited lectures given by Sam around the world are the Willard Josiah Gibbs Lecture of the American Mathematical Society and the Abraham Wald Memorial Lectures of the Institute of Mathematical Statistics, of which he was President in 1978-79 and is a fellow. Currently Sam is an editor or associate editor of ten journals, lecture series, and encyclopedia.

Karlin's 10 books and 330 papers have displayed a wide curiosity and voluminous fund of knowledge. His main work has been in (1) total positivity, mathematical analysis, and approximation theory, (2) probability, statistics, and stochastic processes, (3) operations research and management sciences, (4) genetics and evolution, and (5) epidemiology. He has made many fundamental and original contributions to all of these fields. This volume includes papers in the first three fields; papers in the latter two fields were contributed to an earlier volume, *Mathematical Evolutionary Theory* (Marc Feldman, editor), in honor of Sam.

We want to take this opportunity to thank Erin Nakashima, who has been responsible for the technical editing of this volume.

Theodore W. Anderson
Krishna B. Athreya
Donald L. Iglehart, Editors

May 3, 1989

PUBLICATIONS OF SAMUEL KARLIN

Books

1. *Studies in the Mathematical Theory of Inventory and Production* (with K. Arrow and H. Scarf), Stanford University Press, Stanford, California, 1958.
2. *Mathematical Methods and Theory in Games, Programming, and Economics, I. Matrix Games, Programming, and Mathematical Economics*, Addison–Wesley Publishing Company, 1959.
3. *Mathematical Methods and Theory in Games, Programming, and Economics, II. The Theory of Infinite Games*, Addison–Wesley Publishing Company, 1959.
4. *A First Course in Stochastic Processes*, Academic Press, New York, 1966. Second edition (with H. M. Taylor), 1975.
5. *Tchebycheff Systems: With Applications in Analysis and Statistics* (with W. J. Studden), Interscience, 1966.
6. *Total Positivity*, Stanford University Press, Stanford, California, 1968.
7. *Studies in Spline Functions and Approximation Theory* (with C. A. Micchelli, A. Pinkus, and I. I. Schoenberg), Academic Press, New York, 1976.
8. *A Second Course in Stochastic Processes*, First Edition (with H. M. Taylor), Academic Press, New York, 1980.
9. *An Introduction to Stochastic Modeling* (with H. M. Taylor), Academic Press, New York, 1984.
10. *Theoretical Studies on Sex Ratio Evolution* (with S. Lessard), Princeton University Press, Princeton, New Jersey, 1986.

Edited Volumes

1. *Mathematical Methods in the Social Sciences*, Proceedings of the First Stanford Symposium (edited with K. Arrow and P. Suppes), Stanford University Press, Stanford, California, 1960.
2. *Studies in Applied Probability and Mangement Sciences* (edited with K. Arrow and H. Scarf), Stanford University Press, Stanford, California, 1962.

3. *Population Genetics and Ecology* (edited with E. Nevo), Academic Press, New York, 1976.
4. *Studies in Econometrics, Time Series, and Multivariate Statistics* (edited with T. Amemiya and L. A. Goodman), Academic Press, New York, 1983.
5. *Evolutionary Processes and Theory* (edited with E. Nevo), Academic Press, Florida, 1986.

Papers

1. Unconditional convergence in Banach spaces, *Bulletin of the American Mathematical Society*, **54** (1948), 148-152.
2. Bases in Banach spaces, *Duke Mathematical Journal*, **15** (1948), 971-985.
3. Orthogonal properties of independent functions, *Transactions of the American Mathematical Society*, **66** (1949), 44-64.
4. Geometry of reduced moment spaces (with L. Shapley), *Proceedings of the National Academy of Sciences USA*, **35** (1949), 673-677.
5. Operator treatment of minimax principle, *Contributions to the Theory of Games*, Volume I (1950), 133-154.
6. Solutions of discrete two person games (with H. F. Bohnenblust and L. Shapley), *Contributions to the Theory of Games*, Volume I (1950), 51-72.
7. Games with continuous convex payoff (with H. F. Bohnenblust and L. Shapley), *Contributions to the Theory of Games*, Volume I (1950), 181-192.
8. Some applications of a theorem on convex functions (with L. Shapley), *Annals of Mathematics*, **52** (1950), 148-153.
9. On a theorem of Ville (with H. F. Bohnenblust), *Contributions to the Theory of Games*, Volume I (1950), 155-160.
10. Polynomial Games (with M. Dresher and S. Shapley), *Contributions to the Theory of Games*, Volume I (1950), 161-180.
11. Continuous Games, *Proceedings of the National Academy of Sciences USA*, **37** (1951), 220-223.
12. Reduction of certain classes of games to integral equations, *Contributions to the Theory of Games*, Volume II (1953), 125-158.

13. On a class of games, *Contributions to the Theory of Games*, Volume II (1953), 159-171.

14. Extreme points of vector functions, *Proceedings of the American Mathematical Society*, **4** (1953), 603-610.

15. Solutions of convex games as fixed points (with M. Dresher), *Contributions to the Theory of Games*, Volume II (1953), 75-86.

16. The theory of infinite games, *Annals of Mathematics*, **58** (1953), 371-401.

17. Some random walks arising in learning models I, *Pacific Journal of Mathematics*, **3** (1953), 725-756.

18. Geometry of moment space (with L. Shapley), *Memoirs of the American Mathematical Society*, **12** (1953), 1-93.

19. Contributions to the Theory of Games, Volume II, a review, *Scientific Monthly*, Volume LXXVIII, **4** (1954).

20. Cardano, the gamble scholar, a review, *Scientific Monthly*, Volume LXXVIII, **4** (1954).

21. Theory of games and statistical decisions, a review, *Science*, **120** (1954), 1065.

22. Representation of a class of stochastic processes (with J. McGregor), *Proceedings of the National Academy of Sciences USA*, **41** (1955), 387-391.

23. On the renewal equation, *Pacific Journal of Mathematics*, **5** (1955), 229-257.

24. Geometrical properties of the unit sphere of Banach algebras (with H. F. Bohnenblust), *Annals of Mathematics*, **62** (1955), 217-230.

25. Decision theory for Pólya–type distributions. Case of two actions I, *Proceedings of the Third Berkeley Symposium*, (1955), 115-128.

26. The structure of dynamic programming models, *Naval Research Logistics Quarterly*, **2** (1955), 285-294.

27. The theory of decision procedures for distributions with monotone likelihood ratio (with H. Rubin), *Annals of Mathematical Statistics*, **27** (1956), 272-299.

28. Distribution processing a monotone likelihood ratio (with H. Rubin), *Journal of American Statistical Association*, **51** (1956), 637-643.

29. On the design and comparison of certain dichotomous experiments (with R. Bradt), *Annals of Mathematical Statistics*, **27** (1956), 390-409.

30. On sequential designs for maximizing the sum of n observations (with R. Bradt and S. Johnson), *Annals of Mathematical Statistics*, **27** (1956), 1060-1074.

31. The differential equations of birth and death processes and the Stieltjes moment problem (with J. McGregor), *Transactions of the American Mathematical Society*, **85** (1957), 489-546.

32. Multistage statistical decision procedures (with M. A. Girshick and H. L. Royden), *Annals of Mathematical Statistics*, **28** (1957), 111-125.

33. Pólya–type distributions II, *Annals of Mathematical Statistics*, **28** (1957), 281-308.

34. Steady state inventory policy, *Proceedings of the Conference on Techniques of Operations Research*, Illinois Institute of Technology, June, 1957.

35. An infinite move game with a lag, *Contribution to the Theory of Games*, Volume III (1957), 257-272.

36. Multistage poker models (with R. Restrepo), *Contribution to the Theory of Games*, Volume III (1957), 337-363.

37. On games described by bell–shaped kernels, *Contribution to the Theory of Games*, Volume III (1957), 365-391.

38. The classification of birth and death processes (with J. McGregor), *Transactions of the American Mathematical Society*, **86** (1957), 366-400.

39. Pólya–type distributions, III: Admissibility for multi–action problems, *Annals of Mathematical Statistics*, **28** (1957), 839-860.

40. Pólya–type distributions IV: Some principles of selecting a single procedure from a complete class, *Annals of Mathematical Statistics*, **29** (1958), 1-21.

41. Admissibility for estimation with quadratic loss, *Annals of Mathematical Statistics*, **29** (1958), 406-436.

42. Production over time with increasing marginal costs (with K. J. Arrow), *Studies in the Mathematical Theory of Inventory and Production* (K. J. Arrow, S. Karlin, H. Scarf, eds.), Stanford University Press, Stanford, 1958, 61-69.

43. Smoothed production plans (with K. J. Arrow), *Studies in the Mathematical Theory of Inventory and Production* (K. J. Arrow, S. Karlin, H. Scarf, eds.), Stanford University Press, Stanford, (1958), 70- 85.

44. Production planning without storage (with K. J. Arrow), *Studies in the Mathematical Theory of Inventory and Production* (K. J. Arrow, S. Karlin, H. Scarf, eds.), Stanford University Press, Stanford, (1958), 86-91.

45. The optimal expansion of the capacity of a firm (with K. J. Arrow and M. J. Beckmann), *Studies in the Mathematical Theory of Inventory and Production* (K. J. Arrow, S. Karlin, H. Scarf, eds.), Stanford University Press, Stanford, (1958), 92-108.

46. One stage inventory models with uncertainty, *Studies in the Mathematical Theory of Inventory and Production* (with K. J. Arrow, S. Karlin, H. Scarf, eds.), Stanford University Press, Stanford, (1958), 109-134.

47. Optimal inventory policy for the Arrow–Harris–Marschak dynamic model, *Studies in the Mathematical Theory of Inventory and Production* (K. J. Arrow, S. Karlin, H. Scarf, eds.), Stanford University Press, Stanford, (1958), 135-154.

48. Inventory models of the Arrow–Harris–Marschak type with time lag (with H. Scarf), *Studies in the Mathematical Theory of Inventory and Production* (K. J. Arrow, S. Karlin, H. Scarf, eds.), Stanford University Press, Stanford, (1958), 155-178.

49. Optimal policy for hydroelectric operations (with J. Gessford), *Studies in the Mathematical Theory of Inventory and Production* (K. J. Arrow, S. Karlin, H. Scarf, eds.), Stanford University Press, Stanford, (1958), 179-200.

50. Steady state solutions, *Studies in the Mathematical Theory of Inventory and Production* (K. J. Arrow, S. Karlin, H. Scarf, eds.), Stanford University Press, Stanford, (1958), 223-269.

51. The application of renewal theory to the study of inventory policies, *Studies in the Mathematical Theory of Inventory and Production* (K. J. Arrow, S. Karlin, H. Scarf, eds.), Stanford University Press, Stanford, (1958), 270-297.

52. Inventory models and related stochastic processes (with H. Scarf), *Studies in the Mathematical Theory of Inventory and Production* (K. J. Arrow, S. Karlin, H. Scarf, eds.), Stanford University Press, Stanford, (1958), 319-336.

53. Many server queueing processes with Poisson input and exponential service times (with J. McGregor), *Pacific Journal of Mathematics*, **8** (1958), 87-118.

54. Linear growth, birth and death processes (with J. McGregor), *Journal of Mathematics and Mechanics*, **7** (1958), 643-662.

55. A characterization of birth and death processes (with J. McGregor), *Proceedings of the National Academy of Sciences USA*, **45** (1958), 375-379.

56. Random walks (with J. McGregor), *Illinois Journal of Mathematics*, **3** (1959), 66-81.

57. Note on a moving single server problem (with R. G. Miller, Jr. and N. U. Prabhu), *Annals of Mathematical Statistics*, **30** (1959), 243-246.

58. Price speculation under certainty, chapter in *Studies in Linear and Non-linear Programming* (K. J. Arrow, L. Hurvicz, and H. Uzawa, eds.), 1959, 189-198.

59. Positive operators, *Journal of Mathematics and Mechanics*, **8** (1959), 907-937.

60. Coincidence properties of birth and death processes (with J. McGregor), *Pacific Journal of Mathematics*, **9** (1959), 1109-1140.

61. Coincidence probabilities (with J. McGregor), *Pacific Journal of Mathematics*, **9** (1959), 1141-1164.

62. A stationary inventory model with Markovian demand (with A. J. Fabens), *Mathematical Methods in the Social Sciences*, (1959), 159-175.

63. On certain differential–integral equations (with G. Szegö), *Mathematische Zeitschrift*, **72** (1960), 205-228.

64. Slippage problems (with D. Truax), *Annals of Mathematical Statistics*, **31** (1960), 296-324.

65. Optimal policy for dynamic inventory process with stochastic demands subject to seasonal variations, *Journal of the Society of Industrial and Applied Mathematics*, **8** (1960), 611-629.

66. Dynamic inventory policy with varying stochastic demands, *Management Science*, **6** (1960), 231-258.

67. Pólya–type distributions of convolutions (with F. Proschan), *Annals of Mathematical Statistics*, **31** (1960), 721-736.

68. Classical diffusion processes and total positivity (with J. McGregor), *Journal of Mathematical Analysis and Application*, **1** (1960), 163-183.

69. On certain determinants whose elements are orthogonal polynomials (with G. Szegö), *Journal d'Analyse Mathématique*, **8** (1960-61), 1-157.

70. The mathematical theory of inventory processes, chapter in *Modern Mathematics for Engineers* (E. F. Beckenbach, ed.), 1961, 228-258.

71. Total positivity, absorption probabilities, and applications, *Bulletin of the American Mathematical Society*, **67** (1961), 105-108.

72. The Hahn polynomials, formulas, and an application (with J. McGregor), *Scripta Mathematica*, **26** (1961), 33-46.

73. Moment inequalities of Pólya frequency functions (with F. Proschan and R. E. Barlow) *Pacific Journal of Mathematics*, **11** (1961), 1023-1033.

74. Occupation time laws for birth and death processes (with J. McGregor), *Proceedings of the Fourth Berkeley Symposium in Mathematical Statistics and Probability*, (1961), 249-272.

75. Total positivity of fundamental solutions of parabolic equations (with J. McGregor), *Proceedings of the American Mathematical Society*, **13** (1962), 136-139.

76. Optimal policy for dynamic inventory processes with non–stationary stochastic demands (with D. Iglehart), *Studies in Applied Probability and Management Science* (K. Arrow, S. Karlin, and H. Scarf, eds.), Stanford University Press, Stanford, California, (1962), 127-147.

77. Stochastic models and optimal policy for selling an asset, *Studies in Applied Probability and Management Science* (K. Arrow, S. Karlin, and H. Scarf, eds.), Stanford University Press, Stanford, California, (1962), 148-158.

78. Prices and optimal inventory policy (with C. Carr), *Studies in Applied Probability and Management Science* (K. Arrow, S. Karlin, and H. Scarf, eds.), Stanford University Press, Stanford, California, (1962), 159-172.

79. Determinants of eigenfunctions of Sturm–Liouville equations, *Journal d'Analyse Mathématique*, **9** (1961-62), 365-397.

80. On a genetics model of Moran (with J. McGregor), *Proceedings of the Cambridge Philosophy Society*, **58**, Part 2 (1962), 299-311.

81. Determinants of orthogonal polynomials (with J. McGregor), *Bulletin of the American Mathematical Society*, **68** (1962), 204-209.

82. On some classes of functions associated with exponential polynomials (with C. Loewner), *Studies in Mathematical Analysis and Related Topics*, Stanford University Press, Stanford, California, (1962), 175-182.

83. Generalized renewal functions and stationary inventory models (with A. J. Fabens), *Journal of Mathematical Analysis and Applications*, **5** (1962), 461-487.

84. Representation theorems for positive functions, *Journal of Mathematics and Mechanics*, **12** (1963), 599-618.

85. Total positivity and convexity preserving transformations, *Symposium on Pure Mathematics*, American Mathematical Society, **7** (1963), 329-347.

86. On choosing combinations of weapons (with W. E. Pruitt and W. F. Madow), *Naval Research Logistic Quarterly*, **10** (1963), 95-119.

87. Conditioned limit theorems (with M. Dwass), *Annals of Mathematical Statistics*, **34** (1963), 1147-1167.

88. Generalized convex inequalities (with A. Novikoff), *Pacific Journal of Mathematics*, **13** (1963), 1251-1279.

89. On some stochastic models in genetics (with J. McGregor), *Conference on Stochastic Models in Medicine and Biology*, The University of Wisconsin Press, (1964), 245-279.

90. Total positivity, absorption probabilities and applications, *Transactions of the American Mathematical Society*, **111** (1964), 33-107.

91. Direct product branching processes and related Markov chains (with J. McGregor), *Proceedings of the National Academy of Sciences USA*, **51** (1964), 598-602.

92. Direct product branching processes and related induced Markov chains I. Calculations of rates of approach to homozygosity (with J. McGregor), *Symposium in Memory of Bayes, Bernoulli and Laplace* (J. Neyman, ed.), (1965), 111-145.

93. The existence of eigenvalues for integral operators, *Transactions of the American Mathematical Society*, **113** (1964), 1-17.

94. Oscillation properties of eigenvectors of strictly totally positive matrices, *Journal d'Analyse Mathématique*, **14** (1965), 247-266.

95. Spectral theory of branching processes, I. The case of the discrete spectrum (with J. McGregor), *Zeitschrift für Wahrscheinlichkeitstheorie und Verwandte Gebiete*, **5** (1966), 6-33.

96. Spectral representation of branching processes, II. Case of continuous spectrum (with J. McGregor), *Zeitschrift für Wahrscheinlichkeitstheorie und Verwandte Gebiete*, **5** (1966), 34-54.

97. Ehrenfest urn models (with J. McGregor), *Journal of Applied Probability*, **2** (1965), 352-376.

98. Generalized absolutely monotone functions (with Z. Ziegler), *Israel Journal of Mathematics*, **3** (1965), 173-180.

99. Chebyshevian spline functions (with Z. Ziegler), SIAM Journal of Numerical Analysis, **3** (1966), 514-543.

100. Generalized Bernstein inequalities, *Acta Scientarium Mathematics*, **27** (1966), 1-15.

101. Optimal experimental designs (with W. Studden), *Annals of Mathematical Statistics*, **37** (1966), 783-815.

102. On a theorem of P. Nowosad (with L. Nirenberg), *Mathematical Analysis and Application*, **17** (1967), 61-67.

103. Education in applied mathematics, *Proceedings of Education in Applied Mathematics*, Colorado, 1966, 336-340.

104. Properties of the stationary measure of the critical case simple branching process (with J. McGregor), *Annals of Mathematical Statistics*, **38** (1967), 977-991.

105. Characterization of moment points in terms of Christoffel numbers (with L. Schumaker), *Journal d'Analyse Mathématique*, **20** (1967), 213-231.

106. The fundamental theorem of algebra for Tchebycheffian monosplines (with L. Schumaker), *Journal d'Analyse Mathématique*, **20** (1967), 233-269.

106. Characterization of moment points in terms of Christoffel numbers (with L. Schumaker), *Journal d'Analyse Mathematics*, Volume XX (1967), 213-231.

107. Chebyshevian spline functions (with Z. Ziegler), *Inequalities* (1967), 137-149.

108. Limit theorems for the split times of branching processes (with K. B. Athreya), *Journal of Mathematics and Mechanics*, **17** (1967), 257-278.

109. Central limit theorems for certain infinite urn schemes, *Journal of Mathematics and Mechanics*, **17** (1967), 373-401.

110. The rate of production of recombinants between linked genes in finite populations (with J. McGregor and W. Bodmer), *Proceedings of the Fifth Berkeley Symposium on Mathematical Statistics and Probability*, University of California Press, 1967, 403-414.

111. The number of mutant forms maintained in a population (with J. McGregor), *Proceedings of the Fifth Berkeley Symposium on Mathematical Statistics and Probability*, University of California Press, 1967, 415-438.

112. Rates and probabilities of fixation for two locus random mating finite populations without selection (with J. McGregor), *Genetics*, **58** (1967), 141-159.

113. Uniqueness of stationary measures for branching processes and applications (with J. McGregor), *Proceedings of the Fifth Berkeley Symposium on Mathematical Statistics and Probability*, University of California Press, 1967, 243-254.

114. The role of the Poisson progeny distribution in population genetic models (with J. McGregor), *Mathematical Biosciences*, **2** (1968), 11-17.

115. Embeddability of discrete time simple branching processes into continuous time branching processes (with J. McGregor), *Transactions of the American Mathematical Society*, **132** (1968), 115-136.

116. Embedding interates of analytic functions with two fixed points into continuous groups (with J. McGregor), *Transactions of the American Mathematical Society*, **132** (1968), 137-145.

117. Analysis of models with homozygote—heterozygote matings (with M. W. Feldman), *Genetics*, **59** (1968), 105-116.

118. Further analysis of negative assortative mating (with M. W. Feldman), *Genetics*, **59** (1968), 117-136.

119. Sign regularity properties of classical polynomials, *Proceedings of the Conference on Orthogonal Expansions and Their Continuous Analogues* (D. Tepper Haimo, ed.), Southern Illinois Press, 1968, 55-74.

120. On the spectral representation of branching processes with mean one (with J. McGregor), *Journal of Mathematical Analysis and Applications*, **21** (1968), 485-495.

121. Rates of approach to homozygosity for finite stochastic models with variable population size, *American Naturalist*, **102** (1968), 443-455.

122. Branching processes, *Lectures in Applied Mathematics, Mathematics of the Decision Sciences*, **12** (1968), 195-234.

123. Embedding of urn schemes into continuous time Markov branching processes and related limit theorems (with K. B. Athreya), *Annals of Mathematical Statistics*, **39** (1968), 1801-1817.

124. A variation–diminishing generalized spline approximation method (with J. Karon), *Journal of Approximation Theory*, **1** (1968), 255-268.

125. Equilibrium behavior of population genetic models with non–random mating, Part I: Preliminaries and special mating systems, *Journal of Applied Probability*, **5** (1968), 231-313.

126. Equilibrium behavior of population genetic models with non–random mating, Part II: Pedigrees, homozygosity and stochastic models, *Journal of Applied Probability*, **5** (1968), 487-566.

127. Linkage and selection: New equilibrium properties of the two locus symmetric viability model (with M. W. Feldman), *Proceedings of the National Academy of Sciences USA*, **62** (1969), 70-74.

128. Assortative mating based on phenotype: I. Two alleles with dominance (with F. Scudo), *Genetics*, **63** (1969), 479-498.

129. Assortative mating based on phenotype: II. Two autosomal alleles without dominance (with F. Scudo), *Genetics*, **63** (1969), 499-510.

130. Best quadrature formulas and interpolation by splines satisfying certain boundary conditions and applications to optimal quadrature formulas, *Approximation with Special Emphasis on Spline Functions*, Academic Press (1969), 447-466.

131. The fundamental theorem of algebra for monosplines satisfying certain boundary conditions and applications to optimal quadrature formulas, *Approximation with Special Emphasis on Spline Functions*, Academic Press, 1969, 467-484.

132. Linkage and selection: Two locus symmetric viability model (with M. W. Feldman), *Theoretical Population Biology*, **1** (1970), 39-71.

133. Branching processes with random environments (with K. B. Athreya), *Bulletin of the American Mathematical Society*, **76** (1970), 865-870.

134. Some elementary integrability theorems for special transforms (with R. Askey), *Journal d'Analyse Mathématique*, **23** (1970), 27-38.

135. Iteration of positive approximation operators (with Z. Ziegler), *Journal of Approximation Theory*, **3** (1970), 310-339.

136. Convergence to equilibrium of the two locus additive viability model (with M. W. Feldman), *Journal of Applied Probability*, **7** (1970), 262-271.

137. Periodic boundary–value problems with cyclic totally positive Green's functions with applications to periodic spline theory (with J. Lee), *Journal of Differential Equations*, **8** (1970), 374-396.

138. Iteration of analytic functions of several variables (with J. McGregor), *Symposium in Honor of S. Bochner, Problems in Analysis*, Princeton University Press, (1970), 81-92.

139. Letter to the editor, a remark on B–splines, *Journal of Approximation Theory*, **3** (1970), 455.

140. Total positivity, interpolation by splines and Green's functions of differential operators, *Journal of Approximation Theory*, **4** (1971), 91-112.

141. Best quadrature formulas and splines, *Journal of Approximation Theory*, **4** (1971), 59-90.

142. On mutation selection balance for two–locus haploid and diploid populations (with J. McGregor), *Theoretical Population Biology*, **2** (1971), 60-70.

143. On branching processes with random environments, I: Extinction probabilities (with K. B. Athreya), *Annals of Mathematical Statistics*, **42** (1971), 1499-1520.

144. Branching processes with random environments, II: Limit theorems (with K. B. Athreya), *Annals of Mathematical Statistics*, **42** (1971), 1843-1858.

145. The evolution of dominance: A direct approach through the theory of linkage and selection (with M. W. Feldman), *Theoretical Population Biology*, **2** (1971), 482-492.

146. On the Hermite–Birkhoff interpolation (with J. Karon), Special Volume in Honor of J. Walsh, *Journal of Approximation Theory*, **6** (1972), 90-115.

147. The fundamental theorem of algebra for monosplines satisfying boundary conditions (with C. Micchelli), *Israel Journal of Mathematics*, **11** (1972), 405-451.

148. On a class of best non–linear approximation problems, *Bulletin of the American Mathematical Society*, **78** (1972), 43-49.

149. Polymorphisms for genetic and ecological systems with weak coupling (with J. McGregor), *Theoretical Population Biology*, **3** (1972), 210-238.

150. Application of method of small parameters to multi–niche population genetic models (with J. McGregor), *Theoretical Population Biology*, **3** (1972), 186-209.

151. Some extremal problems for eigenvalues of certain matrix and integral operators, *Advances in Mathematics*, **9** (1972), 93-136.

152. Poised and non-poised Hermite–Birkhoff interpolation (with J. Karon), *Indiana University Mathematics Journal*, **21** (1972), 1131-1170.

153. Mathematical genetics: A hybrid seed for educators to sow (with M. W. Feldman), *International Journal of Mathematical Education in Science and Technology*, **3** (1972), 169-189.

154. The evolutionary development of modifier genes (with J. McGregor), *Proceedings of the National Academy of Sciences USA*, **69** (1972), 3611-3614.

155. Some mathematical models of population genetics, *American Mathematical Monthly*, **79** (1972), 699-739.

156. Addendum to a paper of W. Ewens (with J. McGregor), *Theoretical Population Biology*, **3** (1972), 113-116.

157. Criteria for extinction of certain population growth processes with interacting types (with N. Kaplan), *Advances in Applied Probability*, **5** (1973), 183-199.

158. Some variational problems on certain Sobolev spaces and perfect splines, *Bulletin of the American Mathematical Society*, **79** (1973), 124–128.

159. Equilibria for genetic systems with weak interactions (with J. McGregor), *Proceedings of the Sixth Berkeley Symposium on Mathematical Statistics and Probability* (L. M. LeCam, J. Neyman, E. L. Scott, eds.), University of California Press, 1973, 79-87.

160. To I. J. Schoenberg and his mathematics, *Journal of Approximation Theory*, **8** (1973), vi-ix.

161. Sex and infinity: A mathematical analysis of the advantages and disadvantages of genetic recombination, *Population Dynamics* (M. Bartlett and R. Hiorns, eds.), Academic Press, 1973, 155-194.

162. Inequalities for symmetric sampling plans I, *Annals of Statistics*, **2** (1974), 1065-1094.

163. Oscillation properties of generalized characteristic polynomials for totally positive and positive definite matrices (with A. Pinkus), *Linear Algebra and Its Application*, **8** (1974), 281-312.

164. Toward a theory of the evolution of modifier genes (with J. McGregor), *Theoretical Population Biology*, **5** (1974), 59-103.

165. Some population genetic models combining artificial and natural selection pressures (with D. Carmelli), *Proceedings of the National Academy of Sciences USA*, **71** (1974), 4727-4731.

166. Random temporal variation in selection intensities. Case of large population size (with U. Liberman), *Theoretical Population Biology*, **6** (1974), 355-382.

167. Temporal fluctuations in selection intensities. Case of small population size (with B. Levikson), *Theoretical Population Biology*, **6** (1974), 383-412.

168. Some population genetic models combining artificial and natural selection pressures, I. One–locus theory (with D. Carmelli), *Theoretical Population Biology*, **7** (1975), 94-122.

169. Some population genetic models combining artificial and natural selection pressures, II. Two–locus theory (with D. Carmelli), *Theoretical Population Biology*, **7** (1975), 123-148.

170. Derivation of the eigenvalues of the configuration process induced by a labeled direct product branching process (with H. Avni), *Theoretical Population Biology*, **7** (1975), 221-228.

171. Some inequalities for generalized concave functions (with Z. Ziegler), *Journal of Approximation Theory*, **13** (1975), 276-293.

172. Numerical studies on two loci selection models with general viabilities (with D. Carmelli), *Theoretical Population Biology*, **7** (1975), 399-421.

173. General two–locus selection model: Some objectives, results and interpretations, *Theoretical Population Biology*, **7** (1975), 364-398.

174. Random temporal variation in selection intensities: One–locus two–allele model (with U. Liberman), *Journal of Mathematical Biology*, **2** (1975), 1-17.

175. Interpolation properties of generalized perfect splines and the solutions of certain extremal problems, I, *Transactions of the American Mathematical Society*, **206** (1975), 25-66.

176. Linear growth models with many types and multi–dimensional Hahn polynomials (with J. McGregor), *Theory and Applications of Special Functions*, Academic Press, New York, 1975, 261-288.

177. Some properties of determinants of orthogonal polynomials (with J. McGregor), *Theory and Applications of Special Functions*, Academic Press, New York, 1975, 521-550.

178. Some inequalities for the spectral radius of non–negative matrices and applications (with S. Friedland), *Duke Mathematical Journal*, **42** (1975), 459-490.

179. Random temporal variation in selection intensities acting on infinite diploid populations: Diffusion method analysis (with B. Levikson), *Theoretical Population Biology*, **8** (1975), 292-300.

180. Some theoretical analyses of migration selection interaction in a cline: A generalized two range environment (with N. Richter–Dyn), in *Population Genetics and Ecology*, Academic Press, New York (1976), 659-706.

181. Population subdivision and selection migration interaction, in *Population Genetics and Ecology*, Academic Press, New York (1976), 617-657.

182. On a class of best non–linear approximation problems and extended monosplines, in *Studies in Spline Functions and Approximation Theory* (S. Karlin, C. A. Micchelli, A. Pinkus, and I. J. Schoenberg, eds.), Academic Press, New York, (1976), 19-66.

183. A global improvement theorem for polynomial monosplines, in *Studies in Spline Functions and Approximation Theory* (S. Karlin, C. A. Micchelli, A. Pinkus, and I. J. Schoenberg, eds.), Academic Press, New York, 1976, 67-82.

184. Gaussian quadrature formulae with multiple nodes (with A. Pinkus), in *Studies in Spline Functions and Approximation Theory* (S. Karlin, C. A. Micchelli, A. Pinkus, and I. J. Schoenberg, eds.), Academic Press, New York, 1976, 113-141.

185. An extremal property of multiple Gaussian nodes (with A. Pinkus), in *Studies in Spline Functions and Approximation Theory* (S. Karlin, C. A. Micchelli, A. Pinkus, and I. J. Schoenberg, eds.), Academic Press, New York, 1976, 143-162.

186. Interpolation by splines with mixed boundary conditions (with A. Pinkus), in *Studies in Spline Functions and Approximation Theory* (S. Karlin, C. A. Micchelli, A. Pinkus, and I. J. Schoenberg, eds.), Academic Press, New York, 1976, 305-325.

187. Divided differences and other nonlinear existence problems at extremal points (with A. Pinkus), in *Studies in Spline Functions and Approximation Theory* (S. Karlin, C. A. Micchelli, A. Pinkus, and I. J. Schoenberg, eds.), Academic Press, New York, 1976, 327-352.

188. Oscillatory perfect splines and related extremal problems, in *Studies in Spline Functions and Approximation Theory* (S. Karlin, C. A. Micchelli, A. Pinkus, and I. J. Schoenberg, eds.), Academic Press, New York, 1976, 371-460.

189. Generalized Markov Bernstein type inequalities for spline functions, in *Studies in Spline Functions and Approximation Theory* (S. Karlin, C. A. Micchelli, A. Pinkus, and I. J. Schoenberg, eds.), Academic Press, New York, 1976, 461-484.

190. Some one–sided numerical differentiation formulae and applications, in *Studies in Spline Functions and Approximation Theory* (S. Karlin, C. A. Micchelli, A. Pinkus, and I. J. Schoenberg, eds.), Academic Press, New York, 1976, 485-500.

191. A phenotypic symmetric selection model for three loci, two alleles: The case of tight linkage (with U. Liberman), *Theoretical Population Biology*, **10** (1976), 334-364.

192. Some applications to inequalities of the method of generalized convexity (with Z. Ziegler), *Journal d'Analyse Mathématique*, **30** (1976), 281-303.

193. Selection with many loci and possible relations to quantitative genetics, in *Proceedings of the International Conference on Quantitative Genetics*, Iowa State University Press, Ames, 1977, 207-226.

194. Gene frequency patterns in the Levene subdivided population model, *Theoretical Population Biology*, **11** (1977), 356-385.

195. Variable spatial selection with two stages of migrations and comparisons between different timings (with R. Kenett), *Theoretical Population Biology*, **11** (1977), 386-409.

196. Theoretical aspects of multilocus selection balance I, in *Studies in Mathematical Biology* (S. Levin, ed.), Mathematical Association of America, 1978, 503-587.

197. Protection of recessive and dominant traits in a subdivided population with general migration structure, *American Naturalist*, **111** (1977), 1145-1162.

198. The two–locus multiallele additive viability model (with U. Liberman), *Journal of Mathematical Biology*, **5** (1978), 201-211.

199. Some population genetic models combining sexual selection with assortative mating (with P. O'Donald), *Heredity*, **41** (1978), 165-174.

200. Analysis of central equilibrium configurations for certain multilocus systems in subdivided populations (with R. Campbell), *Genetical Research*, **32** (1978), 151-169.

201. Comparisons of positive assortative mating and sexual selection models, *Theoretical Population Biology*, **14** (1978), 281-312.

202. Analysis of partial assortative mating and sexual selection models for a polygamous species involving a trait based on levels of heterozygosity (with S. Farkash), *Theoretical Population Biology*, **14** (1978), 430-445.

203. Some multiallele partial assortative mating systems for a polygamous species (with S. Farkash), *Theoretical Population Biology* **14** (1978), 197-222.

204. Evolutionary aspects and sensitivity studies of some major gene models (with D. Carmelli), *Journal of Theoretical Biology*, **75** (1978), 197-222.

205. The relationship between HLA antigens and Bermuda grass hayfever (with C. Geller–Bernstein, R. Kenett, R. Zamir, M. Sela), *Immunogenetics*, **7** (1978), 259-264.

206. Simultaneous stability of $D = 0$ and $D \neq 0$ for multiplicative viabilities at two loci (with M. W. Feldman), *Genetics*, **90** (1978), 813-825.

207. Classifications and comparisons of multilocus recombination distributions (with U. Liberman), *Proceedings of National Academy of Sciences USA*, **75** (1978), 6332-6336.

208. Principles of polymorphism and epistasis for multilocus systems, *Proceedings of the National Academy of Sciences USA*, **76** (1979), 541-555.

209. Sexual selection encounter models (with J. Raper), *Theoretical Population Biology*, **15** (1979), 246-256.

210. Index measures for assessing the mode of inheritance of continuously distributed traits. Theory and justifications (with D. Carmelli and R. Williams), *Theoretical Population Biology*, **16** (1979), 81-106.

211. Analysis of genetic data on Jewish populations: I. Historical background, demographic features and genetic markers (with B. Bonné–Tamir and R. Kenett), *American Journal of Human Genetics*, **31** (1979), 324-340.

212. Analysis of biochemical genetic data on Jewish populations: II. Results and interpretations of heterogeneity indices and distance measures with respect to standards (with R. Kenett and B. Bonné–Tamir), *American Journal of Human Genetics*, **31** (1979), 341-365.

213. Models of multifactorial inheritance: I. Multivariate formulations and basic convergence results, *Theoretical Population Biology*, **15** (1979), 308-355.

214. Models for multifactorial inheritance: II. The covariance structure for a scalar phenotype under selective assortative mating and sex–dependent symmetric parental transmission, *Theoretical Population Biology*, **15** (1979), 356-393.

215. Models of multifactorial inheritance: III. Calculation of covariance of relatives under selective assortative mating, *Theoretical Population Biology*, **15** (1979), 394-423.

216. Models of multifactorial inheritance: IV. Asymmetric transmission for a scalar phenotype, *Theoretical Population Biology*, **15** (1979), 424-438.

217. Representation of nonepistatic selection models and analysis of Hardy–Weinberg equilibrium configurations (with U. Liberman), *Journal of Mathematical Biology*, **7** (1979), 353-374.

218. Central equilibria in multilocus systems. I. Generalized nonepistatic selection regimes (with U. Liberman), *Genetics*, **91** (1979), 777-798.

219. Central equilibria in multilocus systems. II. Bisexual and sex–linked nonepistatic selection models (with U. Liberman), *Genetics*, **91** (1979), 799-816.

220. A natural class of multilocus recombination processes and related measures of crossover interference (with U. Liberman), *Advances in Applied Probability*, **11** (1979), 479-501.

221. Applications of a robust measure of mode of inheritance to lipid, height, and weight measurement (with P. Williams, S. Mellen, W. Haskell, C. Villars, S. Jensen), *American Journal of Human Genetics*, **31** (1979), 143A.

222. Monte Carlo study of an index for discriminating monogenic and polygenic modes of inheritance (with D. Carmelli and P. Williams), *American Journal of Human Genetics*, **31** (1979), 137A.

223. Some matrix linear fractional transformations and their properties, *Journal d'Analyse Mathématique*, **36** (1979), 145-155.

224. An assortative mating encounter model (with J. K. Raper, and P. O'Donald), *Heredity*, **43** (1979), 27-34.

225. Approaches in modeling mode of inheritance with distributed traits, in *Genetic Analysis of Common Diseases: Applications to Predictive Factors in Coronary Disease* (C. F. Sing and M. Skolnick, eds.), Alan R. Liss, Inc., New York, 1979, 229-257.

226. Comments on statistical methodology in medical genetics, in *Genetic Analysis of Common Diseases: Applications to Predictive Factors in Coronary Disease* (C. F. Sing and M. Skolnick, eds.), Alan R. Liss, Inc., New York, 1979, 497-520.

227. A class of indices to assess major–gene versus polygenic inheritance of distributed variables (with D. Carmelli and R. Williams), in *Genetic Analysis of Common Diseases: Applications to Predictive Factors in Coronary Disease* (C. F. Sing and M. Skolnick, eds.), Alan R. Liss, Inc., New York, 1979, 259-270.

228. Shapes of velocity curves in multiunit enzyme kinetic systems (with R. Kenett), *Mathematical Biosciences*, **52** (1980), 97-115.

229. Measures of enzyme cooperativity, *Journal of Theoretical Biology*, **87** (1980), 33-54.

230. Selection–migration regimes characterized by a globally stable equilibrium (with R. Campbell), *Genetics*, **94** (1980), 1065-1084.

231. Models of multifactorial inheritance: V. Linear assortative mating as againist selective (nonlinear) assortative mating, *Theoretical Population Biology*, **17** (1980), 255-275.

232. Models of multifactorial inheritance: VI. Formulas and properties of the vector phenotype equilibrium covariance matrix, *Theoretical Population Biology*, **17** (1980), 276-297.

233. Polymorphism in subdivided populations characterized by a major and subordinate demes (with R. Campbell), *Heredity*, **44** (1980), 151-168.

234. The number of stable equilibria for the classical one–locus multiallele selection model, *Journal of Mathematical Biology*, **9** (1980), 189-192.

235. The range of stability of a polymorphic linkage equilibrium state in a general two–locus two–allele selection model, *Journal of Mathematical Biology*, **10** (1980), 189-194.

236. The effects of increased phenotypic variance on the evolutionary outcomes of generalized major–gene models (with D. Carmelli), *Annals of Human Genetics*, **44** (1980), 81-93.

237. Classes of orderings of measures and related correlation inequalities: I. Multivariate totally positive distributions (with Y. Rinott), *Journal of Multivariate Analysis*, **10** (1980), 467-498.

238. Classes of orderings of measures and related correlation inequalities: II. Multivariate reverse rule distributions (with Y. Rinott), *Journal of Multivariate Analysis*, **10** (1980), 499-516.

239. Total positivity properties of absolute value multinormal variables with applications to confidence interval estimates and related probabilistic inequalities (with Y. Rinott), *Annals of Statistics*, **9** (1981), 1035-1049.

240. Entropy inequalities for classes of probability distributions: I. The univariate case (with Y. Rinott), *Advances in Applied Probability*, **13** (1981), 93-112.

241. Entropy inequalities for classes of probability distributions: II. The multivariate case (with Y. Rinott), *Advances in Applied Probability*, **13** (1981), 325-351.

242. Structured exploratory data analysis (SEDA) for determining mode of inheritance of quantitative traits: I. Simulation studies on the effect of background distributions (with P. Williams and D. Carmelli), *American Journal of Human Genetics*, **33** (1981), 262-281.

243. Structured exploratory data analysis (SEDA) for determining mode of inheritance of quantitative traits. II. Simulation studies on the effect of ascertaining families through high-valued probands, *American Journal of Human Genetics*, **33** (1981), 282-292.

244. The existence of a protected polymorphism under conditions of soft as opposed to hard selection in a multideme population system (with R. Campbell), *American Naturalist*, **117** (1981), 262-275.

245. Sexual selection at a multiallelic locus with complete or partial dominance (with P. O'Donald), *Heredity*, **47** (1981), 209-220.

246. Some natural viability systems for a multiallelic locus: A theoretical study, *Genetics*, **97** (1981), 457-473.

247. A theoretical and numerical assessment of genetic variability (with M. W. Feldman), *Genetics*, **97** (1981), 475-493.

248. Preferential mating in symmetric multilocus systems (with J. Raper), *Proceedings of the National Academy of Sciences USA*, **78**, (1981), 3730-3733.

249. The detection of particular genotypes in finite populations. I. Natural selection effects (with S. Tavaré), *Theoretical Population Biology*, **19** (1981), 187-214.

250. The detection of particular genotypes in finite populations. II. The effects of partial penetrance and family structure (with S. Tavaré), *Theoretical Population Biology*, **19** (1981), 215-229.

251. Statistical methods for assessing linkage disequilibrium at the HLA–A,B,C, loci (with A. Piazza), *Annals of Human Genetics*, **45** (1981), 79-94.

252. Genetic analysis of the Stanford LRC family study data. I. Structured exploratory data analysis of height and weight measurements (with P. T. Williams, S. Jensen, and J. W. Farquhar), *American Journal of Epidemiology*, **113** (1981), 307-324.

253. Genetic analysis of the Stanford LRC family study data. II. Structured exploratory data analysis of lipids and lipoproteins (with P. T. Williams. W. L. Haskell, and P. Wood), *American Journal of Epidemiology*, **113** (1981), 325-337.

254. Sibling and parent–offspring correlation estimation with variable family size (with E. C. Cameron and P. T. Williams), *Proceedings of the National Academy of Sciences USA*, **78** (1981), 2664-2668.

255. The detection of a recessive visible gene in finite populations (with S. Tavaré), *Genetical Research*, **37** (1981), 33-46.

256. Univariate and multivariate total positivity, generalized convexity and related inequalities (with Y. Rinott), in *Generalized Concavity in Optimization and Economics* (S. Schaible and W. T. Ziemba, eds.), Academic Press, New York, 1981, 703-718.

257. Analysis of central equilibria in multilocus systems. A generalized symmetric viability regime (with H. Avni), *Theoretical Population Biology*, **20** (1981), 241-280.

258. A class of cumulant plots for assessing mode and asymmetry of transmission of a quantitative trait (with E. Cameron), *Theoretical Population Biology*, **21** (1982), 238-254.

259. Linear birth and death processes with killing (with S. Tavaré), *Journal of Applied Probability*, **19** (1982), 477-487.

260. Somee results on optimal partitioning of variance and monotonicity with truncation level, in *Statistics and Probability Essays in Honor of C. R. Rao* (G. Kallianpur, P. R. Krishnaiah, and J. K. Ghosh, eds.), 1982, 375-382.

261. A diffusion process with killing. The time to formation of recurrent deleterious mutant genes (with S. Tavaré), *Stochastic Processes and Their Applications*, **13** (1982), 249-261.

262. Analysis of biochemical genetic data on Jewish populations. III. The application of individual phenotype measurements for population comparisons (with D. Carmelli and B. Bonné–Tamir), *American Journal of Human Genetics*, **34** (1982), 50-64.

263. Preferential mating in symmetric multilocus systems: Stability conditions of the central equilibrium (with J. Raper), *Genetics*, **100** (1982), 137-147.

264. The reduction property for central polymorphisms in nonepistatic systems (with U. Liberman), *Theoretical Population Biology*, **22** (1982), 69-95.

265. A criterion for stability–instability at fixation states involving an eigenvalue one with applications in population genetics (with S. Lessard), *Theoretical Population Biology*, **22** (1982), 108-126.

266. Applications of ANOVA type decomposition for comparisons of conditional variance statistics including jackknife estimates (with Y. Rinott), *Annals of Statistics*, **10** (1982), 485-501.

267. Detecting particular genotypes in populations under nonrandom mating (with S. Tavaré), *Mathematical Biosciences*, **59** (1982), 57-75.

268. Classifications of selection-migration structures and conditions for a protected polymorphism, in *Evolutionary Biology*, Volume 14 (M. K. Hecht, B. Wallace, and C. T. Prance, eds.), Plenum Publishing Corporation, 1982, 61-204.

269. Association arrays for the study of familial height, weight, lipid and lipoprotein similarity in three West Coast populations (with P. T. Williams, E. Barrett–Connor, J. Hoover, J. W. Farquhar, P. W. Wahl, W. Haskell, R. Bergelin, and L. Suarez), *American Journal of Epidemiology*, **116** (1982), 1001-1021.

270. Association arrays for comparing familial total cholesterol, high density lipoprotein, cholesterol, and triglyceride similarity in the Israeli population by country of origin (with P. T. Williams, D. Carmelli, Y. Friedlander, T. Cohen, and Y. Stein), *American Journal of Epidemiology*, **116** (1982), 1022-1032.

271. Some assortative mating models based on phenotype classes determined by a dominance ordering of multiple alleles, *Theoretical Population Biology*, **22** (1982), 241-257.

272. Unbiasedness in the sense of Lehmann in n–action decision problems, in *A Festschrift for Erich L. Lehmann in Honor of his Sixty–fifth Birthday* (P. J. Bickel, K. Doksum, J. L. Hodges, eds.), Wadsworth Publishing Company, 1982, 261-285.

273. Structured exploratory data analysis applied to mode of inheritance (with E. C. Cameron and P. T. Williams), in *Developments in Statistics*, Volume 4 (P. Krishnaiah, ed.), 1983, 185-277.

274. Association arrays in assessing forms of dependencies between bivariate random variables, *Proceedings of National Academy of Sciences USA*, **80** (1983), 647-651.

275. Letter to the Editor (with E. C. Cameron, D. Carmelli, and P. T. Williams), *American Journal of Human Genetics*, **35** (1983), 534-540.

276. M-matrices as covariance matrices of multinormal distributions (with Y. Rinott), *Linear Algebra and Its Applications*, **52/53** (1983), 419-438.

277. A class of diffusion processes with killing arising in population genetics (with S. Tavaré), *SIAM Journal of Applied Mathematics*, **43** (1983), 31-41.

278. Path analysis in genetic epidemiology: A critique (with E. Cameron and R. Chakraborty), *American Journal of Human Genetics*, **35** (1983), 695-732.

279. New approaches for computer analysis of nucleic acid sequences (with G. Ghandour, F. Ost, S. Tavaré, L. Korn), *Proceedings of the National Academy of Sciences USA*, **80**, (1983), 5660-5664.

280. On the optimal sex ratio (with S. Lessard), *Proceedings of the National Academy of Sciences USA*, **80** (1983), 5931-5935.

281. Measuring interference in the chiasma renewal formation process (with U. Liberman), *Advances in Applied Probability*, **15** (1983), 471-487.

282. Kin selection and altruism (with C. Matessi), *Proceedings of the Royal Society of London, Series B*, **219** (1983), 327-353.

283. Structured exploratory data analysis (SEDA) of finger ridge–count inheritance. I. Major gene index, midparental correlation, and offspring–between–parents function in 125 South Indian families (with R. Chakraborty, P. Williams, and S. Mathew), *American Journal of Physical Anthropology*, **62** (1983), 377-396.

284. Structured exploratory data analysis (SEDA) of finger ridge–count inheritance. II. Association arrays in parent–offspring and sib–sib pairs (with P. Williams, R. Chakraborty, and S. Mathew), *American Journal of Physical Anthropology*, **62** (1983), 397-407.

285. Comparison of measures, multivariate majorization, and applications to statistics (with Y. Rinott), in *Studies in Econometrics, Time Series and Multivariate Statistics* (S. Karlin, L. Goodman, and T. Amemiya, eds.), 1983, Academic Press, New Yor, 465-489.

286. On the optimal sex ratio: A stability analysis based on a characterization for one–locus multiallele viability models (with S. Lessard), *Journal of Mathematical Biology*, **20** (1984), 15-38.

287. Mathematical models, problems, and controversies of evolutionary theory, *Bulletin of the American Mathematical Society*, **10** (1984), 221-273.

288. Theoretical models of genetic map functions (with U. Liberman), *Theoretical Population Biology*, **25** (1984), 331-346.

289. Theoretical aspects of genetic map functions in recombination processes, *The Pittsburgh Symposium Conference in Honor of C. C. Li* (A. Chakravarti, ed.), Van Nostrand, 1984, 209-229.

290. Permutation methods for the structured exploratory data analysis (SEDA) of familial trait values (with P. Williams), *American Journal of Human Genetics*, **36** (1984), 873-898.

291. Comparative analysis of human and bovine papillomaviruses (with G. Ghandour, D. Foulser and L. Korn), *Molecular Biology and Evolution*, **1** (1984), 357-370.

292. On the evolution of altruism by kin selection (with C. Matessi), *Proceedings of the National Academy of Sciences USA*, **81** (1984), 1754-1758.

293. Discussion: Recent developments in SEDA (with P. Williams), *Genetic Epidemiology of Coronary Heart Disease: Past, Present and Future* (D. C. Rao, ed.), Alan Liss, Inc., 1984, 317-323.

294. Random replacement schemes and multivariate majorization, in *Inequalities in Statistics and Probability* (Y. L. Tong, ed.), Institute of Mathematical Statistics Lecture Notes, 1984, 35-41.

295. Statistical approaches in assessing structural relationships in DNA sequences in *Proceedings of the XII International Biometrics Conference, September 2-8, 1984*, Tokyo, Japan, 1985.

296. Some monotonicity properties of Schur powers of matrices and related inequalities (with F. Ost), *Linear Algebra and Its Application*, **68** (1985), 47-65.

297. DNA sequence comparisons of the human, mouse, and rabbit immunoglobulin kappa–gene (with G. Ghandour and D. Foulser), *Molecular Biology and Evolution*, **2** (1985), 35-52.

298. Alignment maps and homology analysis of the $J-C$ intron in human, mouse, and rabbit immunoglobulin kappa–gene (with G. Ghandour), *Molecular Biology and Evolution*, **2** (1985), 53-65.

299. Permutation methods for the structured exploratory data analysis (SEDA) of total cholesterol measured in five Israeli populations (with P. Williams, D. Carmelli, and E. Cameron), *American Journal of Epidemiology*, **122** (1985), 163-186.

300. A study of familial resemblance of two cognitive psychometric tests by permutation analyses (with D. Carmelli), *Behavioral Genetics*, **15** (1985), 223-244.

301. Maximal segmental match length among random sequences from a finite alphabet (with F. Ost), *Proceedings of Berkeley Conference in Honor of Jerzy Neyman and Jack Kiefer* (L. M. LeCam and R. A. Olshen, eds.), Volume 1, Wadsworth, Inc., 1985, 225-243.

302. Some probabilistic aspects in multivariate splines (with C. Micchelli and Y. Rinott), in *Multivariate Analysis–VI* (P. R. Krishnaiah, ed.), (1985), 355-360.

303. The use of multiple alphabets in kappa–gene immunoglobulin DNA sequence comparisons (with G. Ghandour), *EMBO Journal*, **4** (1985), 1217-1223.

304. Comparative statistics for DNA and protein sequences: Single sequence analysis (with G. Ghandour), *Proceedings of the National Academy of Sciences USA* **82**, 1985, 5800-5804.

305. Comparative statistics for DNA and protein sequences: Multiple sequence analysis (with G. Ghandour), *Proceedings of the National Academy of Sciences USA*, **82** (1985), 6186-6190.

306. DNA sequence patterns in human, mouse, and rabbit immunoglobulin kappa–genes (with G. Ghandour), *Journal of Molecular Evolution*, **22** (1985), 195-208.

307. Multiple alphabet amino acid sequence comparisons of the immunoglobulin κ–chain constant domain (with G. Ghandour), *Proceedings of the National Academy of Sciences USA*, **82** (1985), 8597-8601.

308. Some monotonicity properties of Schur powers of matrices and related inequalities (with F. Ost), *Linear Algebra and Its Application*, **68** (1985), 47-65.

309. Comparative analysis of structural relationships in DNA and protein sequences, in *Evolutionary Processes and Theory* (S. Karlin and E. Nevo, eds.), Academic Press, Florida, 1986, 329-363.

310. Altruistic behavior in sibling groups with unrelated intruders (with C. Matessi), in *Evolutionary Processes and Theory* (S. Karlin and E. Nevo, eds.), Academic Press, Florida, 1986, 689-724.

311. Multivariate splines: A probabilistic perspective (with Y. Rinott), *Journal of Multivariate Analysis*, **20** (1986), 69-90.

312. Significant potential secondary structures in the Epstein–Barr virus genome, *Proceedings of the National Academy of Sciences USA*, **83** (1986), 6915-6919.

313. Counts of long aligned word matches among random letter sequences (with F. Ost), *Advances in Applied Probability*, **19** (1987), 293-351.

314. Maximal success durations for a semi–Markov process (with D. E. Foulser), *Stochastic Processes and Their Application*, **24** (1987), 203-224.

315. Models of intergenerational kin altruism (with M. Morris and C. Matessi), *The American Naturalist*, **130** (1987), 544-569.

316. A model for the development of the tandem repeat units in the EBV ori–P region and discussion of their possible function (with B. E. Blaisdell), *Journal of Molecular Evolution*, **25** (1987), 215-229.

317. Path analysis in genetic epidemiology and alternatives, *Journal of Educational Statistics*, **12** (1987), 165-177.

318. Permutation analysis of familial association arrays for lipoprotein concentrations in families of the Stanford Five City Project (with P. J. Williams and J. W. Farquhar), *American Journal of Epidemiology*, **126** (1987), 1126-1140.

319. Efficient algorithms for molecular sequence analysis (with M. Morris, G. Ghandour, and M. Leung), *Proceedings of the National Academy of Sciences USA*, **85** (1988), 841-845.

320. Algorithms for identifying local molecular sequence features (with M. Morris, G. Ghandour, and M. Leung), *CABIOS*, **4** (1988), 41-51.

321. Patterns in DNA and amino acid sequences and their statistical significance (with B. E. Blaisdell and F. Ost), *Mathematical Methods for DNA Sequences* (M. S. Waterman, ed.), CRC Press, Boca Raton, Florida, 1988, 133-157.

322. NonGaussian phenotypic models of quantitative traits, *Proceedings of the Second International Conference on Quantitative Genetics* (B. Weir, E. Eisen, M. Goodman, G. Namkoong, eds.), Sinauer Associates, Inc., Sunderland, Massachusetts, 1988, 123-144.

323. Coincident probabilities and applications in combinatorics, *Journal of Applied Probability*, Special 25th Anniversary Volume, (1988), 185-200.

324. Pólya frequency functions and sequences. In essays accompanying *Selected Works of I. J. Schoenberg*, Birkhäuser, Basel, 1988, 384-387.

325. Total positivity and variation diminishing transformations. In essays accompanying *Selected Works of I. J. Schoenberg*, Birkhäuser, Basel, 1988, 269-273.

326. Maximal length of common words among random letter sequences (with F. Ost), *Annals of Probability*, **16** (1988), 535-563.

327. Distinctive charge configurations in proteins of the Epstein–Barr virus and possible functions (with B. E. Blaisdell), *Proceedings of the National Academy of Sciences USA*, **85** (1988), 6637-6641.

328. A generalized Cauchy–Binet formula and applications to total positivity and majorization (with Y. Rinott), *Journal of Multivariate Analysis*, **27** (1988), 284-299.

329. Charge configurations in viral proteins (with V. Brendel), *Proceedings of the National Academy of Sciences USA*, **85** (1988), 9396-9400.

330. A method to identify distinctive charge configurations in protein sequences, with application to human herpesvirus polypeptides (with B. E. Blaisdell, E. S. Mocarski, and V. Brendel), *Journal of Molecular Biology*, **205** (1989), 165-177.

Ph.D. STUDENTS OF SAMUEL KARLIN

1948	John Denby Wilks: functional analysis
1952	Carlo Comba: Analysis
1954	Russell N. Brandt: statistics James R. McGregor: probability Rodriques Restrepo: game theory
1955	John W. Pratt: statistics Donald R. Truax: statistics
1958	Rupert G. Miller: statistics
1959	Augustus J. Fabens: operations research Frank Proschan: statistics
1960	William E. Pruitt: probability
1961	Richard E. Barlow: statistics Charles Carr: economics John Gessford: operations research Marcel F. Neuts: probability Richard W. Singleton: operations research
1962	Donald L. Iglehart: operations research Charles J. Stone: probability William J. Studden: theory of approximation and statistics
1964	Stephen M. Samuels: probability
1965	Paul R. Milch: probability Zvi Ziegler: theory of approximation
1966	Larry L. Schumaker: theory of approximation
1967	Krishna B. Athreya: probability Burton H. Singer: probability
1968	John M. Karon: theory of approximation
1969	Marcus W. Feldman: mathematical biology John Lee: theory of approximation Thomas M. Liggett: probability Charles A. Micchelli: total positivity Masakazu Takahashi: probability Roy E. Welsch: probability and applied statistics

1970	Ilan Eshel: mathematical biology
	Norman Kaplan: probability theory, mathematical genetics
1974	Dorit Carmelli: statistical genetics
	Benjamin Levikson: mathematical genetics
	Uri Liberman: mathematical genetics
	Allan Pinkus: topics in approximation theory
	Yosef Rinott: multivariate total positivity
1976	Shoshana Farkash: mathematical genetics
1977	Haim Avni: mathematical genetics
	Roni Kenett: statistical genetics
1979	Russell B. Campbell: mathematical genetics
1980	John Raper: mathematical genetics
1981	Elic Yavor: total positivity
1982	Edward Cameron: genetic epidemiology
1984	Marge Foddy: stochastic processes
1985	David Foulser: problems in comparisons of random letter sequences
1986	Paul Williams: epidemiology
1989	Ming Ying Leung: applied stochastic processes
	McDonald Morris: molecular sequence analysis

Second-Order Moments of a Stationary Markov Chain and Some Applications

T. W. Anderson*
Stanford University

1. INTRODUCTION

A finite-state Markov chain is a stochastic process in which the variable takes on one of a finite number of values. Although the values may be numerical, they need not be; they may be simply states or categories. If they are given numerical values, the values are not necessarily the first so many integers, as in the number of customers waiting in a queue. When the chain is described in terms of states, it is convenient for many purposes to treat the chain as a vector-valued process. The vector has 1 in the position corresponding to the given state and 0 in the other positions. Then the vector-valued process is first-order autoregressive in the wide sense when the Markov chain is first-order. Anderson (1979a), (1979b), (1980) pointed out analogies between Gaussian autoregressive processes and Markov chains in terms of moments, sufficient statistics, tests of hypotheses, etc.

AMS 1980 Subject Classifications: Primary 60J10, Secondary 62M05.
Key words and phrases: Markov chain, stationary probabilities, estimation, autoregression.
*Research supported in part by the U.S. Army Research Office Contract No. DAAL03-89-K-0033.

PROBABILITY, STATISTICS, AND MATHEMATICS
Papers in Honor of Samuel Karlin

Copyright © 1989 by Academic Press, Inc.
All rights of reproduction in any form reserved.
ISBN-0-12-058470-0

In this paper the consequences of the autoregressive structure of the vector–valued process are developed further to yield various second–order moments and the spectral density of the process. It is shown that to estimate the stationary probabilities from a finite sequence of observations the mean of these observations is asymptotically equivalent to the maximum likelihood estimator and is asymptotically efficient (Section 4). The numerical–valued Markov chain is considered as a linear function of the vector–valued process, and a simple condition is obtained for it to be a wide–sense first–order autoregressive process (Section 5).

2. A STATIONARY MARKOV CHAIN

A stationary Markov chain $\{x_t\}$ with discrete time parameter and states $1, \ldots, m$ is defined by the transition probabilities

$$\Pr\{x_t = j | x_{t-1} = i, x_{t-2} = k, \ldots\} = p_{ij}, \; i, j, k, \ldots = 1, \ldots, m, \quad t = \ldots, -1, 0, 1, \ldots, \tag{1}$$

where $p_{ij} \geq 0$ and $\sum_{j=1}^{m} p_{ij} = 1$. Let $\Pr\{x_t = i\} = p_i$, $i = 1, \ldots, m$ ($p_i \geq 0$ and $\sum_{i=1}^{p} p_i = 1$). Then

$$\sum_{i=1}^{m} p_i p_{ij} = p_j, \qquad j = 1, \ldots, m. \tag{2}$$

If $\boldsymbol{P} = (p_{ij})$ and $\boldsymbol{p} = (p_i)$, a column vector, then the above properties can be written as

$$\boldsymbol{p}'\boldsymbol{P} = \boldsymbol{p}', \quad \boldsymbol{P}\boldsymbol{\varepsilon} = \boldsymbol{\varepsilon}, \quad \boldsymbol{p}'\boldsymbol{\varepsilon} = 1, \tag{3}$$

where $\boldsymbol{\varepsilon} = (1, 1, \ldots, 1)'$. Let $\boldsymbol{\varepsilon}_i$ be the m–component vector with 1 in the i–th position and 0's elsewhere, and let $\{\boldsymbol{z}_t\}$ be a sequence of m–component random vectors. Then the Markov chain can be written

$$\Pr\{\boldsymbol{z}_t = \boldsymbol{\varepsilon}_j | \boldsymbol{z}_{t-1} = \boldsymbol{\varepsilon}_i, \boldsymbol{z}_{t-2} = \boldsymbol{\varepsilon}_k, \ldots\} = p_{ij}, \; i, j, k, \ldots = 1, \ldots, m, \quad t = \ldots, -1, 0, 1, \ldots. \tag{4}$$

Note that $\boldsymbol{z}_t$ has 1 in one position and 0 in the others; hence $\boldsymbol{\varepsilon}'\boldsymbol{z}_t = 1$.

3. SECOND–ORDER MOMENTS

It follows from the model that

$$\mathcal{E}(z_{jt}|\boldsymbol{z}_{t-1} = \boldsymbol{\varepsilon}_i, \boldsymbol{z}_{t-2} = \boldsymbol{\varepsilon}_k, \ldots) = p_{ij}, \tag{5}$$

where z_{jt} is the j–th component of $\boldsymbol{z}_t$. We can write (5) in vector form as

$$\mathcal{E}(\boldsymbol{z}_t|\boldsymbol{z}_{t-1}, \boldsymbol{z}_{t-2}, \ldots) = \boldsymbol{P}'\boldsymbol{z}_{t-1}. \tag{6}$$

Let

$$\boldsymbol{v}_t = \boldsymbol{z}_t - \boldsymbol{P}'\boldsymbol{z}_{t-1} \tag{7}$$

be the t–th disturbance. Then

$$\begin{aligned} \mathcal{E}\boldsymbol{v}_t &= \mathcal{E}\{\mathcal{E}[(\boldsymbol{z}_t - \boldsymbol{P}'\boldsymbol{z}_{t-1})|\boldsymbol{z}_{t-1}, \boldsymbol{z}_{t-2}, \ldots]\} \\ &= \mathbf{0}. \end{aligned} \tag{8}$$

In (8) the outer expectation is with respect to $\boldsymbol{z}_{t-1}, \boldsymbol{z}_{t-2}, \ldots$. Similarly

$$\begin{aligned} \mathcal{E}\boldsymbol{v}_t\boldsymbol{z}'_{t-s} &= \mathcal{E}[\mathcal{E}(\boldsymbol{v}_t|\boldsymbol{z}_{t-1}, \boldsymbol{z}_{t-2}, \ldots)\boldsymbol{z}_{t-s}] \\ &= \mathbf{0}, \qquad s = 1, 2, \ldots . \end{aligned} \tag{9}$$

Since $\boldsymbol{v}_{t-s} = \boldsymbol{z}_{t-s} - \boldsymbol{P}'\boldsymbol{z}_{t-s-1}$,

$$\mathcal{E}\boldsymbol{v}_t\boldsymbol{v}'_{t-s} = \mathbf{0}, \qquad s = 1, 2, \ldots . \tag{10}$$

Thus $\{\boldsymbol{v}_t\}$ is a sequence of uncorrelated random vectors.

We can iterate $\boldsymbol{z}_t = \boldsymbol{P}'\boldsymbol{z}_{t-1} + \boldsymbol{v}_t$ to obtain

$$\boldsymbol{z}_t = \boldsymbol{v}_t + \boldsymbol{P}'\boldsymbol{v}_{t-1} + \cdots + (\boldsymbol{P}')^{s-1}\boldsymbol{v}_{t-s+1} + (\boldsymbol{P}')^s\boldsymbol{z}_{t-s}. \tag{11}$$

Then

$$\mathcal{E}(\boldsymbol{z}_t|\boldsymbol{z}_{t-s}, \boldsymbol{z}_{t-s-1}, \ldots) = (\boldsymbol{P}')^s\boldsymbol{z}_{t-s}. \tag{12}$$

This conditional expected value could alternatively be obtained from the fact that the transition probabilities from x_{t-s} to x_t are the elements of $\boldsymbol{P}^s$.

Since $\{x_t\}$ is stationary, $\{\boldsymbol{z}_t\}$ is stationary and $\boldsymbol{z}_t$ has a marginal multinomial distribution with probabilities $p_1, \ldots, p_m$. Hence, the expectation of $\boldsymbol{z}_t$ is

$$\mathcal{E}\boldsymbol{z}_t = \boldsymbol{p}, \tag{13}$$

and the covariance matrix is

$$\operatorname{Var}(\boldsymbol{z}_t) = \boldsymbol{D_p} - \boldsymbol{p}\boldsymbol{p}' = \boldsymbol{V}, \tag{14}$$

say, where $\boldsymbol{D_p}$ is a diagonal matrix with i–th diagonal element p_i. From (11) we also find

$$\mathcal{E}\boldsymbol{z}_t\boldsymbol{z}_{t-s}' = (\boldsymbol{P}')^s \mathcal{E}\boldsymbol{z}_t\boldsymbol{z}_t' = (\boldsymbol{P}')^s \boldsymbol{D_p}, \qquad s = 0, 1, \ldots . \tag{15}$$

Since $\mathcal{E}\boldsymbol{z}_t = \mathcal{E}\boldsymbol{z}_{t-s} = \boldsymbol{p}$, and $(\boldsymbol{P}')^s\boldsymbol{p} = \boldsymbol{p}$, the covariance matrix between $\boldsymbol{z}_t$ and $\boldsymbol{z}_{t-s}$ is

$$\operatorname{Cov}(\boldsymbol{z}_t, \boldsymbol{z}_{t-s}') = (\boldsymbol{P}')^s \boldsymbol{V}, \qquad s = 0, 1, \ldots . \tag{16}$$

Thus (14) and (16) determine the second–order moments of $\{\boldsymbol{z}_t\}$.

The conditional covariance matrix of $\boldsymbol{z}_t$ and of $\boldsymbol{v}_t$ is

$$\begin{aligned} \operatorname{Var}(\boldsymbol{z}_t | \boldsymbol{z}_{t-1} = \boldsymbol{\varepsilon}_i, \boldsymbol{z}_{t-2}, \ldots) &= \operatorname{Var}(\boldsymbol{v}_t | \boldsymbol{z}_{t-1} = \boldsymbol{\varepsilon}_i, \boldsymbol{z}_{t-2}, \ldots) \\ &= \boldsymbol{D_{p_i}} - \boldsymbol{p}_i\boldsymbol{p}_i' \\ &= \boldsymbol{V}_i, \end{aligned} \tag{17}$$

say, where $\boldsymbol{D_{p_i}}$ is a diagonal matrix with p_{ij} as the j–th diagonal element and

$$\boldsymbol{P} = \begin{bmatrix} \boldsymbol{p}_1' \\ \boldsymbol{p}_2' \\ \vdots \\ \boldsymbol{p}_m' \end{bmatrix}. \tag{18}$$

From this conditional variance of $\boldsymbol{v}_t$ (which has conditional mean value $\mathbf{0}$) we find the (marginal) covariance matrix of $\boldsymbol{v}_t$ as

$$\begin{aligned} \mathcal{E}\boldsymbol{v}_t\boldsymbol{v}_t' &= \sum_{i=1}^m p_i \boldsymbol{V}_i \\ &= \sum_{i=1}^m p_i(\boldsymbol{D_{p_i}} - \boldsymbol{p}_i\boldsymbol{p}_i') \\ &= \boldsymbol{D_p} - \sum_{i=1}^m p_i\boldsymbol{p}_i\boldsymbol{p}_i'. \end{aligned} \tag{19}$$

by (2). Note that

$$P'VP = P'D_pP - P'pp'P = \sum_{i=1}^{m} p_i p_i p_i' - pp'. \tag{20}$$

Thus

$$V = P'VP + \sum_{i=1}^{m} p_i V_i. \tag{21}$$

The second-order moments of $\{z_t\}$ are the second-order moments of a first-order autoregressive process with coefficient matrix P' and disturbance covariance matrix $\sum_{i=1}^{m} p_i V_i$. (See Anderson (1971), Sections 5.2 and 5.3, for example.) However, the conditional covariance matrix of v_t given z_{t-1} depends on z_{t-1}. This fact shows that v_t and z_{t-1} are dependent, though uncorrelated.

Let $\lambda_1 = 1, \lambda_2, \ldots, \lambda_m$ be the characteristic roots of P, $t_1 = \varepsilon, t_2, \ldots, t_m$ be the corresponding right-sided (column) characteristic vectors, and $q_1' = p', q_2' \ldots, q_m'$ be the corresponding left-sided (row) characteristic vectors. It is assumed that there are m linearly independent right- and left-sided vectors (that is, that the elementary divisors of P are simple). Suppose further that the chain is irreducible and periodic; then $|\lambda_i| < 1$, $i = 2, \ldots, m$. Let

$$\Lambda = \begin{pmatrix} 1 & 0 & \ldots & 0 \\ 0 & \lambda_2 & \ldots & 0 \\ \vdots & \vdots & & \vdots \\ 0 & 0 & \ldots & \lambda_m \end{pmatrix} = \begin{pmatrix} 1 & \mathbf{0} \\ \mathbf{0} & \Lambda_2 \end{pmatrix}, \tag{22}$$

$$Q = (p, q_2, \ldots, q_m) = (p, Q_2), \tag{23}$$

$$T = (\varepsilon, t_2, \ldots, t_m) = (\varepsilon, T_2). \tag{24}$$

Normalize the vectors so $T'Q = I$; that is, $T' = Q^{-1}$ and $Q' = T^{-1}$. Then

$$P = T\Lambda Q' = T\Lambda T^{-1}, \tag{25}$$

$$\boldsymbol{P}^s = \boldsymbol{T}\boldsymbol{\Lambda}^s\boldsymbol{Q}' = \boldsymbol{T}\boldsymbol{\Lambda}^s\boldsymbol{T}^{-1}. \tag{26}$$

Since $\boldsymbol{\varepsilon}'\boldsymbol{V} = \boldsymbol{\varepsilon}'(\boldsymbol{D_p} - \boldsymbol{pp}') = \boldsymbol{0}$,

$$\begin{aligned}(\boldsymbol{P}')^s\boldsymbol{V} &= \boldsymbol{Q}\boldsymbol{\Lambda}^s\boldsymbol{T}'\boldsymbol{V} \\ &= (\boldsymbol{p}, \boldsymbol{Q}_2)\begin{pmatrix} 1 & \boldsymbol{0} \\ \boldsymbol{0} & \boldsymbol{\Lambda}_2^s \end{pmatrix}\begin{pmatrix} \boldsymbol{\varepsilon}' \\ \boldsymbol{T}_2' \end{pmatrix}(\boldsymbol{D_p} - \boldsymbol{pp}') \\ &= \boldsymbol{Q}_2\boldsymbol{\Lambda}_2^s\boldsymbol{T}_2'\boldsymbol{D_p} \\ &= (\boldsymbol{P}' - \boldsymbol{p}\boldsymbol{\varepsilon}')^s\boldsymbol{D_p}, \quad s = 1, 2, \ldots .\end{aligned} \tag{27}$$

As s increases, the covariance function decreases as a linear combination of $\lambda_2^s, \ldots, \lambda_m^s$ since $|\lambda_i| < 1$, $i = 2, \ldots, m$.

We can write $\boldsymbol{z}_t$ in the moving average representation

$$\boldsymbol{z}_t = \sum_{s=0}^{\infty}(\boldsymbol{P}')^s\boldsymbol{v}_{t-s}. \tag{28}$$

Since $(\boldsymbol{P}')^s\boldsymbol{v}_{t-s} = \boldsymbol{Q}_2\boldsymbol{\Lambda}_2^s\boldsymbol{T}_2'\boldsymbol{v}_{t-s}$, $s = 1, 2, \ldots$, (28) converges. The representation (28) is trivial, however, because

$$\begin{aligned}(\boldsymbol{P}')^s\boldsymbol{v}_{t-s} &= (\boldsymbol{P}')^s(\boldsymbol{z}_{t-s} - \boldsymbol{P}'\boldsymbol{z}_{t-s-1}) \\ &= (\boldsymbol{P}')^s\boldsymbol{z}_{t-s} - (\boldsymbol{P}')^{s+1}\boldsymbol{z}_{t-(s+1)}.\end{aligned} \tag{29}$$

The spectral density of $\{\boldsymbol{z}_t\}$ for $\lambda \neq 0$ (Hannan (1970), p. 67, for example) is $1/(2\pi)$ times

$$\begin{aligned}&(\boldsymbol{I} - \boldsymbol{P}'e^{i\lambda})^{-1}\boldsymbol{V}(\boldsymbol{I} - \boldsymbol{P}e^{-i\lambda})^{-1} \\ &\quad = \left[\boldsymbol{Q}(\boldsymbol{I} - \boldsymbol{\Lambda}e^{i\lambda})\boldsymbol{T}'\right]^{-1}\boldsymbol{V}\left[\boldsymbol{T}(\boldsymbol{I} - \boldsymbol{\Lambda}e^{-i\lambda})\boldsymbol{Q}'\right]^{-1} \\ &\quad = (\boldsymbol{T}')^{-1}(\boldsymbol{I} - \boldsymbol{\Lambda}e^{i\lambda})^{-1}\boldsymbol{Q}^{-1}\boldsymbol{V}(\boldsymbol{Q}')^{-1}(\boldsymbol{I} - \boldsymbol{\Lambda}e^{-i\lambda})^{-1}\boldsymbol{T}^{-1} \\ &\quad = \boldsymbol{Q}(\boldsymbol{I} - \boldsymbol{\Lambda}e^{i\lambda})^{-1}\boldsymbol{T}'\boldsymbol{V}\boldsymbol{T}(\boldsymbol{I} - \boldsymbol{\Lambda}e^{-i\lambda})^{-1}\boldsymbol{Q}' \\ &\quad = \boldsymbol{Q}_2(\boldsymbol{I} - \boldsymbol{\Lambda}_2e^{i\lambda})^{-1}\boldsymbol{T}_2'\boldsymbol{D_p}\boldsymbol{T}_2(\boldsymbol{I} - \boldsymbol{\Lambda}_2e^{-i\lambda})^{-1}\boldsymbol{Q}_2'.\end{aligned} \tag{30}$$

Since $|\lambda_i| < 1$, $i = 2, \ldots, m$, $\boldsymbol{I} - \boldsymbol{\Lambda}_2e^{i\lambda}$ and $\boldsymbol{I} - \boldsymbol{\Lambda}_2e^{-i\lambda}$ are nonsingular for all real λ and (30) is the spectral density for all real λ.

4. ESTIMATION OF THE STATIONARY PROBABILITIES

Consider a sequence of observations on the chain, $x_1, \ldots, x_N$. These define a sequence $\boldsymbol{z}_1, \ldots, \boldsymbol{z}_N$ from the process $\{\boldsymbol{z}_t\}$. Let

$$\boldsymbol{S} = \sum_{t=1}^{N} \boldsymbol{z}_t. \tag{31}$$

Then $\mathcal{E}\boldsymbol{S} = N\boldsymbol{p}$. The covariance matrix of $\boldsymbol{S}$ is

$$\begin{aligned}
(32)\quad & \mathcal{E}(\boldsymbol{S} - N\boldsymbol{p})(\boldsymbol{S} - N\boldsymbol{p})' \\
&= \sum_{t,s=1}^{N} \mathcal{E}(\boldsymbol{z}_t - \boldsymbol{p})(\boldsymbol{z}_s - \boldsymbol{p})' \\
&= \sum_{t=1}^{N}\sum_{s=1}^{t-1} \mathcal{E}(\boldsymbol{z}_t - \boldsymbol{p})(\boldsymbol{z}_s - \boldsymbol{p})' + \sum_{s=1}^{N}\sum_{t=1}^{s-1} \mathcal{E}(\boldsymbol{z}_t - \boldsymbol{p})(\boldsymbol{z}_s - \boldsymbol{p})' \\
&\quad + \sum_{t=1}^{N} \mathcal{E}(\boldsymbol{z}_t - \boldsymbol{p})(\boldsymbol{z}_t - \boldsymbol{p})' \\
&= \sum_{t=1}^{N}\sum_{r=1}^{t-1} \left[\mathcal{E}(\boldsymbol{z}_t - \boldsymbol{p})(\boldsymbol{z}_{t-r} - \boldsymbol{p})' + \mathcal{E}(\boldsymbol{z}_{t-r} - \boldsymbol{p})(\boldsymbol{z}_t - \boldsymbol{p})'\right] + N\boldsymbol{V} \\
&= \sum_{t=1}^{N}\sum_{r=1}^{t-1} \left[(\boldsymbol{P}')^r \boldsymbol{V} + \boldsymbol{V}\boldsymbol{P}^r\right] + N\boldsymbol{V} \\
&= \sum_{t=1}^{N}\sum_{r=1}^{t-1} \left[\boldsymbol{Q}_2\boldsymbol{\Lambda}_2^r\boldsymbol{T}_2'\boldsymbol{D}_{\boldsymbol{p}} + \boldsymbol{D}_{\boldsymbol{p}}\boldsymbol{T}_2\boldsymbol{\Lambda}_2^r\boldsymbol{Q}_2'\right] + N\boldsymbol{V} \\
&= \sum_{t=1}^{N} \left[\boldsymbol{Q}_2(\boldsymbol{I} - \boldsymbol{\Lambda}_2)^{-1}\boldsymbol{\Lambda}_2(\boldsymbol{I} - \boldsymbol{\Lambda}_2^{t-1})\boldsymbol{T}_2'\boldsymbol{D}_{\boldsymbol{p}}\right. \\
&\quad \left. + \boldsymbol{D}_{\boldsymbol{p}}\boldsymbol{T}_2(\boldsymbol{I} - \boldsymbol{\Lambda}_2^{t-1})\boldsymbol{\Lambda}_2(\boldsymbol{I} - \boldsymbol{\Lambda}_2)^{-1}\boldsymbol{Q}_2'\right] + N\boldsymbol{V} \\
&= N\left[\boldsymbol{Q}_2(\boldsymbol{I} - \boldsymbol{\Lambda}_2)^{-1}\boldsymbol{\Lambda}_2\boldsymbol{T}_2'\boldsymbol{D}_{\boldsymbol{p}} + \boldsymbol{D}_{\boldsymbol{p}}\boldsymbol{T}_2\boldsymbol{\Lambda}_2(\boldsymbol{I} - \boldsymbol{\Lambda}_2)^{-1}\boldsymbol{Q}_2'\right] \\
&\quad - \boldsymbol{Q}_2(\boldsymbol{I} - \boldsymbol{\Lambda}_2)^{-2}\boldsymbol{\Lambda}_2(\boldsymbol{I} - \boldsymbol{\Lambda}_2^N)\boldsymbol{T}_2'\boldsymbol{D}_{\boldsymbol{p}} \\
&\quad - \boldsymbol{D}_{\boldsymbol{p}}\boldsymbol{T}_2(\boldsymbol{I} - \boldsymbol{\Lambda}_2^N)\boldsymbol{\Lambda}_2(\boldsymbol{I} - \boldsymbol{\Lambda}_2)^{-2}\boldsymbol{Q}_2' + N\boldsymbol{V}.
\end{aligned}$$

Then the covariance matrix of

$$\sqrt{N}\bar{\boldsymbol{z}} = \frac{1}{\sqrt{N}}\boldsymbol{S} \tag{33}$$

is

$$(34) \qquad Q_2(I - \Lambda_2)^{-1}T_2'D_p + D_pT_2(I - \Lambda_2)^{-1}Q_2' - D_p + pp'$$
$$-\frac{1}{N}\left[Q_2(I - \Lambda_2)^{-2}\Lambda_2(I - \Lambda_2^N)T_2'D_p + D_pT_2(I - \Lambda_2^N)\Lambda_2(I - \Lambda_2)^{-2}Q_2'\right].$$

THEOREM.

$$(35) \qquad \sqrt{N}(\bar{z} - p) \xrightarrow{\mathcal{L}} N\left[0, Q_2(I - \Lambda_2)^{-1}T_2'D_p + D_pT_2(I - \Lambda_2)^{-1}Q_2' - D_p + pp'\right].$$

The sum S is the vector of frequencies of the states $1, \ldots, m$ being observed, and $\bar{z}$ is the vector of relative frequencies. Since $\mathcal{E}\bar{z} = p$, the vector $\bar{z}$ is an estimator of p. Grenander (1954), Rosenblatt (1956), and Grenander and Rosenblatt (1957) showed that in the case of a scalar process with expected value a linear function of exogenous series and a stationary covariance function the least squares estimator of this linear function is asymptotically efficient among all linear estimators. (See also Anderson (1971), Section 10.2.) The result holds as well for vector processes. In particular, the mean of a set of observations is an asymptotically efficient linear estimator of the constant mean of a wide-sense stationary process, as is the case here.

An alternative method of estimating p from the data $z_1, \ldots, z_N$ is to estimate P and find its left-sided characteristic vector corresponding to the characteristic root 1. If z_1 is given (that is, fixed), the maximum likelihood estimator of P is

$$(36) \qquad \hat{P} = \left(\sum_{t=2}^{N} z_{t-1}z_{t-1}'\right)^{-1} \sum_{t=2}^{N} z_{t-1}z_t'.$$

(See Anderson and Goodman (1957).) Note that

$$(37) \qquad \sum_{t=2}^{N} z_{t-1}z_{t-1}' = \begin{pmatrix} \sum_{t=1}^{N-1} z_{1t} & 0 & \cdots & 0 \\ 0 & \sum_{t=1}^{N-1} z_{2t} & \cdots & 0 \\ \vdots & \vdots & & \vdots \\ 0 & 0 & \cdots & \sum_{t=1}^{N-1} z_{mt} \end{pmatrix}$$

is diagonal. In fact, the diagonal elements of (37) are the components of $S - z_N$. If there is no absorbing state, every component of p is positive and the probability that (37) is nonsingular approaches 1 as $N \to \infty$. Hence $\hat{\boldsymbol{P}}$ is well-defined. Since $\hat{\boldsymbol{P}}$ is a consistent estimator of $\boldsymbol{P}$, as $N \to \infty$ the probability approaches 1 that $\boldsymbol{\varepsilon}'\hat{\boldsymbol{p}} = 1$ and

$$\hat{\boldsymbol{p}}'\hat{\boldsymbol{P}} = \hat{\boldsymbol{p}} \tag{38}$$

have a unique solution for $\hat{\boldsymbol{p}}$.

We observe that

$$\begin{aligned}(\bar{\boldsymbol{z}} - \frac{1}{N}\boldsymbol{z}_N)'\hat{\boldsymbol{P}} &= \frac{1}{N}\boldsymbol{\varepsilon}'\sum_{t=2}^{N}\boldsymbol{z}_{t-1}\boldsymbol{z}_t' \\ &= \frac{1}{N}\sum_{t=2}^{N}\boldsymbol{z}_t' \\ &= (\bar{\boldsymbol{z}} - \frac{1}{N}\boldsymbol{z}_1)'.\end{aligned} \tag{39}$$

From (38) and (39) we obtain for arbitrary a

$$\begin{aligned}N^{a-1}(\boldsymbol{z}_N'\hat{\boldsymbol{P}} - \boldsymbol{z}_1') &= N^a(\bar{\boldsymbol{z}} - \hat{\boldsymbol{p}})'(\hat{\boldsymbol{P}} - \boldsymbol{I}) \\ &= N^a(\bar{\boldsymbol{z}} - \hat{\boldsymbol{p}})'\hat{\boldsymbol{T}}_2(\hat{\boldsymbol{\Lambda}}_2 - \boldsymbol{I})\hat{\boldsymbol{Q}}_2',\end{aligned} \tag{40}$$

where

$$\hat{\boldsymbol{P}} = (\boldsymbol{\varepsilon}, \hat{\boldsymbol{T}}_2)\begin{pmatrix} 1 & \boldsymbol{0} \\ \boldsymbol{0} & \hat{\boldsymbol{\Lambda}}_2 \end{pmatrix}\begin{pmatrix} \hat{\boldsymbol{p}}' \\ \hat{\boldsymbol{Q}}_2' \end{pmatrix}. \tag{41}$$

Since the left-hand side of (40) approaches $\boldsymbol{0}$ for $a < 1$ as $N \to \infty$ and $(\bar{\boldsymbol{z}} - \hat{\boldsymbol{p}})'\boldsymbol{\varepsilon} = 0$, we deduce that

$$N^a(\bar{\boldsymbol{z}} - \hat{\boldsymbol{p}}) \xrightarrow{\text{p}} \boldsymbol{0}. \tag{42}$$

The two estimators of $\boldsymbol{p}$ are asymptotically equivalent in the sense that $\bar{\boldsymbol{z}} - \hat{\boldsymbol{p}} = o_p(N^{-a})$. See, also, Henry (1970).

In many situations a sample $x_1, \ldots, x_N$ (or equivalently, $\boldsymbol{z}_1, \ldots, \boldsymbol{z}_N$) is not drawn from a stationary process, but from a process starting with some given state. In that case one might discard enough initial observations to ensure that over the remaining period of observation the process is stationary or almost stationary. Thus the use of $\bar{\boldsymbol{z}}$ calculated from the remaining observations as an estimator of $\boldsymbol{p}$ would waste the initial observations. (If one wanted to run a simulation and insisted on *independent* observations, one would start the process with a possibly random state, let it run until stationarity is achieved, and make one observation. Then one would repeat the procedure. Clearly this is expensive in computational resources and is unnecessary.)

To estimate $\boldsymbol{P}$ and then $\boldsymbol{p}$ does not require stationarity; one can start from any state. The estimator of $\boldsymbol{P}$ is a maximum likelihood estimator, and hence the estimator of $\boldsymbol{p}$ (and $\boldsymbol{\Lambda}_2$, $\boldsymbol{W}_2$, and $\boldsymbol{T}_2$) is maximum likelihood and asymptotically efficient. (Several different sequences from the same chain can be aggregated; see Anderson and Goodman (1957).)

The investigator will typically want to estimate the asymptotic covariance matrix of the estimator, given in (35), which involves

$$\boldsymbol{D_p T}_2(\boldsymbol{I}-\boldsymbol{\Lambda}_2)^{-1}\boldsymbol{Q}_2' = \boldsymbol{D_p}\sum_{j=2}^{m}\boldsymbol{t}_j(1-\lambda_j)^{-1}\boldsymbol{q}_j' = \boldsymbol{D_p}\sum_{j=2}^{m}\sum_{s=0}^{\infty}\lambda_j^s\boldsymbol{t}_j\boldsymbol{q}_j'. \tag{43}$$

The rate of convergence is governed by the root that is largest in absolute value. The asymptotic covariance matrix can also be written as

$$\left[\boldsymbol{D_p}\sum_{s=1}^{\infty}(\boldsymbol{P}-\boldsymbol{\varepsilon p}')^s\right]' + \boldsymbol{D_p}\sum_{s=1}^{\infty}(\boldsymbol{P}-\boldsymbol{\varepsilon p}')^s + \boldsymbol{D_p} - \boldsymbol{pp}'. \tag{44}$$

Since

$$(\boldsymbol{P}-\boldsymbol{\varepsilon p}')^s = \boldsymbol{P}^s - \boldsymbol{\varepsilon p}', \quad s = 1, 2, \ldots, \tag{45}$$

the covariance matrix (44) is

$$\left[\boldsymbol{D_p}\sum_{s=1}^{\infty}(\boldsymbol{P}^s-\boldsymbol{\varepsilon p}')\right]' + \boldsymbol{D_p}\sum_{s=1}^{\infty}(\boldsymbol{P}^s-\boldsymbol{\varepsilon p}') + \boldsymbol{D_p} - \boldsymbol{pp}'. \tag{46}$$

The infinite sum can be approximated by a finite sum since the sum is convergent. For an estimator $\boldsymbol{P}$ and $\boldsymbol{p}$ are replaced by $\hat{\boldsymbol{P}}$ and $\hat{\boldsymbol{p}}$, respectively.

A possible computational method is to power $\boldsymbol{P}$ until the rows of $\boldsymbol{P}^s$ are similar enough, that is, until $\boldsymbol{P}^s$ is similar to $\boldsymbol{\varepsilon p}'$. Then (46) follows. Note that at each step only $\boldsymbol{P}^s$ and $\sum_{s=0}^{r} \boldsymbol{P}^s$ need be held in memory.

As a way of simulating a probability distribution of a finite set of outcomes, Persi Diaconis (personal communication) has suggested setting up a Markov chain with this probability distribution as the stationary probability distribution. An example of particular interest is simulating the uniform distribution of matrices with nonnegative integer entries and fixed column and row sums (contingency tables). Diaconis and Efron (1986) have discussed the analysis of two–way tables based on this model.

Diaconis sets up the following procedure to generate a Markov chain in which all possible tables (with assigned marginals) are the states of the chain. At each step select a pair of rows (g, h) and columns (i, j) at random; with probability $\frac{1}{2}$ add 1 to the g, i–th cell and to the h, j–th cell and subtract 1 from the g, j–th cell and from the h, i–th cell. If a step would lead to a negative entry, cancel it. Each step leaves the row and column sums as given and so determines the transition probabilities of a Markov chain with the possible tables as states. Since the matrix of transition probabilities is doubly stochastic (the probability of going from one state to another is the same as the probability of the reverse), the stationary probabilities are uniform; that is, $\boldsymbol{p}$ is proportional to $\boldsymbol{\varepsilon}$.

Diaconis and Efron were interested in the distribution of the χ^2 goodness–of–fit statistic, say $\boldsymbol{T}(\boldsymbol{z})$, where $\boldsymbol{z}$ is the vector representation of the two–way tables; that is, they want $\Pr\{T(\boldsymbol{Z}) \leq \tau\}$ for arbitrary $\tau \in [0, \infty)$. Given $p_i = \Pr\{\boldsymbol{Z} = \boldsymbol{\varepsilon}_i\}$ the probability can be calculated

$$\Pr\{T(\boldsymbol{Z}) \leq \tau\} = \sum_{i=1}^{m} p_i I\big[T(\varepsilon_i) \leq \tau\big], \tag{47}$$

where $I(\cdot)$ is the indicator function; that is, $I\big[T(\varepsilon_i) \leq \tau\big] = 1$ if $T(\varepsilon_i) \leq \tau$, and $= 0$ if $T(\varepsilon_i) > \tau$. Given a sample $\boldsymbol{z}_1, \ldots, \boldsymbol{z}_N$, the probability (47) is estimated by

$$\frac{1}{N} \sum_{t=1}^{N} I\big[T(\boldsymbol{z}_t) \leq \tau\big]. \tag{48}$$

The estimator of the cdf of $T(\boldsymbol{Z})$ is the empirical cdf of $T(\boldsymbol{Z})$ although $\boldsymbol{z}_1, \ldots, \boldsymbol{z}_N$ are not independent. However, the sample variance of $I\big[T(\boldsymbol{z}_t) \leq$

$\tau]$ divided by N is not an estimator of the variance of (48). If a subset of $\boldsymbol{z}_1, \ldots, \boldsymbol{z}_N$ is taken, say, $\boldsymbol{z}_{t_1}, \ldots, \boldsymbol{z}_{t_n}$, so that $\min |t_i - t_j|$ is great enough that $\boldsymbol{z}_{t_1}, \ldots, \boldsymbol{z}_{tn}$ can be considered independent, the sample variance of $I[T(\boldsymbol{z}_{t_i}) \leq \tau]$ provides an estimator of a lower bound to the variance of (48).

If the specified row and column totals are such that each possible value of T can come from only one table, there is a $1-1$ correspondence between the statistic and the table. That is, the value of the statistic is simply another label for the state. Then the generation of the statistic is given by the Markov chain, and its empirical cdf is an asymptotically efficient estimator of the distribution of T. These facts suggest that even if there is not a $1-1$ correspondence between the statistic and the table, the empirical cdf is an asymptotically efficient estimator.

5. SCORING A MARKOV CHAIN

Suppose each state is assigned a numerical value α_i, $i = 1, \ldots, m$. Let $y_t = \alpha_i$ if and only if x_t is in state i. For example, the state of a queue may be indicated by the number of customers waiting. In that case it is convenient to label the states $0, 1, \ldots, N$ and set $\alpha_i = i$; here N is the maximum number of customers who can be waiting. (In many other cases the index of the state may have no numerical meaning.)

The process $\{y_t\}$ can also be defined by

$$y_t = \boldsymbol{\alpha}' \boldsymbol{z}_t, \tag{49}$$

where $\boldsymbol{\alpha}' = (\alpha_1, \ldots, \alpha_m)$. The second–order moments of $\{y_t\}$ can be found from those of $\{\boldsymbol{z}_t\}$. The mean is

$$\mathcal{E} y_t = \boldsymbol{\alpha}' \boldsymbol{p} = \mu; \tag{50}$$

say; the variance is

$$\begin{aligned} \operatorname{Var}(y_t) &= \boldsymbol{\alpha}' \boldsymbol{V} \boldsymbol{\alpha} = \boldsymbol{\alpha}' \boldsymbol{D}_{\boldsymbol{p}} \boldsymbol{\alpha} - (\boldsymbol{\alpha}' \boldsymbol{p})^2 \\ &= \boldsymbol{\alpha}' \boldsymbol{Q} \boldsymbol{T}' \boldsymbol{D}_{\boldsymbol{p}} \boldsymbol{\alpha} - (\boldsymbol{\alpha}' \boldsymbol{p})^2 \\ &= \sum_{i=2}^{m} (\boldsymbol{\alpha}' \boldsymbol{q}_i)(\boldsymbol{\alpha}' \boldsymbol{D}_{\boldsymbol{p}} \boldsymbol{t}_i); \end{aligned} \tag{51}$$

and the covariances are given by

$$
\begin{aligned}
\text{Cov}(y_t, y_{t-s}) &= \boldsymbol{\alpha}' \text{Cov}(\boldsymbol{z}_t, \boldsymbol{z}_{t-s})\boldsymbol{\alpha} \\
&= \boldsymbol{\alpha}'(\boldsymbol{P}')^s \boldsymbol{V}\boldsymbol{\alpha} \\
&= \boldsymbol{\alpha}'\boldsymbol{Q}_2 \boldsymbol{\Lambda}_2^s \boldsymbol{T}_2' \boldsymbol{D_p}\boldsymbol{\alpha} \\
&= \sum_{i=2}^{m} (\boldsymbol{\alpha}'\boldsymbol{q}_i)(\boldsymbol{\alpha}'\boldsymbol{D_p}\boldsymbol{t}_i)\lambda_i^s, \quad s = 1, 2, \ldots .
\end{aligned} \tag{52}
$$

A similar result was derived by Reynolds (1972) in a different way.*

The nature of the process $\{y_t\}$ can be found from the nature of the $\{\boldsymbol{z}_t\}$ process. Let $\mathcal{L}$ be the lag operator; that is, $\mathcal{L}\boldsymbol{z}_t = \boldsymbol{z}_{t-1}$. Then the $\{\boldsymbol{z}_t\}$ process can be written in autoregressive form as

$$(\boldsymbol{I} - \boldsymbol{P}'\mathcal{L})\boldsymbol{z}_t = \boldsymbol{v}_t. \tag{53}$$

Since $\boldsymbol{P} = \boldsymbol{T\Lambda Q}'$ and $\boldsymbol{T}'\boldsymbol{Q} = \boldsymbol{I}$, multiplication of (53) on the left by $\boldsymbol{T}'$ gives

$$(\boldsymbol{I} - \boldsymbol{\Lambda}\mathcal{L})\boldsymbol{w}_t = \boldsymbol{u}_t, \tag{54}$$

where

$$\boldsymbol{w}_t = \boldsymbol{T}'\boldsymbol{z}_t = \begin{pmatrix} \boldsymbol{\varepsilon}' \\ \boldsymbol{T}_2' \end{pmatrix} \boldsymbol{z}_t = \begin{pmatrix} 1 \\ \boldsymbol{w}_t^{(2)} \end{pmatrix} \tag{55}$$

and

$$\boldsymbol{u}_t = \boldsymbol{T}'\boldsymbol{v}_t = \begin{pmatrix} \boldsymbol{\varepsilon}' \\ \boldsymbol{T}_2' \end{pmatrix} \boldsymbol{v}_t = \begin{pmatrix} 0 \\ \boldsymbol{u}_t^{(2)} \end{pmatrix}. \tag{56}$$

We can re-write the last $m-1$ components of (54) as

$$(\boldsymbol{I} - \boldsymbol{\Lambda}_2\mathcal{L})\boldsymbol{w}_t^{(2)} = \boldsymbol{u}_t^{(2)}. \tag{57}$$

The first component of (54) is $(1 - \mathcal{L})1 = 0$. Then

$$\boldsymbol{w}_t^{(2)} = (\boldsymbol{I} - \boldsymbol{\Lambda}_2\mathcal{L})^{-1}\boldsymbol{u}_t^{(2)}. \tag{58}$$

* I am indebted to Jayaram Muthuswamy for calling my attention to this problem and associated literature.

Multiplication of (55) on the left by $\boldsymbol{Q}$ with (58) and (56) yields

$$(59)\quad y_t = \boldsymbol{\alpha}'\boldsymbol{z}_t = \boldsymbol{\alpha}'(\boldsymbol{p}, \boldsymbol{Q}_2)\begin{pmatrix} 1 \\ \boldsymbol{w}_t^{(2)} \end{pmatrix} = \boldsymbol{\alpha}'\boldsymbol{p} + \boldsymbol{\alpha}'\boldsymbol{Q}_2(\boldsymbol{I} - \boldsymbol{\Lambda}_2\mathcal{L})^{-1}\boldsymbol{u}_t^{(2)}$$
$$= \boldsymbol{\alpha}'\boldsymbol{p} + \boldsymbol{\alpha}'\boldsymbol{Q}_2(\boldsymbol{I} - \boldsymbol{\Lambda}_2\mathcal{L})^{-1}\boldsymbol{T}_2'\boldsymbol{v}_t.$$

Multiplication of (59) by $|\boldsymbol{I} - \boldsymbol{\Lambda}_2\mathcal{L}|$ yields

$$(60)\qquad \left[\prod_{j=2}^{m}(1 - \lambda_j\mathcal{L})\right](y_t - \mu) = \sum_{j=2}^{m}(\boldsymbol{\alpha}'\boldsymbol{q}_j)\left[\prod_{h\neq 1,j}(1 - \lambda_h\mathcal{L})\right]\boldsymbol{t}_j'\boldsymbol{v}_t.$$

This equation defines an autoregressive moving average process (in the wide sense) with an autoregressive part of order $m-1$ and a moving average part of order at most $m-2$. Note that if $\alpha_j = 1$ for some j and $\alpha_i = 0$ for $i \neq j$, then (60) defines the marginal process of z_{jt}.

The above development is in terms of real roots and vectors. If a pair of roots are complex conjugate, the corresponding pairs of vectors are complex conjugate and the analysis goes through as before.

The covariance function (52) will be the covariance function of a first–order autoregression if there is only one term in the sum on the right–hand side. Let $\boldsymbol{Q}_2 = (\boldsymbol{q}_2, \boldsymbol{Q}_3)$, $\boldsymbol{T}_2 = (\boldsymbol{t}_2, \boldsymbol{T}_3)$, and

$$(61)\qquad \boldsymbol{\Lambda}_2 = \begin{pmatrix} \lambda_2 & \mathbf{0} \\ \mathbf{0} & \boldsymbol{\Lambda}_3 \end{pmatrix}$$

for a real root λ_2. Then

$$(62)\qquad \operatorname{Cov}(y_t, y_{t-s}) = (\boldsymbol{\alpha}'\boldsymbol{q}_2)(\boldsymbol{\alpha}'\boldsymbol{D_p}\boldsymbol{t}_2)\lambda_2^s + \boldsymbol{\alpha}'\boldsymbol{Q}_3\boldsymbol{\Lambda}_3^s\boldsymbol{T}_3'\boldsymbol{D_p}\boldsymbol{\alpha}.$$

This is the covariance function of a first–order autoregressive process if λ_2 is real and either $\boldsymbol{\alpha}'\boldsymbol{Q}_3 = \mathbf{0}$ or if $\boldsymbol{\alpha}'\boldsymbol{D_p}\boldsymbol{T}_3 = \mathbf{0}$. This fact was given by Lai (1978) in his Theorem 2.3 under the assumption that all of the characteristic roots are real and distinct.

The alternative conditions may be written as

$$(63)\qquad \boldsymbol{\alpha}'\boldsymbol{Q} = (\mu, \boldsymbol{\alpha}'\boldsymbol{q}_2, 0)$$

and

$$(64)\qquad \boldsymbol{\alpha}'\boldsymbol{D_p}\boldsymbol{T} = (\mu, \boldsymbol{\alpha}'\boldsymbol{D_p}\boldsymbol{t}_2, 0).$$

Multiply (63) on the right by $\boldsymbol{Q}^{-1} = \boldsymbol{T}'$ and (64) by $\boldsymbol{T}^{-1}\boldsymbol{D}_p^{-1} = \boldsymbol{Q}'\boldsymbol{D}_{\boldsymbol{p}}^{-1}$ to obtain

$$\boldsymbol{\alpha}' = \mu\boldsymbol{\varepsilon}' + (\boldsymbol{\alpha}'\boldsymbol{q}_2)\boldsymbol{t}_2' \tag{65}$$

and

$$\boldsymbol{\alpha}' = \mu\boldsymbol{\varepsilon}' + (\boldsymbol{\alpha}'\boldsymbol{D}_{\boldsymbol{p}}\boldsymbol{t}_2)\boldsymbol{q}_2'\boldsymbol{D}_{\boldsymbol{p}}^{-1}. \tag{66}$$

The conclusion is that $\{y_t\}$ is a first-order autoregressive process (in the wide sense) if the Markov chain is irreducible and aperiodic, if there are m linearly independent characteristic vectors, and if the vector $\boldsymbol{\alpha}$ is a linear combination of $\boldsymbol{\varepsilon}$ and of either a right-sided characteristic vector of $\boldsymbol{P}$ corresponding to a real root (other than 1) or a left-sided vector corresponding to a real root multiplied by $\boldsymbol{D}_{\boldsymbol{p}}^{-1}$.

REFERENCES

Anderson, T. W. (1971), *The Statistical Analysis of Time Series*, John Wiley and Sons, Inc., New York.

Anderson, T. W. (1979a), Panels and time series analysis: Markov chains and autoregressive processes, *Qualitative and Quantitative Social Research: Papers in Honor of Paul F. Lazarsfeld* (Robert K. Merton, James S. Coleman, and Peter H. Rossi, eds.), The Free Press, New York, 82-97.

Anderson, T. W. (1979b), Some relations between Markov chains and vector autoregressive processes, *International Statistical Institute: Contributed Papers, 42nd Session, December 4-14, 1979*, Manila, 25-28.

Anderson, T. W. (1980), Finite-state Markov chains and vector autoregressive processes, *Proceedings of the Conference on Recent Developments in Statistical Methods and Applications*, Directorate-General of Budget, Accounting and Statistics, Executive Yuan, Taipei, Taiwan, Republic of China, 1-12.

Anderson, T. W., and Leo A. Goodman (1957), Statistical inference about Markov chains, *Annals of Mathematical Statistics*, **28**, 89-110.

Diaconis, Persi, and Bradley Efron (1985), Testing for independence in a two-way table: New interpretations of the chi-square statistic, *Annals of Statistics*, **13**, 845-913.

Grenander, Ulf (1954), On the estimation of regression coefficients in the case of an autocorrelated disturbance, *Annals of Mathematical Statistics,* **25**, 252-272.

Grenander, Ulf, and Murray Rosenblatt (1957), *Statistical Analysis of Stationary Time Series*, John Wiley and Sons, Inc., New York.

Hannan, E. J. (1970), *Multiple Time Series*, John Wiley and Sons, Inc., New York.

Henry, Neil Wylie (1970), *Problems in the Statistical Analysis of Markov Chains*, Doctoral Dissertation, Columbia University Libraries.

Lai, C. D. (1978), First order autoregressive Markov processes, *Stochastic Processes and their Applications,* **7**, 65-72

Reynolds, John F. (1972), Some theorems on the transient covariance of Markov chains, *Journal of Applied Probability*, **9**, 214-218.

Rosenblatt, Murray (1956), Some regression problems in time series analysis, *Proceedings of the Third Berkeley Symposium in Mathematical Statistics and Probability*, University of California Press, Berkeley and Los Angeles, **1**, 165-186.

A "Dynamic" Proof of the Frobenius–Perron Theorem for Metzler Matrices

Kenneth J. Arrow
Stanford University

Matrices with non-negative off-diagonal elements have many applications in mathematical economics and other fields of investigation. Economists have called them "Metzler matrices" because of their study by L. Metzler (1945). An important property, especially for the study of stability of dynamic systems, is that the largest real part of the characteristic roots is itself a characteristic root and has a semi-positive characteristic vector.

There is a less well-known property of linear dynamic systems governed by Metzler matrices: if the forcing term is a non-negative vector and if the system starts in the positive orthant, it will remain there forever. The proposition seems to have been first proved by Samuel Karlin, though never published by him; his proof is referred to by Beckenbach and Bellman (1961), p. 137.

Karlin's result does not appear to be derivable from the standard Frobenius-Perron theorem (Theorem 4 below). Its proof is not very hard, however. The question is then raised, whether the Frobenius-Perron result is

AMS 1980 Subject Classification: Primary 15A48.

Key words and phrases: Metzler matrices, non–negative matrices, dominant roots, stability.

Copyright © 1989 by Academic Press, Inc.
All rights of reproduction in any form reserved.
ISBN-0-12-058470-0

derivable simply from Karlin's theorem. This note shows that the answer is affirmative. The result may very possibly be useful for expository purposes.

I start by demonstrating Karlin's theorem, for completeness. I then derive several familiar properties of Metzler matrices from which the Frobenius-Perron theorem can be derived.

DEF. 1: *A is a* Metzler matrix *if $a_{ij} \geq 0$ for all $i \neq j$.*

There are several ways of proving Karlin's theorem. In the following, I use the concept of the exponential of a matrix (for a more elementary but lengthier proof, see Arrow (1960)).

DEF. 2:

$$\exp A = \Sigma_{n=0}^{\infty} A^n / n!.$$

The infinite series converges absolutely for any matrix A.

If A and B are matrices that commute (i.e., $AB = BA$), then it is easy to see that the binomial theorem is valid for $(A + B)^n$. If we substitute in the infinite series defining exp $(A + B)$ and rearrange terms, it is easy to see that

(1) if A and B commute, then $\exp\,(A + B) = (\exp\ A)(\exp\ B)$.

If A is non-negative, then of course $A^n \geq 0$ for all n. From Def. 2,

(2) $\exp\ A \geq I$ if $A \geq 0$.

Now suppose A is a Metzler matrix. By Def. 1, we can find a scalar, s such that $A - sI \geq 0$. Clearly, sI commutes with any matrix and in particular with $A - sI$. Then,

$$\exp\ A = \exp\ [sI + (A - sI)] = \exp\ (sI) \exp\ (A - sI).$$

But, from Def. 2, $\exp\ (sI) = e^s I$, since $I^n = I$ for all n.

(3) $\exp\ A = p \exp\ (A - sI)$ for some scalar $p > 0$.

From (2), with A replaced by $A - sI$, and (3),

(4) if A is Metzler, exp $A \geq pI$ for some $p > 0$.

Therefore,

If A is Metzler, $(\exp\ A)x \geq px \gg 0$ if $x \gg 0$, and $(\exp\ A)x \geq px \geq 0$ if $x \geq 0$.

(I use the notations $x \geq 0$, $x > 0$, $x \gg 0$, to mean, respectively, $x_i \geq 0$ for all i, $x \geq 0$ and $x_i > 0$ for some i, and $x_i > 0$ for all i.)

LEMMA 1. *If A is Metzler,* $(\exp\ A)x \gg 0$ *if* $x \gg 0, (\exp\ A)x \geq 0$ *if* $x \geq 0$.

Let $x(t)$ be the solution of the differential equation,

$$\dot{x} = Ax + b,$$

with given initial condition $x(0)$. Then $x(t)$ can be written,

$$x(t) = (\exp\ At)x(0) + \int_0^t [\exp\ (Au)b]du.$$

Suppose A Metzler, $b \geq 0$, and $x(0) >> 0$. From Def. 1, At is Metzler for $t \geq 0$. Then, from Lemma 1,

THEOREM 1 (KARLIN). *If A is Metzler and $b \geq 0$ then any solution of the differential equation,*

$$\dot{x} = Ax + b,$$

for which $x(0) \gg 0$ has the property that $x(t) \gg 0$ for all $t \geq 0$.

Now suppose that A is a *stable* Metzler matrix, so that the real parts of all characteristic roots are negative. Take any vector y for which $Ay \leq 0$. Take any solution to the differential equation,

$$\dot{x} = A(x - y),$$

for which $x(0) \gg 0$. Since A is stable, $\lim_{t \to \infty} x(t) = y$. Since $-Ay \geq 0$ and A is Metzler, Theorem 1 implies that $x(t) \gg 0$ for all $t \geq 0$. Therefore, $y \geq 0$.

THEOREM 2. *If A is a stable Metzler matrix and $Ay \leq 0$, then $y \geq 0$.*

Suppose $Ay < 0$. Then clearly $y = 0$ is impossible.

COROLLARY 1. *If A is a stable Metzler matrix and $Ay < 0$, then $y > 0$.*

Now a sort of converse of Theorem 2 can be shown. Suppose A is a Metzler matrix and $Ay \ll 0$ for some $y > 0$. Clearly, by perturbing y, it can be assumed that,

(5) $Ay \ll 0, y \gg 0.$

Let the superscript T denote "transpose." Let $x(t)$ satisfy,

(6) $\dot{x} = A^T x, x(0) \gg 0.$

Then, by Theorem 1, $x(t) \gg 0$ for $t \geq 0$. Define,

(7) $\xi(t) = y^T x(t).$

Since $y \gg 0$ and $x(t) \gg 0$,

(8) $\xi(t) > 0$ for $t \geq 0$.

From (5),

(9) $Ay \leq my$ for some $m < 0$.

From (7) and (6),

$$\dot{\xi} = y^T \dot{x} = y^T A^T x = (Ay)^T x \leq m y^T x = m\xi,$$

By integration and (8),

$$0 < \xi(t) \leq \xi(0)e^{mt} \text{ for all } t \geq 0.$$

Hence, $\xi(t)$ must approach 0 as $t \to \infty$. Since $y \gg 0$ and $x(t) \gg 0$ for all $t \geq 0$, this is possible only if $x(t)$ approaches 0.

Thus, every solution of the differential equation,

$$\dot{x} = A^T x,$$

for which $x(0) \gg 0$, approaches 0. But then every solution approaches 0 so that A^T and therefore A is stable.

THEOREM 3. *If A is Metzler and $Ay \ll 0$ for some $y > 0$, then A is stable.*

We can now deduce the existence of a dominant root and semi-positive dominant vector from Corollary 1 and Theorem 3.

DEF. 3: $\sigma(A)$ = *maximum of real parts of characteristic roots of A.*

For any $s > \sigma(A), A - sI$ is stable. Let A, in particular, be Metzler. Then $A - sI$ is Metzler and stable. It is certainly non-singular. Choose a fixed vector

(10) $\quad c \ll 0.$

For each $s > \sigma(A)$, we can define $y(s)$ as satisfying

(11) $\quad (A - sI)y(s) = c.$

Then by Corollary 1,

(12) $\quad y(s) > 0$ for all $s > \sigma(A)$.

Let e be the vector all of whose components are 1. Then, $e^T y(s) > 0$. Define

(13) $\quad \eta(s) = [e^T y(s)]^{-1}, x(s) = \eta(s)\ y(s),$

so that

(14) $\quad x(s)$ belongs to the unit simplex, S, for all $s > \sigma(A)$.

From (11) and (13),

(15) $(A - sI)x(s) = \eta(s)c$ for $s > \sigma(A)$.

From (14), the function, $x(s)$ has a limit point, say x, as s approaches $\sigma(A) + 0$. Hence, $x \epsilon S$, and, in particular,

(16) $x > 0$.

By definition, there exists a sequence $\{s_n\}$ such that,

$$\lim_{n \to \infty} s_n = \sigma(A) + 0,$$
$$\lim_{n \to \infty} x(s_n) = x.$$

If we substitute s_n for s in (15) and let n approach ∞, then the left-hand side converges, and therefore the right-hand side must also converge, so that,

$$\lim_{n \to \infty} \eta(s_n) = \eta \geq 0.$$

From (15), with s replaced by s_n and n approaching infinity,

(17) $[A - \sigma(A)I]x = \eta c$.

Suppose $\eta > 0$. Then $[A - \sigma(A)I]x \ll 0$. From (16) and Theorem 3,

$$0 > \sigma[A - \sigma(A)I] = \sigma(A) - \sigma(A) = 0,$$

a contradiction. Hence $\eta = 0$, so that, from (17),

$$Ax = \sigma(A)x.$$

THEOREM 4. *If A is a Metzler matrix, there exists a real number, σ and a real vector $x > 0$ such that, (a) $Ax = \sigma x$, and (b) for every characteristic root, λ, of A, $R(\lambda) \leq \sigma$.*

The usual criteria for stability and for existence of non-negative solutions to the equation $Ax = b$ for Metzler matrices are all either all in the theorems already stated or can easily be deduced from them.

APPENDIX

As is well known, Theorem 4 can be strengthened in one way: conclusion (b) can be replaced by,

(b') for every characteristic root, λ, of A, with $\lambda \neq \sigma, R(\lambda) < \sigma$.

In view of (b), this is equivalent to the statement,

(b'') if $\sigma + i\tau$ (τ real) is a characteristic root of A, then $\tau = 0$.

I now prove statement (b'') using Theorems 1 and 4. Without loss of generality, it can be assumed that

(18) $\sigma(A) = 0.$

Suppose (b'') false for some A. Then, under (18),

(19) $i\tau$ is a characteristic root of A for some $\tau \neq 0$.

If A is a matrix of order n, let S be any subset of the integers $1, ..., n$, and S^* its complement.

DEF. 4. *A is* decomposable *if there exists a non-empty set S with non-empty complement such that $A_{SS^*} = 0$. Otherwise A is* indecomposable.

It is easy to see that if (18) and (19) hold for some Metzler matrix, they hold for some indecomposable Metzler matrix. By successive decompositions, we can find a partition of the integers $1, ..., n$, into sets $S(i)(1 = 1, ..., p)$ such that,

$$A_{S(i)S(j)} = 0 \text{ for } i < j.$$

$$A_{S(i)S(i)} \text{ is indecomposable for each } i.$$

Then any characteristic root of A is a characteristic root of $A_{S(i)S(i)}$ for some i, and conversely. From (19), $i\tau$ is a characteristic root of $A_{S(i)S(i)}$

for some i. By Def. 3, $\sigma(A_{S(i)S(i)}) \geq 0$. But, by Theorem 4, $\sigma(A_{S(i)S(i)})$ is a root of $A_{S(i)S(i)}$ and therefore is a root of A, so that,

$$\sigma(A_{S(i)S(i)}) \leq \sigma(A) = 0,$$

and therefore $\sigma(A_{S(i)S(i)}) = 0$. Hence, (18) and (19) hold for $A_{S(i)S(i)}$ which is indecomposable and Metzler, by definition, and we can say that they hold for some matrix A which also satisfies,

(20) A indecomposable.

DEF. 5. *For any vector* $x, Z(x) = \{i | x_i = 0\}$.

We abbreviate $Z(x)$ as Z, when the context makes it clear.

The interesting implication of (20) for the present purpose is,

LEMMA 2. *If A is indecomposable and Metzler, $x > 0$, and $Z(x)$ is non-empty, the* $(Ax)_Z \neq 0$.

PROOF. Since $x > 0$, Z^* is non-empty. By Def. 5, $x_Z = 0, x_{Z^*} \gg 0$. Hence,

$$(Ax)_Z = A_{ZZ}x_Z + A_{ZZ^*}x_{Z^*} = A_{ZZ^*}x_{Z^*},$$

Since A is Metzler, all elements of A_{ZZ^*} are non-negative. Since A is indecomposable, at least one element of A_{ZZ^*} is non-zero and therefore positive. Since $x_{Z^*} \gg 0$, it must be that $(Ax)_Z > 0$.

From (18) and Theorem 4, there exists $x^* > 0$ such that $Ax^* = 0$. If $Z(x^*)$ were non-empty, then certainly $(Ax^*)_Z = 0$, in contradiction to Lemma 2. Hence, $Z(x^*)$ is empty.

(21) There exists $x^* \gg 0$ such that $Ax^* = 0$.

Let $u + iv$ (u, v real) be a characteristic vector of A corresponding to the root $i\tau$. Then it is easy to see that u is any vector satisfying.

(22) $A^2 u = -\tau^2 u, u \neq 0$.

Then the solution of the differential equation.

(23) $\dot{x} = Ax$,

with initial condition $x(0) = u$ is,

(24) $x'(t) = u \cos \tau t + (1/\tau) Au \sin \tau t$.

If $u \geq 0$, then, by Theorem 1, $x'(t) \geq 0$ for all $t \geq 0$, and in particular for $t = \pi/\tau$. But $x'(\pi/\tau) = -u$, so that $u = 0$, a contradiction. Hence, it is impossible that $u \geq 0$. Since $-u$ also satisfies (22), it cannot be that $-u \geq 0$.

(25) If u satisfies (22), then $u_i > 0$ for some i, $u_j < 0$ some j.

For any x^* satisfying (21), the solution to (23) with $x(0) = x^* + u$ is,

(26) $x(t) = x^* + x'(t)$.

From (21) and (25), x^* can be chosen so that $x^* + u \geq 0$, with $(x^* + u)_i = 0$ for at least one i, so that $Z(x^* + u)$ is non-empty. Since $u_i > 0$ for some i, by (25), $(x^* + u)_i > 0$ and therefore $x^* + u > 0$. Then $x(t) \geq 0$ for all $t \geq 0$; but since $x'(t)$ is periodic, in fact, $x(t) \geq 0$ for all t. If $i \epsilon Z(x^* + u)$, then $x_i(0) = 0, x_i(t) \geq 0$ for all t; therefore, $\dot{x}_i(0) = 0$ for all $i \epsilon Z$. From the differential equation (23), $[Az(0)]_Z = 0$, in contradiction to Lemma 2. Hence, the existence of a purely imaginary characteristic root leads to a contradiction.

THEOREM 4′. *If A is Metzler, then there exists a real number σ and a real vector $x > 0$ such that, (a) $Ax = \sigma x$, and (b) $R(\lambda) < \sigma$ if λ is any characteristic root of A with $\lambda \neq \sigma$.*

REFERENCES

Arrow, K.J. (1960), Price-Quantity Adjustments in multiple markets with rising demands, *Mathematical Methods in the Social Sciences, 1959*, (K.J. Arrow, S. Karlin, and P. Suppes, eds.), Stanford University Press, Stanford, 3-15.

Beckenbach, E.F. and R. Bellman (1961), *Inequalities*, Springer, Berlin.

Metzler, L. (1945), Stability of multiple markets: the Hicks conditions. *Econometrica* **13**, 277-292.

Selberg's Second Beta Integral and an Integral of Mehta

Richard Askey* and Donald Richards†
University of Wisconsin and University of Virginia

1. INTRODUCTION

At a meeting in December, 1987 in Sri Lanka, A. Selberg stated a beta integral he had evaluated. The integral is

$$(1.1) \qquad \int_{D_n} \prod_{1 \le i < j \le n} |t_i - t_j|^{2c} (1 - \sum_{i=1}^{n} t_i)^{b-1} \prod_{i=1}^{n} t_i{}^{a-1} dt_i$$

where D_n is the set with $t_i \ge 0$, $\sum_{i=1}^{n} t_i \le 1$. Many decades ago he had evaluated

$$(1.2) \qquad \int_{C_n} \prod_{1 \le i < j \le n} |t_i - t_j|^{2c} \prod_{i=1}^{n} t_i{}^{a-1} (1 - t_i)^{b-1} dt_i$$

AMS 1980 Subject Classifications: Primary 33A75, Secondary 33A15.
Key words and phrases: Selberg's beta integral, gamma integral, hypergeometric function of matrix argument, unitary group, zonal polynomial.
Research supported in part by NSF grants *DMS-8701439 and †DMS-8802929.
Some of the work in this paper was done while the authors were visiting the Institute for Mathematics and its Applications, University of Minnesota. We are happy to acknowledge their support, and thank the organisers of their year on Combinatorics for the interesting program they arranged.

Copyright © 1989 by Academic Press, Inc.
All rights of reproduction in any form reserved.
ISBN-0-12-058470-0

where C_n is the set with $0 \leq t_i \leq 1$, $i = 1, \ldots, n$.

Selberg did not give his evaluation of (1.1) but said that it was different from his earlier evaluation of (1.2), and had the advantage of working in the finite field case. See Evans (1981) for a conjecture of a finite field analogue of (1.2).

New definite integrals that can be evaluated explicitly are interesting, so we worked out an evaluation of (1.1). One of the easiest ways to evaluate integrals like (1.1) and (1.2) is to use an idea of Aomoto (1987). He introduced a new degree of freedom by adding the factors $\prod_{i=1}^{k} t_i$, and so was able to set up a first-order difference equation to determine the dependence of (1.2) in the parameters a and b. A similar method works for (1.1). However, there is an easier way to obtain (1.1) from another integral which can be evaluated by Aomoto's method.

In Section 2, we apply Aomoto's argument to a gamma integral of Selberg. In Section 3, we evaluate Selberg's second beta integral (1.1) using the methods given in Section 2. Finally in Section 4, we use the theory of hypergeometric functions of matrix argument to obtain a direct evaluation of an integral of Mehta (1981). This multidimensional integral over the space of $n \times n$ Hermitian matrices is reduced to an integral over Euclidean space; our method is direct and may be useful in the evaluation of related but more general integrals.

2. SELBERG'S GAMMA INTEGRAL via AOMOTO'S ARGUMENT

If in (1.1) or (1.2) the change of variables $t_i \to t_i/b$ is made and $b \to \infty$, the result is

$$\int_{\mathbf{R}_+^n} \prod_{1 \leq i < j \leq n} |t_i - t_j|^{2c} \prod_{i=1}^{n} t_i^{a-1} e^{-t_i} dt_i \tag{2.1}$$

with $\mathbf{R}_+ = [0, \infty)$. Set

$$\Delta_n(t; c) = \prod_{1 \leq i < j \leq n} |t_i - t_j|^{2c} \tag{2.2}$$

and

$$G_n(k;a,c)=\int_{\mathbf{R}_+^n}\left(\prod_{i=1}^{k}t_i\right)\Delta_n(t;c)\prod_{i=1}^{n}t_i{}^{a-1}e^{-t_i}dt_i. \tag{2.3}$$

Thus for $1\le k\le n$,

$$\begin{aligned}0&=\int_{\mathbf{R}_+^n}\left(\frac{\partial}{\partial t_1}\prod_{i=1}^{k}t_i\right)\Delta_n(t;c)\prod_{i=1}^{n}t_i{}^{a-1}e^{-t_i}dt_i\\&=\ a\int_{\mathbf{R}_+^n}\left(\prod_{i=2}^{k}t_i\right)\Delta_n(t;c)\prod_{i=1}^{n}t_i{}^{a-1}e^{-t_i}dt_i\\&\quad-\int_{\mathbf{R}_+^n}\left(\prod_{i=1}^{k}t_i\right)\Delta_n(t;c)\prod_{i=1}^{n}t_i{}^{a-1}e^{-t_i}dt_i\\&\quad+2c\sum_{j=2}^{n}\int_{\mathbf{R}_+^n}\left(\prod_{i=2}^{k}t_i\right)\frac{\Delta_n(t;c)}{t_1-t_j}\prod_{i=1}^{n}t_i{}^{a-1}e^{-t_i}dt_i.\end{aligned} \tag{2.4}$$

The integrand in $G_n(0;a,c)$ is symmetric in the t_i's, so the integral in (2.4) that is multiplied by a is $G_n(k-1;a,c)$. In the last integral in (2.4), when $2\le j\le k$, the change $t_1\leftrightarrow t_j$ gives

$$\frac{t_1t_j}{t_1-t_j}\quad\to\quad\frac{t_1t_j}{t_j-t_1},$$

so the integrand changes sign. Thus the integral is zero. When $k<j\le n$ the change $t_1\leftrightarrow t_j$ gives

$$\frac{t_1}{t_1-t_j}\quad\to\quad\frac{t_j}{t_j-t_1}\quad=\quad1-\frac{t_1}{t_1-t_j},$$

so the integral is

$$\frac{1}{2}G_n(k-1;a,c).$$

Then (2.4) becomes

$$G_n(k;a,c)=[a+(n-k)c]G_n(k-1;a,c). \tag{2.5}$$

This gives

$$G_n(k;a,c)=\left(\prod_{j=1}^{k}[a+(n-j)c]\right)G_n(0;a,c) \tag{2.6}$$

and

$$G_n(0; a+1, c) = G_n(n; a, c) = \left(\prod_{j=1}^{n}[a + (n-j)c]\right) G_n(0; a, c). \tag{2.7}$$

Iteration of (2.7) gives

$$G_n(0; a, c) = \frac{G_n(0; a+m, c)}{\prod_{j=1}^{n}(a + (n-j)c)_m} \tag{2.8}$$

where

$$(a)_m = \frac{\Gamma(a+m)}{\Gamma(a)}.$$

Formula (2.8) can be rewritten as

$$G_n(0; a, c) = \left(\prod_{j=1}^{n} \frac{\Gamma(a + (n-j)c)}{\Gamma(a + m + (n-j)c}\right) G_n(0; a+m, c). \tag{2.9}$$

We now want to let $m \to \infty$. This can be done as follows. Recall that for $x > 0$,

$$\begin{aligned}
\Gamma(x+1) &= \int_0^\infty t^x e^{-t} dt = \int_0^\infty exp(-t + xlogt)dt \\
&= \left(\frac{x}{e}\right)^x \int_{-x}^\infty exp(-t + xlog(1 + \frac{t}{x}))dt \\
&\cong \left(\frac{x}{e}\right)^x \int_{-x}^\infty exp\left(-t + x\left[t - \frac{t^2}{2x^2}(1 + O(x^{-1}))\right]\right) dt \\
&\cong \left(\frac{x}{e}\right)^x \int_{-x}^\infty exp\left(-\frac{t^2}{2x}\right) dt \\
&= \left(\frac{x}{e}\right)^x x^{1/2} \int_{-x^{1/2}}^\infty exp\left(-\frac{t^2}{2}\right) dt \\
&\cong \left(\frac{x}{e}\right)^x x^{1/2} \int_{-\infty}^\infty exp\left(-\frac{t^2}{2}\right) dt.
\end{aligned} \tag{2.10}$$

This is just one of the standard ways of obtaining the main term in Stirling's approximation of $\Gamma(x)$. See Buck (1965, p.216). Shift the t_i in (2.9) by $a - 1$ and use the argument in (2.10). Also use (2.10) in the form

$$\Gamma(x) \cong \left(\frac{x}{e}\right)^x \left(\frac{2\pi}{x}\right)^{1/2}$$

to get

(2.11) $G_n(0;a,c)$

$$= \frac{\prod_{j=1}^{n} \Gamma(a+(n-j)c)}{(2\pi)^{n/2}} \int_{\mathbf{R}_+^n} \Delta_n(t;c) \prod_{i=1}^{n} exp\left(-\frac{t_i{}^2}{2}\right) dt_i.$$

Then by the symmetry of the integrand in $G_n(0;a,c)$ we have
(2.12)

$$a \int_{\mathbf{R}_+^n} \Delta_n(t;c) \prod_{i=1}^{n} t_i{}^{a-1} e^{-t_i} dt_i$$

$$= n!a \underset{0<t_n<\cdots<t_1<\infty}{\int \cdots \int} \Delta_n(t;c) \prod_{i=1}^{n} t_i{}^{a-1} e^{-t_i} dt_i$$

$$= M_n(c)\Gamma(a+1) \prod_{j=1}^{n-1} \Gamma(a+(n-j)c)$$

with

$$M_n(c) = \frac{1}{(2\pi)^{n/2}} \int_{\mathbf{R}^n} \Delta_n(t;c) \prod_{i=1}^{n} exp\left(-\frac{t_i{}^2}{2}\right) dt_i. \tag{2.13}$$

Let $f(t,a)$ be a continuous function for $0 \le t < \infty$, $0 \le a < \epsilon$, $\epsilon > 0$, which is differentiable in t for $t > 0$. Then

$$a \int_0^\infty f(t,a) t^{a-1} dt = \int_0^\infty f(t,a) dt^a$$

$$= t^a f(t,a)\Big|_0^\infty - \int_0^\infty t^a df(t,a)$$

$$= -\int_0^\infty t^a df(t,a)$$

and so

$$\lim_{a\to 0} a \int_0^\infty f(t,a) t^{a-1} dt = -\int_0^\infty df(t,0) = f(0,0).$$

Using this on (2.12) gives

$$nG_{n-1}(0;2c,c) = \left(\prod_{j=1}^{n-1} \Gamma((n-j)c)\right) M_n(c)$$

and (2.11) used directly gives

$$nG_{n-1}(0;2c,c) = n\left(\prod_{j=1}^{n-1}\Gamma(2c+(n-1-j)c)\right)M_{n-1}(c).$$

These combine to give

$$M_n(c) = \frac{n\Gamma(nc)}{\Gamma(c)}M_{n-1}(c) = \frac{\Gamma(nc+1)}{\Gamma(c+1)}M_{n-1}(c).$$

This gives

$$M_n(c) = \prod_{j=1}^{n}\frac{\Gamma(jc+1)}{\Gamma(c+1)} \tag{2.14}$$

since $M_1(c) = 1$.

Combining (2.14), (2.11) and (2.6) gives

$$G_n(k;a,c) = \prod_{j=1}^{k}[a+(n-j)c]\cdot\prod_{j=1}^{n}\frac{\Gamma(a+(n-j)c)\Gamma(jc+1)}{\Gamma(c+1)}. \tag{2.15}$$

Of course, this follows from Aomoto's extension of Selberg's first beta integral, but the argument to obtain $G_n(k;a,c)$ from $G_n(0;a,c)$ is slightly easier than the beta function argument (Aomoto, 1987) which twice uses the fundamental theorem of calculus, and it is interesting to see how the step from G_n to G_{n-1} arises.

3. SELBERG'S SECOND BETA INTEGRAL

To evaluate Selberg's second beta integral, first consider

$$G(\lambda) = \int_{\mathbf{R}_+^n}\left(\prod_{i=1}^{k}t_i\right)\Delta_n(t;c)\prod_{i=1}^{n}t_i{}^{a-1}e^{-\lambda t_i}dt_i. \tag{3.1}$$

A change of variables gives

$$G(\lambda) = G_n(k)\lambda^{-k-na-(n-1)nc}. \tag{3.2}$$

Multiply $G(\lambda)$ by $\lambda^{b+k+na+(n-1)nc-1}e^{-\lambda}$ and integrate on $[0,\infty)$. The result is

$$(3.3)\quad \int_{\mathbf{R}_+^n} \frac{\left(\prod_{i=1}^k t_i\right)\Delta_n(t;c)}{[1+\sum_{i=1}^n t_i]^{b+k+na+(n-1)nc}} \prod_{i=1}^n t_i{}^{a-1}dt_i$$
$$= \frac{\Gamma(b)\prod_{j=1}^k[a+(n-j)c]}{\Gamma(b+k+na+(n-1)nc)} \prod_{j=1}^n \frac{\Gamma(a+(n-j)c)\Gamma(jc+1)}{\Gamma(c+1)}.$$

Then the change of variables

$$t_i = s_i\left(1-\sum_{i=1}^n s_i\right)^{-1}$$

gives

$$(3.4)\quad \int_{D_n}\left(\prod_{i=1}^k s_i\right)\Delta_n(s;c)\left(1-\sum_{i=1}^n s_i\right)^{b+n} |J|\prod_{i=1}^n s_i{}^{a-1}ds_i$$
$$= \frac{\Gamma(b)\prod_{j=1}^k[a+(n-j)c]}{\Gamma(b+k+na+(n-1)nc)} \prod_{j=1}^n \frac{\Gamma(a+(n-j)c)\Gamma(jc+1)}{\Gamma(c+1)}$$

with

$$J = det\left[\frac{\partial t_i}{\partial s_j}\right].$$

If $u=[1-\sum_{i=1}^n s_i]^{-1}$ the Jacobian is

$$J = \det\begin{vmatrix} u+u^2s_1 & u^2s_1 & \dots & u^2s_1 \\ u^2s_2 & u+u^2s_2 & \dots & u^2s_2 \\ \vdots & \vdots & & \vdots \\ a^2s_n & a^2s_n & \dots & a+a^2s_n \end{vmatrix}.$$

Subtract the last column from the others, then add each of the first $n-1$ rows to the last one. The result is an upper triangular determinant, so

$$J = u^n\left(1+u\sum_{i=1}^n s_i\right) = u^{n+1} = \left[1-\sum_{i=1}^n s_i\right]^{-n-1}.$$

Then

$$
(3.5)\quad \int_{D_n} \left(\prod_{i=1}^{k} s_i\right) \Delta_n(s;c) \left(1-\sum_{i=1}^{n} s_i\right)^{b-1} \prod_{i=1}^{n} s_i{}^{a-1} ds_i
= \frac{\Gamma(b)}{\Gamma(b+k+na+(n-1)nc)} \prod_{j=1}^{n} \frac{\Gamma(a+(n-j)c)\Gamma(jc+1)}{\Gamma(c+1)}.
$$

It is possible to use Aomoto's argument to evaluate (3.5) directly by getting the recurrence relation to determine the factors depending on b as b increases by one and then letting b approach infinity to reduce the unknown part to one known from Selberg's first beta integral. The proof given in Sections 2 and 3 seems to us to better explain why one can evaluate these integrals. However this proof does not work in the finite field case, so there is still great interest in seeing Selberg's proof.

One of the very few applications of Selberg's first beta integral before its reappearance in the late 1970's was by Karlin and Shapley (1953). Also see Karlin and Studden (1966, Chapter IV, §6).

4. SOME INTEGRALS OF MEHTA

Let $\mathcal{H}_n$ denote the space of all Hermitian $n \times n$ matrices, and $U(n)$ be the group of $n \times n$ unitary matrices. If $S, T \in \mathcal{H}_n$, denote their eigenvalues by $s_1, \ldots, s_n$ and $t_1, \ldots, t_n$ respectively. Let $f : \mathcal{H}_n \to \mathbf{C}$ be unitarily invariant; that is, $f(S) = f(gSg^{-1})$, $S \in \mathcal{H}_n$, $g \in U(n)$. Since $f(S)$ is unitarily invariant, then $f(S)$ depends only on the eigenvalues $s_1, \ldots, s_n$ of S; so with no possibility of confusion, we can write $f(s)$ in place of $f(S)$, where $s = (s_1, \ldots, s_n)$.

In work in quantum field theory, Mehta (1981) has used the method of diffusion equations to prove the following result: If $V(S) = tr(S^2 + cn^{-1}S^4)$, c constant, and dS denotes the Lebesgue measure on $\mathcal{H}_n$, then

$$
(4.1)\quad \int_{\mathcal{H}_n}\int_{\mathcal{H}_n} exp[-V(S) - V(T) + tr(ST)]dSdT
= \frac{\pi^{n(n-1)}}{1!2!\cdots n!} \int_{\mathbf{R}^n}\int_{\mathbf{R}^n} exp[-V(s) - V(t) + \sum_{i=1}^{n} s_i t_i]\Delta_n(s)\Delta_n(t)dsdt
$$

where $\Delta_n(s) = \prod_{1 \leq i < j \leq n}(s_i - s_j)$.

Here, we give a direct proof of Mehta's formula (4.1). Our proof uses techniques which were recently applied by Richards (1989) to prove another formula of Mehta (1986): If $f(S)$ is unitarily invariant, then

$$(4.2) \quad \int_{\mathcal{H}_n} f(S) exp(tr\ ST) dS = \pi^{n(n-1)/2} \int_{\mathbf{R}^n} f(s) exp(\sum_{i=1}^{n} s_i t_i) \frac{\Delta_n(s)}{\Delta_n(t)} ds.$$

As before, we also use the theory of generalized hypergeometric functions of matrix argument to extend (4.1). The terms $exp(-V(S))$ and $exp(-V(T))$ will be replaced by unitarily invariant functions $f(S)$ and $g(T)$, respectively, while the term $exp(tr\ ST)$ will be replaced by a hypergeometric function of matrix argument.

Let $m = (m_1, \ldots, m_n)$ be a partition; that is, the m_i are nonnegative integers such that $m_1 \geq \cdots \geq m_n \geq 0$. Let $(a)_j = a(a+1)\cdots(a+j-1)$, $j = 0, 1, \ldots$, the shifted factorial; and define the partitional shifted factorial $[a]_m = (a)_{m_1}(a-1)_{m_2}\cdots(a-n+1)_{m_n}$ for any partition m.

For $S \in \mathcal{H}_n$ let $Z_m(S)$ be the zonal polynomial corresponding to the partition m, (James, 1964). These zonal polynomials are unitarily invariant and are normalized so that

$$(4.3) \qquad \sum_{|m|=j} Z_m(S) = (tr\ S)^j,$$

where $|m| = m_1 + \cdots + m_n$. The polynomials $Z_m(S)$ also have the property (James, 1964) that

$$(4.4) \qquad \int_{U(n)} Z_m(SgTg^{-1}) dg = \frac{Z_m(S) Z_m(T)}{Z_m(I_n)},$$

where I_n denotes the $n \times n$ identity matrix; and dg is the Haar measure on $U(n)$, normalized to have volume one.

Define the generalized hypergeometric function of two matrix arguments,

$$(4.5) \quad {}_pF_q(a_1, \ldots, a_p; b_1, \ldots, b_q; S, T) = \sum_{j=0}^{\infty} \sum_{|m|=j} \frac{[a_1]_m \cdots [a_p]_m}{[b_1]_m \cdots [b_q]_m} \frac{Z_m(S) Z_m(T)}{j!\ Z_m(I_n)},$$

where $-b_i + j - 1$ $(1 \leq i \leq q, 1 \leq j \leq n)$ is not a nonnegative integer. If we set $T = I_n$, we get the hypergeometric function of one matrix argument

$$(4.6)\quad {}_pF_q(a_1, \dots, a_p; b_1, \dots, b_q; S) = \sum_{j=0}^{\infty} \sum_{|m|=j} \frac{[a_1]_m \cdots [a_p]_m}{[b_1]_m \cdots [b_q]_m} \frac{Z_m(S)}{j!}.$$

By (4.4), (4.5) and (4.6), it follows that

$$(4.7)\quad {}_pF_q(a_1, \dots, a_p; b_1, \dots, b_q; S, T) = \int_{U(n)} {}_pF_q(a_1, \dots, a_p; b_1, \dots, b_q; SgTg^{-1}) dg.$$

In particular, by (4.3) and (4.6), ${}_0F_0(S) = exp(tr\ S)$. Hence by (4.7),

$$(4.8)\quad {}_0F_0(S, T) = \int_{U(n)} exp(tr\ SgTg^{-1}) dg.$$

We will also need the result, independently proved by Khatri (1970) and Gross and Richards (1989), that

$$(4.9)\quad {}_pF_q(a_1, \dots, a_p; b_1, \dots, b_q; S, T) = \frac{det({}_pF_q(a_1 - n + 1, \dots, a_p - n + 1; b_1 - n + 1, \dots, b_q - n + 1; s_i t_j))}{c_{p,q} \Delta_n(s) \Delta_n(t)}$$

where

$$(4.10)\quad c_{p,q} = \frac{\prod_{i=1}^{n} \prod_{j=1}^{p} (a_j - n + 1)_{n-i}}{\prod_{i=1}^{n} (n-i)! \prod_{j=1}^{q} (b_j - n + 1)_{n-i}}.$$

It is worth noting that both proofs of (4.9) rely on the basic composition formula (Karlin, 1968).

We will now prove the following extension of Mehta's formula (4.1): If $f(S)$ and $g(T)$ are unitarily invariant, then

$$(4.11)\quad \int_{\mathcal{H}_n} \int_{\mathcal{H}_n} f(S) g(T) {}_pF_q(a_1, \dots, a_p; b_1, \dots, b_q; ST) dS dT = d_n \int_{\mathbf{R}^n} \int_{\mathbf{R}^n} f(s) g(t) \Delta_n(s) \Delta_n(t) \cdot$$

$$\prod_{i=1}^{n} {}_pF_q(a_1 - n + 1, \dots, a_p - n + 1; b_1 - n + 1, \dots, b_q - n + 1; s_i t_i) ds_i dt_i,$$

where

$$d_n = \frac{\pi^{n(n-1)} \prod_{i=1}^n \prod_{j=1}^q (b_j - n + 1)_{n-i}}{1!2!\cdots n! \prod_{i=1}^n \prod_{j=1}^p (a_j - n + 1)_{n-i}}.$$

To prove (4.11), pass to polar coordinates on $\mathcal{H}_n$: $S = g_1 S_1 {g_1}^{-1}$, $T = g_2 T_1 {g_2}^{-1}$, where $S_1 = diag(s_1, \ldots, s_n)$, $T_1 = diag(t_1, \ldots, t_n)$ and $g_1, g_2 \in U(n)$. Then $dS = c_n \Delta_n(s)^2 ds dg_1$ where

$$c_n = \pi^{n(n-1)/2}/1!2!\cdots n!,$$

and similarly for dT. Then in (4.11), the integral with respect to g_1 and g_2 is

$$(4.12) \quad \int_{U(n)} \int_{U(n)} {}_pF_q(a_1, \ldots, a_p; b_1, \ldots, b_q; g_1 S_1 {g_1}^{-1} g_2 T_1 {g_2}^{-1}) dg_1 dg_2.$$

Replacing g_2 by $g_1 g_2$, applying the unitary invariance of the ${}_pF_q$ functions, and integrating over g_1, then (4.12) becomes

$$(4.13) \quad \int_{U(n)} {}_pF_q(a_1, \ldots, a_p; b_1, \ldots, b_q; S_1 g_2 T_1 {g_2}^{-1}) dg_2 = {}_pF_q(a_1, \ldots, a_p; b_1, \ldots, b_q; S_1, T_1).$$

On applying (4.9), (4.12) and (4.13), we see that the left-hand side of (4.11) equals

$$(4.14) \quad \frac{{c_n}^2}{c_{p,q}} \int_{\mathbf{R}^n} \int_{\mathbf{R}^n} f(s) g(t) \Delta_n(s) \Delta_n(t) \cdot det({}_pF_q(a_1 - n + 1, \ldots, a_p - n + 1; b_1 - n + 1, \ldots, b_q - n + 1; s_i t_j)) ds dt.$$

Note that the integrand in (4.14) is symmetric in $s_1, \ldots, s_n$ and in $t_1, \ldots, t_n$. Expanding the determinant as an alternating sum over the symmetric group and using the symmetry in $s_1, \ldots, s_n$, then (4.14) becomes

$$\frac{n!{c_n}^2}{c_{p,q}} \int_{\mathbf{R}^n} \int_{\mathbf{R}^n} f(s) g(t) \Delta_n(s) \Delta_n(t) \cdot \prod_{i=1}^n {}_pF_q(a_1 - n + 1, \ldots, a_p - n + 1; b_1 - n + 1, \ldots, b_q - n + 1; s_i t_i) ds_i dt_i,$$

which is the right-hand side of (4.11).

Finally, we note that (4.11) can also be indirectly proved by repeatedly applying the appropriate extension (Richards, 1989, Integral 5.2) of (4.2) involving the matrix argument hypergeometric functions. Our interest in the direct proof given above is that Mehta (1981) (cf. Chadha, Mahoux and Mehta, 1981) has raised the problem of evaluating integrals more general than – but similar to – (4.1). In that situation, the indirect approach using (4.2) does not appear to be sufficiently powerful.

REFERENCES

Aomoto, K. (1987), Jacobi polynomials associated with Selberg's integrals, *SIAM Journal on Mathematical Analysis*, **18**, 545-549.

Buck, R. C. (1965), *Advanced Calculus*, 2nd ed., McGraw Hill, New York.

Chadha, S., Mahoux, G. and Mehta, M. L. (1981), A method of integration over matrix variables: II, *Journal of Physics, A*, **14**, 579-586.

Evans, R. (1981), Identities for products of Gauss sums over finite fields, *L'Enseignement Mathematique*, **27**, 197-209.

Gross, K. I. and Richards, D. St. P. (1989), Special functions of matrix argument. II: The complex case, *in preparation.*

James, A. T. (1964), Distributions of matrix variates and latent roots derived from normal samples, *Annals of Mathematical Statistics*, **35**, 475-501.

Karlin, S. (1968), *Total Positivity*, Volume 1, Stanford University Press, Stanford.

Karlin, S. and Shapley, L. S. (1953), *Geometry of Moment Spaces*, Memoirs of the American Mathematical Society, Vol. 12, American Mathematical Society, Providence, RI.

Karlin, S. and Studden, W. J. (1966), *Tchebycheff Systems: With Applications in Analysis and Statistics*, Interscience-Wiley, New York.

Khatri, C. G. (1970), On the moments of traces of two matrices in three situations for complex multivariate normal populations, *Sankhyā, Ser. A*, **32**, 65-80.

Mehta, M. L. (1981), A method of integration over matrix variables, *Communications in Mathematical Physics*, **79**, 327-340.

Mehta, M. L. (1986), Random matrices in nuclear physics and number theory, in: *Random Matrices* (J. E. Cohen *et al.*, eds.), *Contemporary Mathematics*, Volume 50, 295-309, American Mathematical Society, Providence, RI.

Richards, D. St. P. (1989), Analogs and extensions of Selberg's integral, to appear in: *Proceedings of the Workshop on q-Series and Partitions* (D. Stanton, ed.), Institute for Mathematics and its Applications, University of Minnesota, Springer, New York.

Selberg, A. (1944), Bemerkninger om et multipelt integral, *Norsk Matematisk Tidsskrift*, **26**, 71-78.

Selberg, A. (1987), private communication.

Exponentiality of the Local Time at Hitting Times for Reflecting Diffusions and an Application

K. B. Athreya* and A. P. N. Weerasinghe*
Iowa State University

1. INTRODUCTION

Consider the following sequence of gambling problems. For each n, $\{X_n(k)\,;\ \ k = 0, 1, 2, \ldots\}$ is a Markov chain with state space $\{jn^{-1}\ :, -1 \leq j \leq n\}$ and such that if $X_n(k) = -n^{-1}$ then $X_n(k+1) = 0$ with probability one while if $X_n(k) = jn^{-1}$, $j \neq -1$, then $X_n(k+1) = (j+1)n^{-1}$ or $(j-1)n^{-1}$ with probability one half each. The state 1 is absorbing for the chain $\{X_n\}$ Let N_n be the number of visits by $X_n(\cdot)$ to $-n^{-1}$ before hitting 1. Then for $X_n(0) = 0$, N_n has a geometric distribution with mean $(n+1)$, and hence if a penalty of n^{-1} dollars were to be charged for each visit to -1, then the total amount spent before reaching the target of 1 is the random variable $n^{-1}N_n$ which tends to the exponential random variable with mean one as $n \to \infty$. This suggests the following continuous time analog to the above gambling problem.

AMS 1980 Subject Classifications: Primary 60H60J.
Keywords and phrases: Reflecting diffusion, local times, stochastic control, Itô processes.
*Research supported in part by National Science Foundation Grant DMS8803639.

Copyright © 1989 by Academic Press, Inc.
All rights of reproduction in any form reserved.
ISBN-0-12-058470-0

Let $\{X(t) : t \geq 0\}$ be a reflecting Brownian motion (reflecting at the origin) and with $X(0) = 0$. For the process X let $L(t)$ denote the local time at 0 and for $a > 0$, T_a the hitting time of level a. Now think of $X(t)$ as the position of the gambler at time t whose goal is to reach a. Suppose that each time the gambler's position reaches zero an infinitesimal penality is charged to continue the game. Then, the local time at the origin, $L(t)$, could be viewed as the penalty charged during $[0, t]$. The discussion in the previous paragraph suggests that $L(T_a)$, the penality for hitting 0, should be exponentially distributed.

We show this to be the case in the next section and proceed in Section 3 to the more general case of reflecting diffusions. In Section 4 we indicate the changes necessary for a general initial condition, while in Section 5 we study the process $\{L(T_a)\}$ where $L(\cdot)$ is the local time and T_a is the hitting time of the target a. Section 6 applies this theory to a stochastic control problem.

We are grateful to Professor William Sudderth of the University of Minnesota for several useful and illuminating discussions.

2. REFLECTING BROWNIAN MOTION

Let $\{B(t, \omega); t \geq 0\}$ be standard one dimensional Brownian motion. Let

$$\begin{cases} L(t,\omega) = -\min_{0 \leq s \leq t} B(s,\omega) \quad \text{and} \\ X(t,\omega) = B(t,\omega) + L(t,\omega). \end{cases} \tag{2.1}$$

Clearly, $X(\cdot, \omega)$ is nonnegative and $L(\cdot, \omega)$ is nondecreasing. Further, $L(0, \omega) = 0$ and

$$\int_0^t I\left(X(t,\omega) > 0\right)) \, dL(t,\omega) = 0$$

for all $t \geq 0$.

The process X is known as reflecting Brownian motion (reflecting at the origin) and the above construction is called the Skorohod decomposition. (See Chung and Williams, 1983). Clearly, both L and X are adapted to the Brownian filtration $\mathcal{F}_t^B = \sigma\{B(u,\omega)\,;\ u \leq t\}$. It is also known that X is a diffusion.

For $a > 0$ define stopping times

$$(2.2) \qquad \begin{cases} T_a = \inf\{t : t \geq 0, \quad X(t) = a\} & \text{and} \\ \mathcal{T}_a = \inf\{t : t \geq 0, \quad L(t) = a\}. \end{cases}$$

The problem mentioned in the introduction can now be formulated and solved by establishing the following

THEOREM 1. *For each $a > 0$, $L(T_a)$ is exponentially distributed with mean a.*

We prove this theorem in via the following three lemmas.

LEMMA 1. *The process L is adapted to the filtration $\mathcal{F}_t \equiv \overline{\sigma\{X(u,\omega)\,;\ u \leq t\}}$, where $\overline{}$ stands for closure with respect to the probability measure of B.*

PROOF: It can be shown (see Chung and Williams, 1983) that for X and L defined by (2.1) there is a set A of probability one such that for each ω in A and for all $t > 0$

$$\frac{1}{2\varepsilon} \int_0^t I_{[0,\varepsilon)}\left(X(s,\omega)\right) ds \to L(t,\omega)$$

as $\varepsilon \to 0$. This is known as Tanaka's formula. This shows that $L(t,\omega)$ is measurable with respect to $\mathcal{F}_t$. ∎

REMARK 1. Since $B(t,\omega) = X(t,\omega) - L(t,\omega)$ this shows that $B(t,\omega)$ is also measurable with respect to the same σ-algebra. Thus $\mathcal{G}_t \equiv \overline{\sigma\left(B(u,\omega) : u \leq t\right)}$ is contained in $\mathcal{F}_t$. That $\mathcal{F}_t \subset \mathcal{G}_t$ is a consequence of our construction. Thus X and B generate the same filtrations (up to completion).

LEMMA 2. *Let $G(x) \equiv P\left(L(T_a) > x\right)$ for $x > 0$. Then $G(x+y) = G(x) \cdot G(y)$ for all $x, y > 0$.*

PROOF: By Lemma 1 for any $x > 0$, $a > 0$, the event $\{\mathcal{T}_x < T_a\}$ is $\mathcal{F}_{\mathcal{T}_x}$ measurable. Since with probability 1, $L(t,\omega) \uparrow \infty$ as $t \uparrow \infty$ for all $a > 0$, the event $\{T_x < T_a\}$ has positive probability. Conditioning on this event yields

$$\begin{aligned} P\,(L(T_a) > x+y|\{T_x < T_a\}) \\ &= E\,(P\,(L(T_a) > x+y|\mathcal{F}_{T_x})\,|\{T_x < T_a\}) \\ &= E\,(P\,(T_a > T_{x+y}|\mathcal{F}_{T_x})\,|\{T_x < T_a\})\,. \end{aligned}$$

Next by the strong Markov property of the process X

$$P\,(T_a > T_{x+y}|\mathcal{F}_{T_x}) = P_{X(T_x)}\,(T_a\,(\theta_{T_x}\omega) > T_y\,(\theta_{T_x}\omega))$$

where $\theta_t\omega$ is the shift operator and P_y refers to the probability measure when the process $X(\cdot)$ satisfies $X(0) = y$. By the continuity of the trajectories of X and L it follows that $X(T_x) = 0$ and $P_{X(T_x)}$ is the same as P_0 which is what we have denoted as P. Thus

$$P\,(L(T_a) > x+y|\{T_x < T_a\}) = P(T_a > T_y)$$

implying

$$P\,(L(T_a) > x+y) = P(T_a > T_y)\cdot P(T_a > T_x),$$

which is the same as

$$G(x+y) = G(x)\cdot G(y). \qquad \blacksquare$$

Since $P(T_a > 0) = 1$, $G(0+)$ must be one and so the above lemma proves that $L(T_a)$ is exponentially distributed. The next step identifies its mean.

LEMMA 3. $E\,L(T_a) = a$.

PROOF: From (2.1), $X(T_a \wedge t) = B(T_a \wedge t) + L(T_a \wedge t)$. But $B(\cdot)$ is a martingale and hence $E\,(B(T_a \wedge t)) = 0$ for all $t > 0$. Also $0 \le X(T_a \wedge t) \le a$ and as $t \uparrow \infty$, $X(T_a \wedge t) \to X(T_a) = a$ and $L(T_a \wedge t) \uparrow L(T_a)$.

Now, by the bounded and monotone convergence theorems,

$$\begin{aligned} E\,L(T_a) &= \lim_t E\,(L(T_a \wedge t)) = \lim_t E\,(X(T_a \wedge t)) \\ &= E\,(X(T_a)) = a. \end{aligned} \qquad \blacksquare$$

REMARK 2. If $\{X_x(t)\,;\, t > 0\}$ denotes reflecting Brownian motion (still reflecting at 0) but starting at x, i.e., $X_x(0) = x$, then for $a > x$, $L(T_a)$

has an atom at 0 of size $\frac{x}{a}$ and conditioned on being strictly positive has an exponential distribution on $(0, \infty)$ with mean a. This suggests that for $0 < a < b$, $L(T_b) - L(T_a)$ has an atom of size $\frac{a}{b}$ at zero with the conditional distribution on $(0, \in fty)$ being exponential with mean b. These will be made precise in Section 4.

3. REFLECTING DIFFUSIONS

By a reflecting diffusion in R_1 (reflecting at the origin) with drift $b(\cdot)$ and diffusion $\sigma(\cdot)$ we mean a probability measure P on the space $\Omega = C([0, \infty), R^+)$ of all nonnegative valued continuous functions with Borel σ-algebra $\mathcal{F}$ such that for all $f \in C_0^{1,2}([0, \infty) \times R)$, (the space of all functions $f(t, x)$ that are continuously differentiable once with respect to t and twice with respect to x and have compact support), the process

$$(3.1) \qquad f(t, X(t, \omega)) - \int_0^t \left(\frac{\partial f}{\partial u} + Af\right)(u, X(u, \omega))\, I_{[0,\infty)}(X(u, \omega))\, du$$

is a submartingale provided f satisfies

$$\lim_{x \downarrow 0} \frac{\partial f}{\partial x}(t, x) \geq 0.$$

where $X(t, \omega) = \omega(t)$ is the coordinate projection and $(Af)(t, x) = \frac{1}{2}\sigma^2(x)\frac{\partial^2 f}{\partial x^2} + b(x)\frac{\partial f}{\partial x}$. For the submartingale the filtration $\mathcal{F}_t$ is the completion with respect to P of the σ-algebra generated by $\{X(u, \omega)\,;\, u \leq t\}$.

By the accepted abuse of notation we shall refer to the above process X under P as a reflecting diffusion. The existence and uniqueness of such a process (i.e., such a P) assuming that $\sigma(\cdot)$ and $b(\cdot)$ are continuous and bounded with $\sigma(x) > 0$ for all $x \geq 0$ follows from the general theory of diffusions with boundary conditions developed by Stroock and Varadhan (1971). It is also shown there that the process X is strong Markov.

It is shown in the appendix that there are processes $B(\cdot, \omega)$ and $L(\d ot, \omega)$ on the above probability space $(\Omega, \mathcal{F}, P)$ such that

i) they are adapted to the filtration $\{\mathcal{F}_t\}$ and

ii) $B(\cdot,\omega)$ is a standard Brownian motion and $L(\cdot,\omega)$ is a nondecreasing continuous process with $L(0,\omega) = 0$ and

iii)

$$X(t,\omega) = X(0,\omega) + \int_0^t \sigma\,(X(u,\omega))\,dB(u) + \int_0^t (X(u,\omega))\,du + L(t,\omega) \tag{3.1$'$}$$

with $\int_0^t I\,(X(u,\omega) > 0)\,dL(u,\omega) = 0$ for all $t \geq 0$.

For $a \geq 0$ define stopping times

$$\begin{cases} T_a = \inf\{t : t \geq 0, \quad X(t) = a\} & \text{and} \\ \mathcal{T}_a = \inf\{t : t \geq 0, \quad L(t) = a\}. \end{cases} \tag{3.2}$$

The generalization of Theorem 1 of the previous section can now be stated.

THEOREM 2. *Let $X(0) = 0$. Then for each $a > 0$, $L(T_a)$ is exponentially distributed and*

$$E\,(L(T_a)) = \int_0^a e^{-A(r)}dr \qquad \textit{where} \tag{3.3}$$

$$A(r) = \int_0^r \frac{2b(s)}{\sigma^2(s)}ds.$$

The proof follows that of Theorem 1.

The extension of Lemma 1 to the present case is a consequence of the Stroock-Varadhan theory as indicated in the appendix. Lemma 2 carries over here without any change. These two yield the assertion that $L(T_a)$ is exponentially distributed. In order to compute the mean of $L(T_a)$ the following proposition is needed.

PROPOSITION 1. *Let $g : [0,\infty) \to R$ be continuous. Then,*

$$E\left(\int_0^{T_a} g\left(X(s)\right) ds\right)$$

$$= \int_0^a \left(\int_0^x \frac{2g(r)}{\sigma^2(r)} e^{-(A(x)-A(r))} dr\right) dx \tag{3.4}$$

where $A(r)$ is as in (3.3).

PROOF: Let $u(\cdot)$ be solution to the differential equation

$$\frac{1}{2}\sigma^2(x)u''(x) + b(x)u'(x) = g(x) \tag{3.5}$$

with $u(0) = 0$ and $u'(0) = 0$. Setting $u'(x) = v(x)$ and reducing (3.5) to a first order equation we can solve for $v(\cdot)$ using the condition $u'(0) = 0$ and then for $u(\cdot)$ using $u(0) = 0$. It can be checked that $u(a)$ is precisely the rightside of (3.4). Now applying Itô's formula to $u\left(X(t)\right)$ we get using (3.1)

(3.6)

$$\begin{aligned} u\left(X(t)\right) = u(0) &+ \int_0^t u'\left(X(s)\right)\sigma\left(X(s)\right) dB(s) \\ &+ \int_0^t u'\left(X(s)\right) dL(s) \\ &+ \int_0^t \left(\frac{1}{2}\sigma^2\left(X(s)\right)u''\left(X(s)\right) + b\left(X(s)\right)u'\left(X(s)\right)\right) ds \\ = u(0) &+ \int_0^t u'\left(X(s)\right)\sigma\left(X(s)\right) dB(s) + u'(0)L(t) \\ &+ \int_0^t g\left(X(s)\right) ds, \end{aligned}$$

yielding

(3.7)

$$u(X(t)) = \int_0^t g(X(s))\,ds + \int_0^t u'(X(s))\,\sigma(X(s))\,dB(s).$$

since $u(0) = 0 = u'(0)$. Thus,

$$E\,u(X(t \wedge T_a)) = E\left(\int_0^{t\wedge T_a} g(X(s))\,ds\right) \tag{3.8}$$

because the second term on the right in (3.7) is a martingale. Taking $g \equiv 1$ and letting $t \uparrow \infty$ and using the monotone convergence theorem on the right and the bounded convergence theorem on the left we get from (3.8) that $E(T_a)$ is finite and is given by

$$E(T_a) = \int_0^a \left(\int_0^x \frac{2}{\sigma^2(r)} e^{-(A(x)-A(r))} dr\right) dx. \tag{3.9}$$

Now returning to (3.8) for a general g continuous on $[0,\infty)$ using dominated convergence theorem on both sides, we get (3.4) on letting $t \uparrow \infty$, by noting that $X(T_a) = a$. ∎

COROLLARY 1.

$$E\left(\int_0^{T_a} b(X(s))\,ds\right) = \int_0^a \left(1 - e^{-A(r)}\right) dr.$$

PROOF: Take $g(\cdot) = b(\cdot)$ in (3.4) and note that

$$A'(r) = 2\frac{b(r)}{\sigma^2(r)}.$$

We now compute $E\,L(T_a)$. From (3.1) on taking expectation we get

$$E(X(t \wedge T_a)) = E\left(\int_0^{t\wedge T_a} \sigma(X(s))\,dB(s)\right) + E\left(\int_0^{t\wedge T_a} b(X_s)ds\right) + E(L(t \wedge T_a))$$

Since $\int_0^{t\wedge T_a} \sigma(X(s))\,dB(s)$ is a martingale, we have

$$E(X(t\wedge T_a)) = E\left(\int_0^{t\wedge T_a} b(X(s))\,ds\right) + E(L(t\wedge T_a)).$$

Letting $t\uparrow\infty$ and noting that $ET_a<\infty$ and $b(\cdot)$ is bounded on $[0,a]$ we find that

$$a = E\left(\int_0^{T_a} b(X(s))\,ds\right) + E\,L(T_a).$$

Now Corollary 1 yields (3.3) and completes the proof of Theorem 2.

REMARK 3. The argument used in Proposition 1 suggests yet another application of Itô's formula to the process $e^{-\theta X(t)}$ where $\theta>0$. This yields (using (3.3))

$$e^{-\theta a} + \theta\int_0^a e^{-A(r)}dr = 1 - \theta E\left(\int_0^{T_a} e^{-\theta X(s)} b(X(s))\,ds\right) + \frac{\theta^2}{2}E\left(\int_0^{T_a} e^{-\theta X(s)}\sigma^2(X(s))\,ds\right).$$

Expanding in powers of θ and equating coefficients on both sides we get for $n\ge 2$

$$\frac{a^n}{n} = E\left(\int_0^{T_a}\left((X(s))^{n-1}\,b(X(s)) + \left(\frac{n-1}{2}\right)(X(s))^{n-2}\,\sigma^2(X(s))\right)ds\right).$$

We have not yet found any use for this formula.

4. GENERAL INITIAL CONDITIONS

In both of the previous two sections it was assumed that the reflecting process

$\{X(t);\, t \geq 0\}$ starts at $X(0) = 0$. In this section we consider the general case $X(0) = x$ for $x \geq 0$. We consider the diffusion case directly and not treat the Brownian motion case separately. We prove the following

THEOREM 3. *Let the process $\{X(t);\, t \geq 0\}$ satisfy (3. 1) with $X(0) = x \geq 0$. Then, for all $y \geq 0$ and $0 \leq x \leq a$,*

$$P_x\left(L(T_a) \leq y\right) = \frac{S(x)}{S(a)} + \left(1 - \frac{S(x)}{S(a)}\right)\left(1 - e^{-(1/S(a))y}\right) \tag{4.1}$$

where

$$S(a) = E_0\left(L(T_a)\right) = \int_0^a e^{-2\int_0^r \left(b(s)/\sigma^2(s)\right)ds}\, dr$$

P_x and E_x denote the probability measure and expectation corresponding to $X(0) = x$.

PROOF: Writing $P_x\left(L(T_a) \leq y\right)$ as an expectation yields

$$\begin{aligned} P_x\left(L(T_a) \leq y\right) &= E_x\left(I_{(L(T_a) \leq y)}\right) \\ &= E_x\left(I_{(L(T_a) \leq y)}; T_a < T_0\right) + E_x\left(I_{(L(T_a) \leq y)};\, T_a > T_0\right) \\ &= P_x(T_a < T_0) + E_x\left(P_0\left(L(T_a) \leq y\right);\, T_a > T_0\right), \end{aligned}$$

where we have invoked the strong Markov property for the second term.

If $x > 0$, then in the interval $[0, T_0 \wedge T_a)$ the process $\{X(t);\, t \geq q0\}$ behaves like an ordinary (nonreflecting) diffusion with scale function $S(\cdot)$ and hence $P_x(T_a < T_0) = \frac{S(x)}{S(a)}$. By Theorem 2 we know that $P_0\left(L(T_a) > y\right) = e^{-y/S(a)}$. ∎

5. THE PROCESS $\{L(T_a);\, a \geq 0\}$

We have seen in Theorems 2 and 3 that the marginal distribution of $L(T_a)$ is exponential when $X(0) = 0$ and is a mixture of an atom at zero and an exponential distribution when $X(0) > 0$. In this section we investigate the finite dimensional distributions and the sample path properties of the process $\{L(T_a) : a \geq 0\}$. We prove

THEOREM 4. *For any initial condition* $X(0) = x \geq 0$, *the process* $\{L(T_a) : a \geq 0\}$ *has independent but not stationary increments. In particular,*

$$P_x\left(L(T_y) = 0\right) = 1 \qquad \textit{for} \quad 0 \leq y \leq x,$$

$P_x\left(L(T_a) \leq y\right)$ *is given by (4.1) and for any* $x < a_1 < a_2$,

$$P_x\left(L(T_{a_2}) - L(T_{a_1}) \leq y\right) = P_{a_1}\left(L(T_{a_2}) \leq y\right). \tag{5.1}$$

Further, the map $a \to L(T_a)$ *is almost surely nondecreasing, left continuous and has right limits.*

PROOF: From (3.1) we have for $s \geq 0$,

$$\begin{aligned} X(T_{a_1}+s) = X(T_{a_1}) + \int_{T_{a_1}}^{T_{a_1}+s} \sigma\left(X(u)\right) dB(u) + \int_{T_{a_1}}^{T_{a_1}+s} b\left(X(u)\right) du \\ + L(T_{a_1}+s) - L(T_{a_1}). \end{aligned} \tag{5.2}$$

Let $\tilde{X}(s) = X\left(T_{a_1}+s\right)$, $\tilde{B}(s) = B\left(T_{a_1}+s\right) - B(T_{a_1})$ and $\tilde{L}(s) = L\left(T_{a_1}+s\right) - L(T_{a_1})$. Then (5.2) becomes

$$\tilde{X}(s) = a_1 + \int_0^s \sigma\left(\tilde{X}(u)\right) d\tilde{B}(u) + \int_0^s b\left(\tilde{X}(u)\right) du + \tilde{L}(s). \tag{5.3}$$

By the strong Markov property $\tilde{B}(\cdot)$ is independent of $\mathcal{F}_{T_{a_1}}$ and is also a Brownian motion. Further, $\tilde{X}$ and $\tilde{L}$ are nonanticipating with respect to $\tilde{B}_s$ and $\tilde{L}$ is nondecreasing and satisfies $\int_0^t I_{(X(s)>0)} d\tilde{L}(s) = 0$. By uniqueness of solutions to (3.1) (see [3]), we conclude that $\{\tilde{X}(s);\ s \geq 0\}$ is a reflecting diffusion with $\tilde{X}(0) = a_1$ and σ and b as diffusion and drift coefficients. Also $L(T_{a_2}) - L(T_{a_1}) = \tilde{L}(\tilde{T}_{a_2})$, where $\tilde{T}_{a_2} = \inf\{s;\ s \geq 0, \quad \tilde{X}(s) = a_2\}$. Therefore the conditional distribution of $L(T_{a_1}) - L(T_{a_1})$ given $\mathcal{F}_{T_{a_1}}$ is the same as that of $L(T_{a_2})$ under P_{a_1} which is given by (4.1). Since this is nonrandom and $L(T_{a_1})$ is $\mathcal{F}_{T_{a_1}}$ measurable, it follows that $L(T_{a_2}) - L(T_{a_1})$ is independent of $\{L(T_a); a \leq a_1\}$ and has the distribution given by (5.1).

Finally, since the map $a \to T_a$ is almost surely left continuous (this is true for any process $X(\cdot)$ with continuous paths) and the map $t \to L(t)$ is almost surely continuous and nondecreasing, the left continuity of map $a \to L(T_a)$ follows. Also since T_a is almost surely nondecreasing, $L(T_a)$ has right limits. ∎

REMARK 4. For fixed a, we can claim that $L(T_a+) = L(T_a)$ with probability 1, since $L(T_{a+h}) - L(T_a) \to 0$ in distribution as $h \downarrow 0$.

6. AN APPLICATION TO A STOCHASTIC CONTROL PROBLEM

Consider a gambler who travels in $[0, \infty)$ according to a stochastic process $\{Y(t);\, t \geq 0\}$. Starting from $Y(0) = y$ in $[0, a)$ his goal is to reach a. Assume that the process $Y(\cdot)$ is represented by an Itô process with instantaneous reflection at the origin. That is, $\{Y(t) : t \geq 0\}$ is a nonnegative valued stochastic process that satisfies

$$Y(0) = y \geq 0, \qquad Y(t) = Z(t) + L^Y(t), \tag{6.1}$$

where

$$Z(t) = y + \int_0^t b(u)du + \int_0^t \sigma(u)dB(u),$$
$$L^Y(t) = -\inf_{0 \leq u \leq t} Z(u),$$

and $B(t)$ is a given standard Brownian motion with filtration $\mathcal{F}_t$, $\sigma(t)$ and $b(t)$ are random functions of t that are progressively measurable with respect to $\{\mathcal{F}_t\}$. If σ and b happen to be functions of only $Y(t)$, then we get the reflecting diffusions treated earlier.

For the Itô process Y, in order to ensure existence we shall assume that for each $t > 0$

$$\int_0^t \left(\sigma^2(s) + |b(s)|\right) ds < \infty \tag{6.2}$$

with probability 1. Notice that L^Y is $\mathcal{F}_t$ adapted nondecreasing continuous process such that

$$\int_0^t I_{(Y(s)>0)} dL^Y(s) = 0 \qquad \text{for all} \quad t \geq 0$$

(i.e., $L^Y(\cdot)$ increases only when Y is zero).

The choice of σ and b for the Itô process $Y(\cdot)$ are restricted so that for each $t \geq 0$,

$$(\sigma(t), b(t)) \in A(Y(t)) \tag{6.3}$$

where $\{A(y);\ 0 \leq y < a\}$ is a given family of control sets such that $A(y) \subset (0, \infty) \times (-\infty, +\infty)$.

As explained in the introduction, the quantity $L^Y(T_a^Y)$ could be viewed as the amount of penalty paid by the gambler before the process reaches the level a. An appropriate optimality criterion is the mean value of this penalty. Thus the gambler's objective is to choose $\sigma(\cdot)$ and $b(\cdot)$ to minimize $E_Y L^Y(T_a^Y)$ where

$$T_a^Y = \inf\{t : t \geq 0, \quad Y(t) = a\} \tag{6.4}$$

and the minimum is taken over the class $\Sigma(y)$

$$\Sigma(y) \equiv \{Y : Y \quad \text{an Itô process satisfying (6.1)-(6.3) and} \quad P_y\left(T_a^Y < \infty\right) = 1\} \tag{6.5}$$

We assume that our only data, i.e. the family $\{A(y);\ y \geq 0\}$, is such that $\Sigma(y)$ is nonempty for all y in $[0, a)$.

We shall show that an optimal choice for the gambler is actually a reflecting diffusion with coefficients $\sigma_0(\cdot)$ and $b_0(\cdot)$ that satisfy

$$\frac{b_0(x)}{\sigma_0^2(x)} = \rho(x) \equiv \sup\{\frac{b}{\sigma^2}, (\sigma, b) \in A(x)\} \qquad \text{and} \tag{6.6}$$

$$b_0(\cdot) \quad \text{and} \quad \sigma_0(\cdot) \quad \text{are continuous and for each} \quad x \quad (\sigma_0(x), b_0(x)) \in A(x). \tag{6.7}$$

It is interesting to note that the above optimal strategy is also optimal for a class of continuous time gambling problems where there is no reflection.

See for e.g. Pestien and Sudderth (1985) and Sudderth and Weerasinghe (1988).

Since $\sigma_0(\cdot)$ and $b_0(\cdot)$ above are continuous in $[0, a]$ and $\sigma_0(x) > 0$ for all x it follows that $\sigma_0(\cdot)$ and $b_0(\cdot)$ are bounded in $[0, a]$ and satisfy $\inf_{[0,a]} \sigma_0(x) > 0$. Extend $\sigma_0(\cdot)$ and $b_0(\cdot)$ for $x > a$ by defining $\sigma_0(x) = \sigma_0(a)$ and $b_0(x) = b_0(a)$ for $x > a$. Let X be the reflecting diffusion for this pair $(\sigma_0(\cdot), b_0(\cdot))$ as defined in Section 3. Recall that this X satisfies

$$X(t) = X(0) + \int_0^t b_0(X(s))\, ds + \int_0^t \sigma_0(X(s))\, dB(s) + L^X(t) \tag{6. 8}$$

where $\{B(t), t \geq 0\}$ is standard Brownian motion and $L^X(t)$ is a nondecreasing and continuous process such that

$$\int_0^t I_{(X(s)>0)} dL^X(s) = 0 \qquad \text{for all} \quad t \geq 0,$$

and both X and L are progressively measurable with respect to the filtration $\mathcal{F}_t$ of the Brownian motion.

It was shown in Section 4 that $E_y T_a^X < \infty$ and that

$$V_y(a) \equiv -E_y\left(L^X(T_a^X)\right) = S(y) - S(a) \tag{6.9}$$

for $0 \leq y < a$, where $S(\cdot)$ is as in (4.1) (with σ and b replaced by σ_0 and b_0 respectively) and is twice differentiable in a (for each fixed y). We now state and prove the main result of this section.

THEOREM 5. *Suppose there exist $\sigma_0(\cdot)$ and $b_0(\cdot)$ satisfying (6.6) and (6.7). Let X be the process defined by (6.8). Then, for all y in $[0, a)$ and $a > 0$,*

$$E_y\left(L^X(T_a^X)\right) = \inf_{Y \in \Sigma(y)} E_y\left(L^Y(T_a^Y)\right). \tag{6.10}$$

REMARK 5. Since $X \in \Sigma(y)$, X is an optimal process for our control problem.

PROOF: Fix y in $[0, a)$. Let

$$V(x) = S(y) - S(x) \qquad \text{for} \quad 0 \le x \le a \tag{6.11}$$

where $S(\cdot)$ is as in (4.1). Note that $V(a) = V_y(a)$ of (6.9). It is easily verified that $V(\cdot)$ satisfies

$$V''(x) + 2\rho(x)V'(x) = 0 \qquad \text{in} \quad [0, a) \tag{6.12}$$

with $V'(0) = -1$ and $V(y) = 0$.

Let $Y \in \Sigma(y)$. Define stopping times $\alpha_n = \inf\{s : s \ge 0, \quad |b(s)| + |\sigma(s)| \ge n\}$, $n = 1, 2, \ldots$. By Itô's formula

$$\begin{aligned} &V\left(Y(t \wedge \alpha_n \wedge T_a^Y)\right) \\ &= V(y) + \int_0^{t\wedge\alpha_n\wedge T_a^Y} V'\left(Y(s)\right)\sigma(s)dB(s) + \int_0^{t\wedge\alpha_n\wedge T_a^Y} V'\left(Y(s)\right)dL^Y(s) \\ &\quad + 7frac12 \int_0^{t\wedge\alpha_n\wedge T_a^Y} \left(\frac{1}{2}\sigma^2 V''\left(Y(s)\right) + b(s)V'\left(Y(s)\right)\right) ds. \end{aligned}$$

Note that on the right side, the first term $V(y) = 0$, the second term is a martingale and the third term is $-L^y(t \wedge \alpha_n \wedge T_a^Y)$ since $V'(0) = -1$ and $\int_0^t I_{((Y(s))>0)} dL^Y(s) = 0$ for all t. Taking expectations, we get

$$\begin{aligned} &E\left(V\left(Y(t \wedge \alpha_n \wedge T_a^Y)\right)\right) \\ &= -E\left(L^Y(t \wedge \alpha_n \wedge T_a^Y)\right) \\ &\quad + E\left(\int_0^{t\wedge\alpha_n\wedge T_a^Y} \frac{1}{2}\sigma^2(s)\left(V''\left(y(s)\right) + \frac{2b(s)}{\sigma^2(s)}V'\left(Y(s)\right)\right) ds\right). \end{aligned} \tag{6.13}$$

By (6.6), $\frac{2b(s)}{\sigma^2(s)} \le 2\rho\left(Y(s)\right)$, and since $V'(x) = -S'(x) \le 0$, we have

$$V''\left(Y(s)\right) + 2\frac{b(s)}{\sigma^2(s)}V'\left(Y(s)\right) \ge V''\left(Y(s)\right) + 2\rho\left(Y(s)\right)V'\left(Y(s)\right)$$

which is 0 by (6.12). Now (6.13) implies $EV\left(Y(t \wedge \alpha_n \wedge T_a^Y)\right) \geq -E\left(L^Y(t \wedge \alpha_n \wedge T_a^Y)\right)$ for all n. Letting $n \uparrow \infty$ and then $t \uparrow \infty$ and using the bounded convergence theorem on the left side and the monotone convergence theorem on the right yields

$$E\,V\left(Y(T_a^Y)\right) \geq -E\,L^Y(T_a^Y).$$

Since $P_y(T_a^Y < \infty) = 1$, $Y(T_a^Y) = a$ with probability 1 and we conclude that

$$V(a) \geq -E\,L^Y(T_a^Y).$$

But $V(a) = -E_y L^X(T_a^X)$ and we are done. ∎

APPENDIX

We show that (3.1) implies (3.1)$'$. This fact is implicit in Stroock-Varadha n (1971).

Taking $a = \sigma^2$, $b = b$, $\rho = 0$, $\gamma = 1$ and $L(t) = \xi_0(t)$ in Theorem 2.5 of (Stroock-Varadhan 1971), we have $L(0) = 0$, $L(\cdot)$ is nondecreasing, continuous and

$$\int_0^t I\left(X(u) > 0\right) dL(u) = 0 \qquad \text{for} \quad t \geq 0.$$

Further, L is adapted to $\mathcal{F}_t$, and that for all bounded nonanticipating θ,

$$X_\theta(t) \equiv \exp\left(\int_0^t \theta(u) dX(u) - \frac{1}{2}\int_0^t I\left(X(u) > 0\right)\sigma^2\left(X(u)\right)\theta^2(u)du\right.$$
$$-\int_0^t I\left(X(u) > 0\right) b\left(X(u)\right)\theta(u)du$$
$$-\int_{[0,t]} \theta(u)dL(u)$$

is a $\mathcal{F}_t$ martingale. Now let

$$(*) \qquad B(t) = \int_0^t \frac{1}{\sigma(X(s))} dX(s) - \int_0^t \frac{b(X(s))}{\sigma(X(s))} ds - \int_{[0,t]} \frac{1}{\sigma(X(s))} dL(s).$$

Clearly, B is a $\mathcal{F}_t$ adapted continuous process. The first term on the right is a stochastic integral with respect to the submartingale process X and is defined in [3]. Now choose $\theta(u) = \frac{\lambda}{\sigma(X(u))}$ where $\lambda \in R$. Since $\sigma(\cdot)$ is bounded away from zero, $\theta(\cdot)$ is bounded. For this θ

$$\begin{aligned} X_\theta(t) &= \exp\left(\lambda \int_0^t \frac{1}{\sigma(X(u))} dX(u) - \frac{\lambda^2}{2} \int_0^t I(X(u) > 0)\, du \right. \\ &\qquad \left. -\lambda \int_0^t \frac{b(X(u))}{\sigma(X(u))} du - \lambda \int_{[0,t]} \frac{1}{\sigma(X(u))} dL(u)\right) \\ &= \exp\left(\lambda B(t) - \frac{\lambda^2}{2} t\right). \end{aligned}$$

This being a martingale for all $\lambda \in R$ implies that B is a standard Brownian motion. Now $(3.1)'$ follows from $(*)$.

It is possible to recover L from X by using a formula similar to the one given in the proof of Lemma 1 in Section 2. This is based upon the fact that

$$\sup_{0 \leq t \leq T} \left| \frac{1}{2\varepsilon} \int_0^t \sigma^2(X(s))\, I_{[0,\varepsilon)}(X(s))\, ds - L(t) \right|$$

goes to zero in mean square as $\varepsilon \to 0$.

REMARK 6. After this paper was completed we learned from E. Perkins (through W. Sudderth) that Theorem 1 is known in the literature. Indeed, Theorem 1 is equivalent to (4.16) on p. 429 of I. Karatzas and S. E. Shreve's book "Brownian Motion and Stochastic Calculus," Springer Verlag, 1988, who deduce this by an application of the Feynman-Kac formula. We came upon Theorem 1 from the gambling example and our proof is somewhat more elementary and direct.

REFERENCES

Chung, K. L. and Williams, R. J. (1983), An introduction to stochastic integration. *Progress in Probability and Statistics*, Volume 4, Birkhäuser.

Pestien, V. and Sudderth, W. (1985), Continuous-time red and black: How to control a diffusion to a goal, *Mathematics of Opererations Research*, **10**, 599–611.

Stoock, D. W. and Varadhan, S. R. S. (1971), Diffusion processes with boundary conditions, *Communications on Pure and Applied Mathematics*, Volume 23, 147–225.

Sudderth, W. and Weerasinghe, A. (1988), Controlling a process to a goal in finite time, *Mathematics of Operations Research*, to appear.

A Normal Approximation for the Number of Local Maxima of a Random Function on a Graph

P. Baldi[*,**]
University of California, San Diego
Y. Rinott[*]
Hebrew University, Jerusalem
University of California, San Diego
C. Stein[*]
Stanford University

1. INTRODUCTION

We shall study conditions for the approximate normality of the distribution of the number of local maxima of a random function on the set

AMS 1980 Subject Classifications: Primary 62E99, Secondary 60C05.
Key words and phrases: approximation to distributions, central limit theorem, random local maxima, graphs.
*Research supported by NSF grant DMS-8800323 to P. B., in part by NIH grant IROIGM 39907-01 to Y. R. and by NSF grant MCS80-24649 to C. S.
**Present address: Jet Propulsion Laboratory, California Institute of Technology.

PROBABILITY, STATISTICS, AND MATHEMATICS
Papers in Honor of Samuel Karlin

Copyright © 1989 by Academic Press, Inc.
All rights of reproduction in any form reserved.
ISBN-0-12-058470-0

of vertices of a graph when the values of the random function are independently identically distributed with a continuous distribution function. For a regular graph, the distribution of the number of local maxima is approximately normal if and only if its variance is large.

A precise statement of this result is given in Section 3. This result covers a number of interesting special cases that were not covered by an earlier paper of Baldi and Rinott (1988). Section 2 contains a basic lemma on normal approximation for sums of indicator random variables. Section 3 contains the main result. Section 4 contains a number of examples. In particular, Example 4 shows that, without the conditions of regularity, asymptotic normality is not implied by a large variance.

Before writing down expressions for the mean and variance of W, the number of local maxima of the random function, we recall some terminology and introduce some notation. A graph $(\mathcal{V}, \mathcal{E})$ consists of a finite set $\mathcal{V}$ of vertices and a set $\mathcal{E}$ of edges, which may be thought of as two-element subsets of $\mathcal{V}$. If $\{v_1, v_2\} \in \mathcal{E}$, the vertices v_1 and v_2 are said to be neighbors. The distance $\delta(v, v')$ is the smallest number n for which there exist $v_0, \ldots, v_n$ with $v_0 = v$ and $v_n = v'$, such that each pair $\{v_i, v_{i+1}\}$ belongs to $\mathcal{E}$. The degree $d(v)$ of a vertex v is the number of edges to which it belongs. A graph is regular if all vertices have the same degree, which we denote by d. A triangle is a set of three vertices v_1, v_2, v_3 with $\delta(v_1, v_2) = \delta(v_2, v_3) = \delta(v_3, v_1) = 1$.

For a regular graph, the mean and variance of W are given by

$$\lambda = EW = \frac{|\mathcal{V}|}{d+1}, \tag{1.1}$$

where $|\mathcal{V}|$ is the number of elements in the set $\mathcal{V}$ and

$$\sigma^2 = \operatorname{Var} W = \sum_{\substack{u,v \\ \delta(u,v)=2}} s(u,v)(2d+2-s(u,v))^{-1}(d+1)^{-2}, \tag{1.2}$$

where $s(u, v)$ is the number of common neighbors of u and v. Observe that

$$\frac{\mathcal{S}}{2(d+1)^3} \leq \sigma^2 \leq \frac{\mathcal{S}}{(d+1)^3}, \tag{1.3}$$

where

$$\mathcal{S} = \sum_{\substack{u,v \\ \delta(u,v)=2}} s(u,v) = |\mathcal{V}|d(d-1) - 6T \tag{1.4}$$

and T is the number of triangles in the graph. From (1.1), (1.3), and (1.4) it follows that the variance is always less than the mean. More precisely

$$(1.5)\quad \frac{1}{2}\frac{d(d-1)}{(d+1)^2}\left(1-6\frac{T}{|\mathcal{V}|d(d-1)}\right) \le \frac{\sigma^2}{\lambda} \le \frac{d(d-1)}{(d+1)^2}\left(1-6\frac{T}{|\mathcal{V}|d(d-1)}\right).$$

Finally note that by (1.5), the variance σ^2 is large if $\lambda = |\mathcal{V}|/(d+1)$ is large and the ratio of T to the maximal possible number of triangles in a regular graph of degree d, which is of the order of $\frac{1}{6}|\mathcal{V}|d(d-1)$, is suitably bounded away from 1.

2. A NORMAL APPROXIMATION THEOREM FOR SUMS OF INDICATOR RANDOM VARIABLES

In this section we develop a normal approximation theorem that will be used, in section 3, to study the distribution of the number of local maxima of a random function on a regular graph. The basic idea of this proof was developed by Stein (1972) and other versions are treated in Stein (1986) and elsewhere. The present version was influenced by a paper by Barbour (1982). It exploits the fact that we are interested in a sum of indicator random variables. The first lemma, related to work of Chen (1975) is an identity applicable to any sum W of indicator random variables. The second lemma gives an expression for the error of the normal approximation to the expectation of any reasonable function of W. With the aid of a third lemma, of a technical nature, which is stated without proof, this leads to the main theorem, which gives a bound for the error of the normal approximation to the distribution of W. We shall write $E^Q Z$ for the conditional expectation of the random variable Z given the random variable Q (not necessarily real valued).

LEMMA 2.1. *Let $\mathcal{V}$ be a finite set and X and X^* random functions on $\mathcal{V}$ taking on only the values 0 and 1. Also let V be a random variable uniformly distributed in $\mathcal{V}$, independent of X. Let the unconditional distribution of X^* be the same as the conditional distribution of X given that $X_V = 1$. Define*

$$(2.1)\qquad W = \sum_{x\in V} X_v = |\mathcal{V}|E^X X_V$$

$$\lambda = EW = |\mathcal{V}|EX_V \tag{2.2}$$

and

$$W^* = \sum_{v\in\mathcal{V}} X_v^*. \tag{2.3}$$

Then, for all f: $\mathbf{Z}^+ \to \mathbf{R}$

$$EWf(W) = \lambda Ef(W^*). \tag{2.4}$$

PROOF.

$$\begin{aligned} EWf(W) &= E(|\mathcal{V}|E^X X_V)f(W) \\ &= |\mathcal{V}|EX_V f(W) \\ &= |\mathcal{V}|EX_V E^{X_V} f(W) \\ &= |\mathcal{V}|EX_V E\,[f(W)|X_V = 1] = \lambda Ef(W^*). \end{aligned} \tag{2.5}$$

∎

This lemma is useful if we have a good approximation to the conditional distribution of W^* given W. Starting on p. 90 of Stein (1986) it was used in order to show that, if $E|W+1-W^*|$ is small, then W has approximately a Poisson distribution. The argument is essentially equivalent to an earlier argument of Chen (1975).

We shall now study the normal approximation to the distribution of W. For this purpose the following notation will be convenient. For h: $\mathbf{R} \to \mathbf{R}$ of bounded variation, let

$$Nh = \frac{1}{\sqrt{2\pi}} \int_{-\infty}^{+\infty} h(x)e^{-\frac{1}{2}x^2}\,dx \tag{2.6}$$

and

$$(U_N h)(y) = e^{\frac{1}{2}y^2} \int_{-\infty}^{y} [h(x) - Nh]\,e^{-\frac{1}{2}x^2}\,dx. \tag{2.7}$$

LEMMA 2.2. *Suppose the hypotheses of Lemma 2.1 are satisfied and let*

$$\sigma^2 = E(W - \lambda)^2. \tag{2.8}$$

Then

$$\sigma^2 = \lambda E(W^* - W), \tag{2.9}$$

and, for any h: $\mathbf{R} \to \mathbf{R}$ *of bounded variation,*

$$\begin{aligned} Eh(\frac{W-\lambda}{\sigma}) &\\ = Nh - \frac{\lambda}{\sigma^2} E\{[E^W(W^*-W) - E(W^*-W)](U_N h)'(\frac{W-\lambda}{\sigma})\} &\\ - \frac{\lambda}{\sigma^2} E \int_W^{W^*} (W^*-w) d(U_N h)'(\frac{w-\lambda}{\sigma}). & \end{aligned} \tag{2.10}$$

For $W^* < W$, it is understood that

$$\int_{W^*}^{W} (W^*-w)d(U_N h)'(\frac{w-\lambda}{\sigma}) = -\int_W^{W^*} (W^*-w)d(U_N h)'(\frac{w-\lambda}{\sigma}). \tag{2.11}$$

PROOF. The second expression for σ^2 follows from Lemma 2.1 by setting $f(w) = w - \lambda$. In order to prove (2.10) we first rewrite (2.4), assuming f to be differentiable with derivative of bounded variation, as

$$\begin{aligned} EWf(W) &= \lambda Ef(W^*) \\ &= \lambda E\big[f(W) + (W^*-W)f'(W) + \int_W^{W^*} (W^*-w)df'(w)\big] \\ &= \lambda Ef(W) + \sigma^2 Ef'(W) \\ &\quad + \lambda E\{[E^W(W^*-W) - E(W^*-W)]\, f'(W)\} \\ &\quad + \lambda E \int_W^{W^*} (W^*-w)df'(w). \end{aligned} \tag{2.12}$$

At the second equality sign we have used Taylor's theorem with remainder and the last equality follows from (2.9). We can rewrite (2.12) as

$$\begin{aligned} &E\big[f'(W) - \frac{W-\lambda}{\sigma^2} f(W)\big] \\ &= -\frac{\lambda}{\sigma^2} E\{[E^W(W^*-W) - E(W^*-W)]f'(W) + \int_W^{W^*} (W^*-w)df'(w)\}. \end{aligned} \tag{2.13}$$

Next, observing that the function $U_N h$, defined by (2.7), satisfies the differential equation

$$(U_N h)'(y) - y(U_N h)(y) = h(y) - Nh, \tag{2.14}$$

we substitute

$$f(w) = \sigma(U_N h)(\frac{w-\lambda}{\sigma}) \tag{2.15}$$

in (2.13) to obtain

$$\begin{aligned} (2.16)\; E\big[h(\frac{W-\lambda}{\sigma}) - Nh\big] &= E\big[(U_N h)'(\frac{W-\lambda}{\sigma}) - \frac{W-\lambda}{\sigma}(U_N h)(\frac{W-\lambda}{\sigma})\big] \\ &= -\frac{\lambda}{\sigma^2} E\{[E^W(W^* - W) - E(W^* - W)](U_N h)'(\frac{W-\lambda}{\sigma})\} \\ &\quad - \frac{\lambda}{\sigma^2} E \int_W^{W^*} (W^* - w) d(U_N h)'(\frac{w-\lambda}{\sigma}), \end{aligned}$$

which is (2.10). ∎

The following lemma will be used for the purpose of bounding the remainder in (2.10). It is proved as Lemma 3, formulas (46) and (47) on p. 25 of Stein (1986). A trivial improvement has been made in (2.17).

LEMMA 2.3. *If h: $\mathbf{R} \to \mathbf{R}$ is bounded and piecewise continuously differentiable with bounded derivative, then*

$$\sup|(U_N h)'| \le \sup h - \inf h \tag{2.17}$$

and

$$\sup|(U_N h)''| \le 2\sup|h'|. \tag{2.18}$$

By combining Lemmas 2.2 and 2.3 we obtain a result that asserts, qualitatively, that, under the hypotheses of Lemma 2.1, W is approximately normally distributed with mean λ and variance σ^2 given by (2.9) if

$$\frac{\sqrt{\operatorname{Var} E^W(W^* - W)}}{E(W^* - W)} \tag{2.19}$$

and

$$\frac{1}{\sigma}\frac{E(W^* - W)^2}{E(W^* - W)} \tag{2.20}$$

are small.

LEMMA 2.4. *Under the hypotheses of Lemma 2.1, for arbitrary piecewise continuously differentiable* h: $\mathbf{R} \to \mathbf{R}$,

$$\begin{aligned} |Eh(\frac{W-\lambda}{\sigma}) - Nh| \le (\sup h - \inf h)&\frac{\sqrt{\operatorname{Var} E^W(W^* - W)}}{E(W^* - W)} \\ &+ \sup|h'|\frac{E(W^* - W)^2}{\sigma E(W^* - W)}. \end{aligned} \tag{2.21}$$

PROOF. This follows immediately from the identity (2.10) with the aid of (2.9) and Lemma 2.3. ∎

Finally, we obtain a bound for the error in the normal approximation for the distribution of W.

THEOREM 2.1. *Under the hypothesis of Lemma 2.1, we have for all* $w \in \mathbf{R}$

$$\begin{aligned} |P(W \le w) - \Phi(\frac{w-\lambda}{\sigma})| \le &\frac{\sqrt{\operatorname{Var} E^W(W^* - W)}}{E(W^* - W)} \\ &+ (\frac{2}{\pi})^{1/4}\sqrt{\frac{1}{\sigma}\frac{E(W^* - W)^2}{E(W^* - W)}}. \end{aligned} \tag{2.22}$$

PROOF. We apply (2.21) with

$$h(x) = \begin{cases} 1 & \text{if } x \le \frac{w-\lambda}{\sigma}; \\ 1 - \frac{1}{\epsilon}(x - \frac{w-\lambda}{\sigma}) & \text{if } \frac{w-\lambda}{\sigma} \le x \le \frac{w-\lambda}{\sigma} + \epsilon; \\ 0 & \text{otherwise.} \end{cases}$$

to obtain

$$
\begin{aligned}
(2.23)\quad P(W \le w) &\le Eh\left(\frac{W-\lambda}{\sigma}\right) \\
&\le Nh + \frac{\sqrt{\operatorname{Var} E^W(W^* - W)}}{E(W^* - W)} + \frac{1}{\epsilon}\frac{E(W^* - W)^2}{\sigma E(W^* - W)} \\
&\le \Phi\left(\frac{w-\lambda}{\sigma}\right) + \frac{\epsilon}{2\sqrt{2\pi}} + \frac{\sqrt{\operatorname{Var} E^W(W^* - W)}}{E(W^* - W)} \\
&\quad + \frac{1}{\epsilon}\frac{E(W^* - W)^2}{\sigma E(W^* - W)}.
\end{aligned}
$$

The latter expression is minimized by

$$
(2.24)\qquad \epsilon = \left[\frac{2\sqrt{2\pi}}{\sigma}\frac{E(W^* - W)^2}{E(W^* - W)}\right]^{\frac{1}{2}}.
$$

The resulting upper bound for $P(W \le w)$ and the symmetric lower bound yield (2.22). ∎

REMARK: The classical De Moivre-Laplace CLT can be obtained from Theorem 2.1 as follows. Let $X_1, \ldots, X_n$ be i.i.d. Bernoulli random variables with $P(X_i = 1) = p$, $W = \sum_{i=1}^n X_i$. Let I^* be a uniform random variable on $\{1, \ldots, n\}$ independent of $X_1, \ldots, X_n$ and set $X^*_{I^*} = 1$, $X^*_i = X_i$ for $i \ne I^*$ and $W^* = \sum_{i=1}^n X^*_i$. It is readily seen that X^* satisfies the conditions of Lemma 2.1. Now the relations

$$
E(W^* - W)^2 = E(W^* - W) = E\left(1 - \frac{W}{n}\right) = 1 - p
$$

$$
\operatorname{Var} E^W(W^* - W) = \operatorname{Var}\left(1 - \frac{W}{n}\right) = p(1-p)/n,
$$

combined with Theorem 2.1, yield the asymptotic normality of W.

3. A NORMAL APPROXIMATION FOR THE DISTRIBUTION OF THE NUMBER OF LOCAL MAXIMA

Returning to the problem described in Section 1, we consider a graph $\mathcal{G} = (\mathcal{V}, \mathcal{E})$ and independently identically distributed continuous random variables $Y = \{Y_v, v \in \mathcal{V}\}$. We shall see in Theorem 3.1, that the distribution of W, the number of local maxima of the random function Y, is approximately normal with mean λ and variance σ^2 given in section 1, with error bounded by $C\sigma^{-1/2}$, where C is an absolute constant. A more careful argument in Section 2 presumably would yield an error of the form $C\sigma^{-1}$ as in the treatment of a similar problem by Barbour (1982).

Define $X = \{X_v, v \in \mathcal{V}\}$ by

$$X_v = \begin{cases} 1 & \text{if } Y_v > Y_u \text{ for all } u \in N(v); \\ 0 & \text{otherwise} \end{cases} \tag{3.1}$$

where $N(v) = \{u \in \mathcal{V} : \delta(u, v) = 1\}$, the neighborhood of v. Let $d(v) = |N(v)|$ denote the degree of v. In order to apply Theorem 2.1 we shall construct a random function X^* which will be seen, in Lemma 3.1, to satisfy the condition required of X^* in Section 2. Let V^* be a random variable, independent of Y, taking values in $\mathcal{V}$ and satisfying

$$P(V^* = v) = (d(v) + 1)^{-1} / \sum_{u \in \mathcal{V}} (d(u) + 1)^{-1}. \tag{3.2}$$

For $v \in \mathcal{V}$ define $Z(v)$ to be the vertex in $N(v) \cup \{v\}$ satisfying

$$Y_{Z(v)} = \max_{w \in N(v) \cup \{v\}} Y_w. \tag{3.3}$$

Note that $Z(v)$ is a random variable, depending on Y. Next define

$$Y_u^* = \begin{cases} Y_{Z(V^*)} & \text{if } u = V^*; \\ Y_{V^*} & \text{if } u = Z(V^*); \\ Y_u & \text{otherwise.} \end{cases} \tag{3.4}$$

In words, if V^* is a vertex at which Y has a local maximum, then $Y^* = Y$. Otherwise, to obtain Y^* from Y, interchange the Y-values of V^* and the vertex $Z(V^*)$ having the largest Y-value in $N(V^*)$, leaving all other values of Y unchanged. Now define X^* from Y^* by analogy to (3.1) and let

$$W = \sum_{v \in \mathcal{V}} X_v, \quad W^* = \sum_{v \in \mathcal{V}} X_v^* \tag{3.5}$$

count the number of local maxima of Y and Y^*, respectively.

LEMMA 3.1. *Let V be a uniform random variable taking values in $\mathcal{V}$, independently of Y, and let Y^* be defined in (3.4). Then, for any measurable set $A \subset \mathbf{R}^{|\mathcal{V}|}$,*

$$P(Y^* \in A) = P(Y \in A | X_V = 1).$$

PROOF. The key observation is that under the present assumptions

$$P(Y^* \in A | V^* = v) = P(Y \in A | X_v = 1). \tag{3.6}$$

Note simply that, conditioned on $V^* = v$, Y_v^* is a local maximum (i.e., $X_v^* = 1$), and the remaining Y_v^* have the conditional distribution given a maximum at v. Similarly, $X_v = 1$ indicates a maximum of Y at v and the remaining Y's as above. From (3.2) and (3.6) we have

$$\begin{aligned} P(Y^* \in A) &= \sum_{v \in \mathcal{V}} P(Y^* \in A | V^* = v) P(V^* = v) \\ &= \sum_{v \in \mathcal{V}} P(Y \in A | X_v = 1)(d(v)+1)^{-1} / \sum_{u \in \mathcal{V}} (d(u)+1)^{-1}. \end{aligned} \tag{3.7}$$

Also

$$\begin{aligned} P(X_V = 1) &= \sum_{u \in \mathcal{V}} P(X_u = 1 | V = u) P(V = u) \\ &= |\mathcal{V}|^{-1} \sum_{u \in \mathcal{V}} (d(u)+1)^{-1}, \end{aligned} \tag{3.8}$$

and

$$\begin{aligned} P(Y \in A | X_V = 1) &= \sum_{v \in \mathcal{V}} P(Y \in A, X_v = 1, V = v) / P(X_V = 1) \\ &= \sum_{v \in \mathcal{V}} P(Y \in A | X_v = 1) P(X_v = 1) P(V = v) / P(X_V = 1). \end{aligned}$$

The desired result now follows from $P(V = v) = |\mathcal{V}|^{-1}$, $P(X_v = 1) = (d(v)+1)^{-1}$, (3.7) and (3.8). ∎

Before we state the main Theorem, we recall that a regular graph is one in which all vertices have the same degree, which will be denoted by d. Also, for $u, v \in \mathcal{V}$ let $s(u,v) = |N(u) \cap N(v)|$.

THEOREM 3.1. *Let $(\mathcal{V}, \mathcal{E})$ be a regular graph and Y a random function on $\mathcal{V}$ whose values are independently distributed with a common continuous distribution and let W be the number of of local maxima of Y. Then the mean and variance of W are given by*

$$\lambda = EW = \frac{|\mathcal{V}|}{d+1}$$

and

$$\sigma^2 = \operatorname{Var} W = \sum_{\substack{u,v \\ \delta(u,v)=2}} s(u,v)(2d+2-s(u,v))^{-1}(d+1)^{-2},$$

and, for all $w \in \mathbf{R}^+$,

$$|P(W \le w) - \Phi(\frac{w-\lambda}{\sigma})| \le C\sigma^{-\frac{1}{2}}, \tag{3.9}$$

where C is an absolute constant.

For the proof of Theorem 3.1 we need the following lemma, where we use the notation $a \vee b = \max(a,b)$.

LEMMA 3.2. *Let u_i and S_j denote elements and subsets of $\mathcal{V}$, respectively, and assume that for all i and j, $u_i \notin S_j$, and let $Y_S = \max_{v \in S} Y_v$, for $S \subset \mathcal{V}$. Then*

(a) $P(\{Y_{u_2} > Y_{u_1} \vee Y_{S_2}\} \cap \{Y_{u_1} > Y_{S_1}\}) = (|S_1 \cup S_2| + 2)^{-1}(|S_1| + 1)^{-1}$.

In general

(b) $P(\cap_{i=2}^k \{Y_{u_i} > Y_{u_{i-1}} \vee Y_{S_i}\} \cap \{Y_{u_1} > Y_{S_1}\}) = \prod_{i=1}^{k} (|\cup_{j=1}^i S_i| + i)^{-1}$.

PROOF. The left hand side of (a) equals

$$P(Y_{u_2} > Y_{u_1} \vee Y_{S_2} \vee Y_{S_1}) P(Y_{u_1} > Y_{S_1} | Y_{u_2} > Y_{u_1} \vee Y_{S_2} \vee Y_{S_1}).$$

Since $|\{u_1\} \cup \{u_2\} \cup S_1 \cup S_2| = |S_1 \cup S_2| + 2$ and the Y-values are i. i. d. (hence exchangeable, in fact Y's exchangeable would suffice) we have

$$P(Y_{u_2} > Y_{u_1} \vee Y_{S_2} \vee Y_{S_1}) = (|S_1 \cup S_2| + 2)^{-1}.$$

Conditionally on the event $Y_{u_2} > Y_{u_1} \vee Y_{S_2} \vee Y_{S_1}$, the Y-values in S_1 and Y_{u_1} are again i.i.d. and exchangeable and Y_{u_1} is the largest among $\{Y_{u_1}, Y_{S_1}\}$ with conditional probability $|u_1 \cup S_1|^{-1} = (|S_1| + 1)^{-1}$ and (a) follows. A similar argument and induction lead to (b). ∎

PROOF OF THEOREM 3.1. First note that

$$E^W(W^* - W) = E^W E^Y (W^* - W),$$

implying

$$\operatorname{Var} E^W(W^* - W) \leq \operatorname{Var} E^Y(W^* - W). \tag{3.10}$$

Thus, in order to apply Theorem 2.1 we need only bound the quantities $E(W^* - W)^2/E(W^* - W)$ and $(\operatorname{Var} E^Y(W^* - W))^{1/2}/E(W^* - W)$. In order to write down an expression for $W^* - W$, which is needed for the computation of the bounds, we observe that

$$X_u^* = \begin{cases} X_u & \text{if } \delta(u, V^*) \geq 3; \\ X_u + A(u, V^*) & \text{if } \delta(u, V^*) = 2; \\ 0 & \text{if } \delta(u, V^*) = 1; \\ 1 & \text{if } u = V^*; \end{cases} \tag{3.11}$$

where $A(u, v)$ is the event that $\delta(u, v) = 2$ and Y does not have a local maximum at u but, if the values of Y at v and $Z(v)$ of (3.3) are interchanged, then the value at u will be a local maximum for the new function. In the sequel, we identify an event with its indicator function. It follows that

$$\begin{aligned} W^* - W &= \sum_u (X_u^* - X_u) \\ &= [1 - (X_{V^*} + \sum_{u \in N(V^*)} X_u)] + \sum_u A(u, V^*). \end{aligned} \tag{3.12}$$

In order to compute $E^Y(W^* - W)$ we observe that, because the graph is regular and V^* is independent of Y, $P^Y(V^* = v) = P(V^* = v) = |\mathcal{V}|^{-1}$. Thus it follows from (3.12) that

$$E^Y(W^* - W) = (1 - \frac{d+1}{|\mathcal{V}|} W) + \frac{1}{|\mathcal{V}|} \sum_u \sum_v A(u, v). \tag{3.13}$$

Next let us verify the formulas for the mean and variance of W. Clearly

$$EW = E\sum_v X_v = \frac{|\mathcal{V}|}{d+1}$$

and thus, by (3.13),

$$E(W^* - W) = \frac{1}{|\mathcal{V}|}\sum_u \sum_v P(A(u,v)). \tag{3.14}$$

Let $B(u,v,z)$ be the event that $A(u,v)$ occurs and $Z(v) = z$. Observe that this can only occur with $z \in N(u) \cap N(v)$ and $\delta(u,v) = 2$. In order to express the $B(u,v,z)$ explicitely in terms of Y, let

$$C = \{v\} \cup N(u) - \{z\} \tag{3.15}$$

and

$$D = \{v\} \cup N(v) - \{z\}. \tag{3.16}$$

Then

$$B(u,v,z) = \{Y_z > Y_u \vee Y_D\} \cap \{Y_u > Y_C\}. \tag{3.17}$$

Thus, by Lemma 3.2

$$\begin{aligned} P(B(u,v,z)) &= (|C \cup D| + 2)^{-1}(|C| + 1)^{-1} \\ &= (2d + 2 - s(u,v))^{-1}(d+1)^{-1} \end{aligned} \tag{3.18}$$

when $z \in N(u) \cap N(v)$ and $\delta(u,v) = 2$, otherwise 0. Finally, it follows from (3.14) that

$$\begin{aligned} &E(W^* - W) \\ &= \frac{1}{|\mathcal{V}|} \sum_{\substack{u,v \\ \delta(u,v)=2}} \sum_{z \in N(u)\cap N(v)} (2d + 2 - s(u,v))^{-1}(d+1)^{-1} \\ &= \frac{1}{|\mathcal{V}|} \sum_{\substack{u,v \\ \delta(u,v)=2}} s(u,v)(2d + 2 - s(u,v))^{-1}(d+1)^{-1}. \end{aligned} \tag{3.19}$$

Thus, by (2.9)

$$\sigma^2 = \lambda E(W^* - W) = (d+1)^{-2} \sum_{\substack{u,v \\ \delta(u,v)=2}} s(u,v)(2d+2-s(u,v))^{-1}$$

A straightforward calculation which appears in Baldi and Rinott (1988) shows that, for any graph (not necessarily regular),

$$\text{(3.20)} \qquad \operatorname{Var} W = \frac{1}{2} \sum_{\substack{u,v \\ \delta(u,v)=1}} [(d(u)+1)^{-1} - (d(v)+1)^{-1}]^2$$

$$+ \sum_{\substack{u,v \\ \delta(u,v)=2}} s(u,v)[d(u)+1]^{-1}[d(v)+1]^{-1}[d(u)+d(v)+2-s(u,v)]^{-1}.$$

The verification of the formula for the variance in our case was included because it is similar to, but simpler than calculations that will be needed later.

We now return to the bounds required for the application of Theorem 2.1 and prove first that

$$\text{(3.21)} \qquad E(W^* - W)^2 / E(W^* - W) \leq 10.$$

By (3.12),

$$\text{(3.22)} \qquad E(W^* - W)^2 \leq 2\Big\{E[1 - (X_{V^*} + \sum_{u \in N(V^*)} X_u)]^2 + E[\sum_{\substack{u \\ \delta(u,V^*)=2}} A(u,V^*)]^2\Big\}.$$

Since $X_{V^*} X_u = 0$ when $u \in N(V^*)$, and $E(X_{V^*} + \sum_{u \in N(V^*)} X_u) = 1$, we have

$$\text{(3.23)} \qquad E[1 - (X_{V^*} + \sum_{u \in N(V^*)} X_u)]^2$$

$$= E(X_{V^*} + \sum_{u \in N(V^*)} X_u)^2 - E(X_{V^*} + \sum_{u \in N(V^*)} X_u)$$

$$= E \sum_{u_1 \in N(V^*)} \sum_{\substack{u_2 \neq u_1 \\ u_2 \in N(V^*)}} X_{u_1} X_{u_2} = \frac{1}{|\mathcal{V}|} \sum_{v} \sum_{u_1 \in N(v)} \sum_{\substack{u_2 \neq u_1 \\ u_2 \in N(v)}} E X_{u_1} X_{u_2}.$$

If $\delta(u_1, u_2) = 1$, then $X_{u_1} X_{u_2} = 0$. If $\delta(u_1, u_2) = 2$, then, by Lemma 3.2,

$$\begin{aligned}
(3.24)\quad EX_{u_1}X_{u_2} &= P\{Y \text{ has local maxima at } u_1 \text{and } u_2\} \\
&= P(\{Y_{u_1} > Y_{u_2} \vee Y_{N(u_1)}\} \cap \{Y_{u_2} > Y_{N(u_2)}\}) \\
&\quad + P(\{Y_{u_2} > Y_{u_1} \vee Y_{N(u_2)}\} \cap \{Y_{u_1} > Y_{N(u_1)}\}) \\
&= 2[|N(u_1) \cup N(u_2)| + 2]^{-1}(d+1)^{-1} \\
&= 2[2d + 2 - s(u_1, u_2)]^{-1}(d+1)^{-1}.
\end{aligned}$$

It follows from (3.23) and (3.24) that

$$\begin{aligned}
(3.25)\qquad & E[1 - (X_{V^*} + \sum_{u \in N(V^*)} X_u)]^2 \\
&= \frac{1}{|\mathcal{V}|} \sum_{u_1} \sum_{\substack{u_2 \\ \delta(u_1,u_2)=2}} \sum_{v \in N(u_1) \cap N(u_2)} 2[2d + 2 - s(u_1, u_2)]^{-1}(d+1)^{-1} \\
&= 2E(W^* - W).
\end{aligned}$$

by (3.19).

In order to bound the second term on the right hand side of (3.22) we recall the formula (3.17) for $B(u, v, z)$ and write

$$\begin{aligned}
(3.26)\quad \Delta &= E[\sum_{\delta(u,V^*)=2} A(u, V^*)]^2 - E \sum_{\delta(u,V^*)=2} A(u, V^*) \\
&= \sum_{u_1} \sum_{\substack{u_2 \\ u_2 \neq u_1}} \sum_{z_1} \sum_{z_2} P(B(u_1, V^*, z_1)B(u_2, V^*, z_2)) \\
&= \frac{1}{|\mathcal{V}|} \sum_{v} \sum_{u_1} \sum_{\substack{u_2 \\ u_2 \neq u_1}} \sum_{z_1} \sum_{z_2} P(B(u_1, v, z_1)B(u_2, v, z_2)).
\end{aligned}$$

But for both $B(u_1, v, z_1)$ and $B(u_2, v, z_2)$ to occur we must have $z_1 = z_2 \in N(v) \cap N(u_1) \cap N(u_2)$ and $\delta(u_1, u_2) = 2$. In fact $z_1 = z_2$ because both are the vertices in $N(V^*)$ where Y attains its maximum, $z_1 \in N(u_1)$ because a local maximum is created at u_1 by exchanging the value of Y at V^* and the larger value at z_1 and $\delta(u_1, u_2) \neq 1$ because there cannot be local maxima at two neighboring vertices. With D as defined in (3.16) with $z = z_1 = z_2$ and C_i defined for $i \in \{1, 2\}$ by

$$C_i = \{v\} \cup N(u_i) - \{z\},$$

we use (3.17) and Lemma 3.2 to conclude that, under these conditions

$$
\begin{aligned}
(3.27)\quad & P(B(u_1, v, z)B(u_2, v, z)) \\
&= P(\{Y_z > Y_{u_1} \vee Y_{u_2} \vee Y_D\} \cap \{Y_{u_1} > Y_{C_1}\} \cap \{Y_{u_2} > Y_{C_2}\}) \\
&= (|C_1 \cup C_2 \cup D| + 3)^{-1}(|C_1 \cup C_2| + 2)^{-1}(|C_1| + 1)^{-1} \\
&\quad + (|C_1 \cup C_2 \cup D| + 3)^{-1}(|C_1 \cup C_2| + 2)^{-1}(|C_2| + 1)^{-1} \\
&\leq 2(2d + 2 - s(u_1, u_2))^{-1}(d + 1)^{-2}.
\end{aligned}
$$

It follows from (3.26) and (3.27) and the associated remarks that

$$
\begin{aligned}
(3.28)\qquad & \Delta \\
&\leq \frac{1}{|\mathcal{V}|}\sum_{u_1} \sum_{\substack{u_2 \\ \delta(u_2,u_1)=2}} \sum_{z \in N(u_1) \cap N(u_2)} \sum_{v \in N(z)} 2(2d + 2 - s(u_1, u_2))^{-1}(d + 1)^{-2} \\
&\leq 2\frac{1}{|\mathcal{V}|} \sum_{\substack{u_1, u_2 \\ \delta(u_1,u_2)=2}} s(u_1, u_2)(2d + 2 - s(u_1, u_2))^{-1}(d + 1)^{-1} \\
&= 2E(W^* - W)
\end{aligned}
$$

since, after u_1 and u_2 have been chosen, z can be chosen in $s(u_1, u_2)$ ways and then v in at most d ways for each choice of z. Combining (3.22), (3.25), (3.28) and the fact that, by (3.14)

$$
(3.29)\qquad E(W^* - W) = E \sum_{\delta(u,V^*)=2} A(u, V^*),
$$

we obtain (3.21).

Next we bound $\operatorname{Var} E^Y(W^* - W)$. In view of (3.13), we start with a bound for $\operatorname{Var}((d + 1)W/|\mathcal{V}|)$. We obtain

$$
\begin{aligned}
(3.30)\quad \operatorname{Var}\left(\frac{d + 1}{|\mathcal{V}|}W\right) &= \frac{(d + 1)^2}{|\mathcal{V}|^2}\sigma^2 \\
&= |\mathcal{V}|^{-2} \sum_{\substack{u,v \\ \delta(u,v)=2}} s(u, v)(2d + 2 - s(u, v))^{-1}.
\end{aligned}
$$

Next we deal with

$$
\begin{aligned}
(3.31)\quad \operatorname{Var}\Big(|\mathcal{V}|^{-1} \sum_{u,v} A(u, v)\Big) &= |\mathcal{V}|^{-2} \sum_{\substack{u,v \\ \delta(u,v)=2}} \operatorname{Var} A(u, v) \\
&\quad + |\mathcal{V}|^{-2} \sum_{u_1,v_1} \sum_{u_2,v_2} \operatorname{Cov}(A(u_1, v_1), A(u_2, v_2)),
\end{aligned}
$$

where in the last double sum u_1, v_1, u_2, v_2 satisfy $\delta(u_1, v_1) = \delta(u_2, v_2) = 2$ and $(u_1, v_1) \neq (u_2, v_2)$. Recalling the fact that

$$A(u,v) = \sum_{z \in N(u) \cap N(v)} B(u,v,z) \tag{3.32}$$

and henceforth suppressing the condition $z \in N(u) \cap N(v)$ in the summations, we have

$$\begin{aligned} \operatorname{Var} A(u,v) &= \operatorname{Var} \sum_z B(u,v,z) \\ &\leq \sum_z \operatorname{Var} B(u,v,z) \end{aligned} \tag{3.33}$$

the inequality following from $E(B(u,v,z_1)B(u,v,z_2)) = 0$ for $z_1 \neq z_2$. Clearly $\operatorname{Var} B(u,v,z) \leq P(B(u,v,z))$, and with (3.18) and (3.33) we obtain that the first term on the right hand side of (3.31) is bounded by

$$|\mathcal{V}|^{-2} \sum_{\substack{u,v \\ \delta(u,v)=2}} s(u,v)(2d+2-s(u,v))^{-1}(d+1)^{-1}. \tag{3.34}$$

We now consider the sum of the covariances on the right hand side of (3.31), which by (3.32) equals

$$|\mathcal{V}|^{-2} \sum_{u_1,v_1,z_1} \sum_{u_2,v_2,z_2} \operatorname{Cov}(B(u_1,v_1,z_1), B(u_2,v_2,z_2)), \tag{3.35}$$

where again u_1, v_1, u_2, v_2 satisfy $\delta(u_1, v_1) = \delta(u_2, v_2) = 2$ and $(u_1, v_1) \neq (u_2, v_2)$. In order to obtain a bound for this expression, we shall consider different cases depending on the distances $\delta(u_1, u_2)$, $\delta(v_1, v_2)$, etc. For each case we calculate a bound for $\operatorname{Cov}(B(u_1, v_1, z_1), B(u_2, v_2, z_2))$ and a bound for the number of (u_1, v_1, z_1), (u_2, v_2, z_2) of this case, thus obtaining a bound for the sum of all covariances. We consider the following cases, with subcases where $z_1 = z_2$ and $z_1 \neq z_2$.

Case 1: all $\delta(u_1, u_2), \delta(u_1, v_2), \delta(v_1, u_2), \delta(v_1, v_2) \geq 3$
Case 2: all the above distances are ≥ 2 and at least one is equal to 2.
Case 3: all the above distances are ≥ 1 and at least one is equal to 1.
Case 4: at least one of them is equal to 0.

In case 1, the corresponding covariances of (3.35) vanish because in this case the indicated events $B(u_1, v_1, z_1)$, $B(u_2, v_2, z_2)$ are independent.

Next we consider case 2, with $z_1 \neq z_2$. In order to compute a bound on $\mathrm{Cov}(B(u_1, v_1, z_1), B(u_2, v_2, z_2))$, define $C_i = \{v_i\} \cup N(u_i) - \{z_i\}$, $D_i = \{v_i\} \cup N(v_i) - \{z_i\}$ and for brevity write B_i for $B(u_i, v_i, z_i)$, $i = 1, 2$. Then we can write the probability of $B_1 \cap B_2$ as a sum of six terms, each of which can be computed with the aid of Lemma 3.2. We have

$$\begin{aligned} P_1 &= P(\{Y_{z_1} > Y_{z_2} > Y_{u_2} > Y_{u_1}\} \cap B_1 \cap B_2) \\ &= (|C_1 \cup C_2 \cup D_1 \cup D_2| + 4)^{-1}(|C_1 \cup C_2 \cup D_2| + 3)^{-1} \\ &\quad (|C_1 \cup C_2| + 2)^{-1}(|C_1| + 1)^{-1} \end{aligned}$$

$$\begin{aligned} P_2 &= P(\{Y_{z_1} > Y_{z_2} > Y_{u_1} > Y_{u_2}\} \cap B_1 \cap B_2) \\ &= (|C_1 \cup C_2 \cup D_1 \cup D_2| + 4)^{-1}(|C_1 \cup C_2 \cup D_2| + 3)^{-1} \\ &\quad (|C_1 \cup C_2| + 2)^{-1}(|C_2| + 1)^{-1} \end{aligned}$$

$$\begin{aligned} P_3 &= P(\{Y_{z_1} > Y_{u_1} > Y_{z_2} > Y_{u_2}\} \cap B_1 \cap B_2) \\ &= (|C_1 \cup C_2 \cup D_1 \cup C_2| + 4)^{-1}(|C_1 \cup C_2 \cup D_2| + 3)^{-1} \\ &\quad (|C_2 \cup D_2| + 2)^{-1}(|C_2| + 1)^{-1} \end{aligned}$$

and we define P_4, P_5, P_6 similarly with the indices 1 and 2 interchanged. It follows that

$$\begin{aligned} P_1 + P_2 = {} & (|C_1 \cup C_2 \cup D_1 \cup D_2| + 4)^{-1}(|C_1 \cup C_2 \cup D_2| + 3)^{-1} \\ & (|C_1| + 1)^{-1}(|C_2| + 1)^{-1}(1 + \frac{|C_1 \cap C_2|}{|C_1 \cup C_2| + 2}) \end{aligned}$$

and consequently

$$P_1 + P_2 + P_3 = (|C_1 \cup C_2 \cup D_1 \cup D_2| + 4)^{-1}(|C_2 \cup D_2| + 2)^{-1}(|C_1| + 1)^{-1}$$

$$(|C_2| + 1)^{-1}\left[1 + \frac{|C_1 \cap C_2|}{|C_1 \cup C_2| + 2}\frac{|C_2 \cup D_2| + 2}{|C_1 \cup C_2 \cup D_2| + 3} + \frac{|C_1 \cap (C_2 \cup D_2)|}{|C_1 \cup C_2 \cup D_2| + 3}\right]$$

adding this to the quantity obtained by interchanging the indices and using (3.18) we obtain

$$\mathrm{Cov}(B_1, B_2) = P(B_1 \cap B_2) - P(B_1)P(B_2) \tag{3.36}$$

$$= \sum_{i=1}^{6} P_i - (|C_1 \cup D_1| + 2)^{-1}(|C_1| + 1)^{-1}(|C_2 \cup D_2| + 2)^{-1}(|C_2| + 1)^{-1}$$

$$= (|C_1 \cup C_2 \cup D_1 \cup D_2| + 4)^{-1}(|C_1| + 1)^{-1}(|C_2| + 1)^{-1}$$

$$\Big[(|C_1 \cap C_2|)(|C_1 \cup C_2| + 2)^{-1}((|C_1 \cup C_2 \cup D_2| + 3)^{-1} + (|C_1 \cup C_2 \cup D_1| + 3)^{-1})$$

$$+ (|C_1 \cap (C_2 \cup D_2)|)(|C_2 \cup D_2| + 2)^{-1}(|C_1 \cup C_2 \cup D_2| + 3)^{-1}$$

$$+ (|C_2 \cap (C_1 \cup D_1)|)(|C_1 \cup D_1| + 2)^{-1}(|C_1 \cup C_2 \cup D_1| + 3)^{-1}$$

$$+ (|(C_1 \cup D_1) \cap (C_2 \cup D_2)|)(|C_1 \cup D_1| + 2)^{-1}(|C_2 \cup D_2| + 2)^{-1}\Big]$$

$$\leq \frac{1}{(d+1)^4}\Big[\frac{5s(u_1, u_2)}{2d + 2 - s(u_1, u_2)} + \frac{2s(u_1, v_2)}{2d + 2 - s(u_1, v_2)}$$

$$+ \frac{2s(u_2, v_1)}{2d + 2 - s(u_2, v_1)} + \frac{s(v_1, v_2)}{2d + 2 - s(v_1, v_2)}\Big].$$

We now sum over all (u_1, v_1, z_1) and (u_2, v_2, z_2) of case 2. The first term on the right hand side of (3.36) gives

$$\frac{5}{(d+1)^4} \sum_{u_1} \sum_{u_2} s(u_1, u_2)(2d + 2 - s(u_1, u_2))^{-1}[d(d-1)]^2$$

where the term $[d(d-1)]^2$ arises from the fact that given u_1, u_2 there are at most $[d(d-1)]^2$ possible choices of z_1, z_2, v_1, v_2. With a similar argument for the other terms it follows that the sum of covariances in (3.35) of case 2 and $z_1 \neq z_2$ is bounded by

$$10|\mathcal{V}|^{-2} \sum \sum s(u_1, u_2)(2d + 2 - s(u_1, u_2))^{-1}. \tag{3.37}$$

Next, consider $z_1 = z_2 = z$. Here and in all remaining cases it suffices to use the bound

$$\mathrm{Cov}(B_1, B_2) \leq E(B_1 B_2) \tag{3.38}$$

and the calculations become considerably simpler. In the present case, Lemma 3.2 yields

$$\begin{aligned} E(B_1 B_2) &= 2(|C_1 \cup C_2 \cup D_1 \cup D_2| + 4)^{-1} \\ &\quad (|C_1 \cup C_2| + 2)^{-1}(d+1)^{-1} \\ &\leq (2d + 2 - s(u_1, u_2))^{-1}(d+1)^{-1}. \end{aligned} \tag{3.39}$$

To obtain a bound for the sum of covariances in (3.35), we sum the right hand side of (3.39) over $u_1, v_1, z_1, u_2, v_2, z_2$ of this case. Observe that after choosing u_1, v_1, the vertex $z = z_1 = z_2$ can be chosen in $s(u_1, v_1)$ ways and given z, $u_2, v_2 \in N(z)$ can be chosen in $(d-2)(d-3)$ ways. Combining (3.38) and (3.39) we conclude that an upper bound on the sum of covariances in case 2 when $z_1 = z_2$ is

$$2|\mathcal{V}|^{-2} \sum \sum s(u_1, u_2)(2d + 2 - s(u_1, v_1))^{-1}. \tag{3.40}$$

The remaining cases are treated similarly. For example in the case that $\delta(u_1, u_2) = 2$, $\delta(v_1, v_2) = 1$ and $z_1 \neq z_2$ we obtain

$$E(B_1 B_2) \leq 6(2d + 2 - s(u_1, u_2))^{-1}(d + 1)^{-3} \tag{3.41}$$

(with the factor 6 corresponding to the six arrangements described in case 2). The number of terms in this case is counted as follows. Given u_1 and v_1, z_1 can be chosen in $s(u_1, v_1)$ ways, then v_2 in $d - 1$ ways since $\delta(v_1, v_2) = 1$, and z_2, u_2 in $d(d - 1)$ ways. This shows that an upper bound on covariances in this case is given by

$$6|\mathcal{V}|^{-2} \sum_{\substack{u_1, u_2 \\ \delta(u_1, u_2) = 2}} s(u_1, u_2)(2d + 2 - s(u_1, u_2))^{-1}. \tag{3.42}$$

In other cases, the bound on $E(B_1, B_2)$ may be of order $(2d+2-s(u_1, u_2))^{-1}$ $(d + 1)^{-i}$ $i = 1, 2$ or 3 but in each case the count of the associated terms leads to a bound similar to (3.37), (3.40), (3.42). In conclusion of this discussion, we obtain

$$\begin{aligned} \operatorname{Var} E^Y(W^* - W) & \\ \leq \alpha |\mathcal{V}|^{-2} & \sum_{\substack{u_1, u_2 \\ \delta(u_1, u_2) = 2}} s(u_1, u_2)(2d + 2 - s(u_1, u_2))^{-1}. \end{aligned} \tag{3.43}$$

where α is an absolute constant, independent of the particular graph. Recalling also (3.19) and (1.2) we have

$$\begin{aligned} & \frac{\operatorname{Var} E^Y(W^* - W)}{[E(W^* - W)]^2} \\ & \leq \alpha(d + 1)^2 \Big[\sum_{u_1} \sum_{u_2} s(u_1, u_2)(2d + 2 - s(u_1, u_2))^{-1} \Big]^{-1} \\ & \leq \alpha \sigma^{-2}. \end{aligned} \tag{3.44}$$

Theorem 3.1 now follows from Theorem 2.1, with (3.21) and (3.44). ∎

4. EXAMPLES

The emphasis in this section is on the asymptotics; the bounds of section 3 will be computed only to the extent needed to verify asymptotic normality.

EXAMPLE 1. The graph $\{0,\ldots,m-1\}^n$. Set $\mathcal{V}=\{0,\ldots,m-1\}^n$ and for $u=(u_1,\ldots,u_n)$, $v=(v_1,\ldots,v_n)$ in $\mathcal{V}$ define $\delta(u,v)=|\{i\in\{1,\ldots,n\}: u_i\neq v_i\}|$. Thus two vertices are connected by an edge if and only if they differ in exactly one coordinate. Here $|\mathcal{V}|=m^n$, $d=n(m-1)$, $s(u,v)=2$ for all $u,v\in\mathcal{V}$ satisfying $\delta(u,v)=2$, so that

$$\sigma^2=|\mathcal{V}|\binom{n}{2}(m-1)^2(d+1)^{-2}d^{-1}. \tag{4.1}$$

It is readily seen that σ^2 is of the order of m^{n-1}/n, diverging to ∞ as $m\to\infty$ for any $n\geq 2$, or as $n\to\infty$ for any $m\geq 2$, and asymptotic normality of W follows from Theorem 3.1.

REMARK. In the case $n=2$, the exact distribution of W can be identified as the hypergeometric distribution. We claim

$$P(W=k)=\frac{\binom{m}{k}\binom{m-1}{m-k}}{\binom{2m-1}{m}}\qquad k=1,2,\ldots,m. \tag{4.2}$$

To see this, consider the graph $\{0,\ldots,m-1\}^2$ as a two dimensional array of m rows and m columns. Each row (or column) may contain at most one local maximum. Let $I_j=I\{$ there exists a maximum in row $j\}$. Then $W=\sum_{j=1}^m I_j$. The I_j's are exchangeable and by Lemma 3.2 accounting also for the $m(m-1)\ldots(m-k+1)$ possible locations of the maxima at rows $1,\ldots,k$ we obtain

$$\begin{aligned}E(I_1\ldots I_k)&=\frac{k!\,m(m-1)\ldots(m-k+1)}{\prod_{l=1}^k(ld+l-2\frac{l(l-1)}{2})}\\&=\frac{k!\,m(m-1)\ldots(m-k+1)}{\prod_{l=1}^k l(2m-l)}=\frac{\binom{m}{k}}{\binom{2m-1}{k}}.\end{aligned} \tag{4.3}$$

This determines (4.2). Thus the distribution of the number of local maxima is the same as that of the number of white balls obtained when m balls are drawn at random without replacement from an urn containing m white balls and $m-1$ black balls.

EXAMPLE 2. Again the graph $\{0,\ldots,m-1\}^n$ but with distance δ defined for $u=(u_1,\ldots,u_n)$, $v=(v_1,\ldots,v_n)\in\mathcal{V}$ by

$$\delta(u,v)=\begin{cases}0 & \text{if } u=v;\\ 1 & \text{if } u\neq v \text{ and for some } i,\ u_i=v_i;\\ 2 & \text{otherwise.}\end{cases}$$

Thus the neighborhood of a vertex consists of the union of all $n-1$-dimensional hyperplanes containing it. It is not hard to see that in this case

$$d=m^n-(m-1)^n-1\approx nm^{n-1},$$

$$s=s(u,v)=m^n-2(m-1)^n+(m-2)^n\approx n(n-1)m^{n-2},$$

and

$$\mathcal{S}=[m(m-1)]^n s\approx n(n-1)m^{3n-2}$$

where the asymptotic expressions are valid if $m\to\infty$, $n=o(m)$ (with $n\to\infty$ allowed). Therefore $\mathcal{S}(d+1)^{-3}$ is of the order of m/n. We obtain asymptotic normality if $m/n\to\infty$. If m/n is bounded, then $\sigma^2=\operatorname{Var}W$ is bounded and aymptotic normality is impossible.

EXAMPLE 3. The complete bipartite graph $K_{d,d}$. In this graph $\mathcal{V}=\mathcal{V}_1\cup\mathcal{V}_2$ with $|\mathcal{V}_i|=d$ and $\delta(v_1,v_2)=1$ if and only if $v_1\in\mathcal{V}_1$ and $v_2\in\mathcal{V}_2$ or the reverse. This graph is regular with degree d. Also

$$s(u,v)=d,\quad \mathcal{S}=4\binom{d}{2}d,\quad |\mathcal{V}|=2d.$$

In this case $\mathcal{S}d^{-3}\approx 2$ as $d\to\infty$ and by (1.3) asymptotic normality is obviously ruled out. In fact for $k=1,2,\ldots,d$

$$P(W=k)=2\frac{d(d-1)\ldots(d-k+1)d}{2d(2d-1)\ldots(2d-k)}. \tag{4.4}$$

Note that as $d\to\infty$, $P(W=k)\to 2^{-k}$ and $W\xrightarrow{\mathcal{D}}$ *Geometric* $(\frac{1}{2})$.

EXAMPLE 4. The complete bipartite graph $K_{1,d}$ (star). Here $|\mathcal{V}_1| = 1$, $|\mathcal{V}_2| = d$, with notation as in Example 3. This graph is not regular. From (3.20) one can readily show that $\sigma^2 = \operatorname{Var} W \to \infty$ as $d \to \infty$. However W is not asymptotically normal. In fact

$$P(W = k) = \begin{cases} 2/(d+1), & \text{if } k = 1; \\ 1/(d+1), & \text{if } k = 2, \dots, d. \end{cases} \tag{4.5}$$

So the distribution is nearly uniform and $\frac{W-\lambda}{\sigma} \xrightarrow{\mathcal{D}} \mathit{Uniform}\ (-\sqrt{3}, \sqrt{3})$. This example indicates that without some regularity conditions on the graph (perhaps local regularity) the problem is much more complicated and obviously $\operatorname{Var} W \to \infty$ is not a sufficient condition for asymptotic normality.

REFERENCES

Baldi, P and Rinott, Y. (1988), On normal approximation of distributions in terms of dependency graphs, to appear in *Annals of Probability.*

Barbour, A. D. (1982), Poisson convergence and random graphs, *Mathematical Proceedings Cambridge Philosophical Society,* **92**, 349-359.

Chen, L. H. Y. (1975), Poisson approximation for dependent trials, *Annals of Probability,* **3**, 534-545.

Stein, C. (1972), A bound for the error in the normal approximation to the distribution of a sum of dependent random variables, *Proceedings Sixth Berkeley Symposium on Mathematical Statistics and Probability,* **2**, 583-602.

Stein, Charles (1986), *Approximate Computation of Expectations* , Institute of Mathematical Statistics Lecture Notes, S. S. Gupta Series Editor, Volume **7**.

Operator Solution of Infinite G_δ Games of Imperfect Information

David Blackwell
University of California, Berkeley

1. INTRODUCTION

Let I, J be non-empty finite sets and put $Z = I \times J$. A *position of length* k ($k = 0, 1, 2, \ldots$) is a sequence $p = (z_1, \ldots, z_k)$ with $z_i \in Z$ for $i = 1, \ldots, k$. We denote by e the position of length 0. A *play* is an infinite sequence $w = (z_1, z_2, \ldots)$ with $z_i \in Z$ for all i. Any bounded Borel function f on the set W of plays defines a two-person zero-sum game, denoted by $G(f)$, played as follows. Player A chooses an element $i_1 \in I$ and, simultaneously, Player B chooses an element $j_1 \in J$. Then both players are told $z_1 = (i_1, j_1)$. Then A chooses $i_2 \in I$ and, simultaneously, B chooses $j_2 \in J$. Then both are told $z_2 = (i_2, j_2)$, etc. They thus produce a play $w = (z_1, z_2, \ldots)$. Then B pays A the amount $f(w)$, ending the game.

It is not known whether all games $G(f)$ have a value, i.e. whether there is a number v such that, for every $\epsilon > 0$, A can guarantee expected income at least $v - \epsilon$ and B can restrict A's expected income to at most $v + \epsilon$. For f the indicator of a G_δ set of plays, I showed (1969) that $G(f)$ does have a value. My proof was very non-constructive, based on calculations with

AMS 1980 Subject Classifications: Primary 90D15, Secondary 04A15.
Key words and phrases: G_δ, imperfect information.

PROBABILITY, STATISTICS, AND MATHEMATICS
Papers in Honor of Samuel Karlin

Copyright © 1989 by Academic Press, Inc.
All rights of reproduction in any form reserved.
ISBN-0-12-058470-0

the upper value of $G(f)$. I present here a somewhat more constructive (less non- constructive?) proof, based on transfinite iteration of a simple operator T.

2. THE OPERATOR

Let H be any set of positions, fixed throughout this section. The set of all plays w with infinitely many positions in H, i.e. with $(z_1, \ldots, z_k) \in H$ for infinitely many k, is a G_δ set, and every G_δ set has this representation for some H (Wolfe (1953)).

Let U be the set of all functions u defined on the set P of positions with $0 \leq u \leq 1$. We associate with each position p of P and each function u a game $g(u,p)$, played as follows. Starting from position p, A and B play as in Section 1 until a position p' strictly after p in H is reached. Then play stops, and B pays A the amount $u(p')$. If no such p' is ever reached, play continues forever and B pays A 0. The game is essentially the same as the game $G(f)$, with $f(w)$ defined as follows. If w has a position p' of length ≥ 1 with $pp' \in H$, let p'' be the shortest such position p'. Then $f(w) = u(pp'')$. If w has no such position then $f(w) = 0$. It follows from a minimax theorem of Karlin (1950) that the game $g(u,p)$ has a value, which we denote by $v(u,p)$.

We now define the operator T, which associates with each function $u \in U$ another function in U, denoted by Tu, defined as follows.

$$Tu(p) = v(u,p).$$

3. THE RESULT

THEOREM. *Define for each countable ordinal a the element $u(a)$ of U as follows: $u(0) \equiv 1$; $u(a+1) = Tu(a)$; if a is a limit ordinal then $u(a) = \inf u(b)$, $b < a$. Then*

(1) $u(b) \geq u(a)$ for $a > b$;

(2) there is an a^* with $u(a^*+1) = u(a^*)$; i.e. with $u^* = u(a^*)$, u^* is a fixed point of T;

(3) the value of the game $G(f)$, where f is the indicator of the set of all plays that hit H infinitely often, is $u^*(e)$. PROOF. The basic monotoneity

property of T that insures (1) is that $u1 \geq u2$ implies $Tu1 \geq Tu2$. Since $u(0)$ is the largest element of U, (1) holds for $b = 0$. Suppose (1) holds for all $b < b0$. If $b0$ is a limit ordinal, taking the inf in (1) over $b < b0$ for fixed $a > b0$ gives (1) for $b0$. If $b0 = b + 1$, applying T to the inequality (1) gives $u(b0) \geq u(a + 1)$ for $a \geq b0$, and (1) for $b0$ follows easily. (2) is a well-known consequence of (1): a non-increasing function on the countable ordinals is ultimately constant. ∎

Now for (3). Put $v = u^*(e)$ and fix $\epsilon > 0$. We first describe a strategy for A that gives him expected income at least $v - \epsilon$ against any strategy B. Let $\epsilon_0, \epsilon_1, \ldots$ be positive numbers whose sum is ϵ. Starting from e, A plays in $g(u^*, e)$ to get at least $Tu^*(e) - \epsilon_0 = u^*(e) - \epsilon_0 = v - \epsilon_0$. If H is ever hit strictly after e, say at $p1$, A then plays in $g(u^*, p1)$ to get at least $Tu^*(p1) - \epsilon_1 = u^*(p1) - \epsilon_1$. If H is hit again, say at $p2$, A then plays in $g(u^*, p2)$ to get at least $u^*(p2) - \epsilon_2$, etc. We have now described a strategy for A, and must show that it gets him at least $v - \epsilon$ against any strategy for B. Fix a strategy for B. Define $x(0) = u^*(e) = v$ and, for $k > 0$, put $x(k) = u^*(pk)$ if pk is defined, i.e. if w hits H at least k times after e, and put $x(k) = 0$ otherwise. Then

$$(1) \qquad E(x(k+1)|x(1), \ldots, x(k)) \geq x(k) - \epsilon_k.$$

(1) is clear if $x(k) = 0$. If $x(k) > 0$, then $x(k+1)$ is A's income in $g(u^*, pk)$. Since the value of this game is $x(k)$ and A played ϵ_k-optimally, (1) follows. Taking expected values in (1) gives $E(x(k + 1) \geq E(x(k)) - \epsilon_k$, so that $E(x(k)) > E(x(0)) - \epsilon = v - \epsilon$ for all k. But $x(k) \leq I(k)$, the indicator of the event: w hits H at least k times after e, so that $E(I(k)) \geq v - \epsilon$. Letting k become infinite gives $E(f) \geq v - \epsilon$: A's strategy gives him at least $v - \epsilon$.

Now for B. We shall prove, inductively on the countable ordinal a, that, for all positions p and any $\epsilon > 0$, in the game $G(f)$ starting from p, B can restrict A's expected income to $u(a)(p) + \epsilon$. The case we need is $a = a^*$, $p = e$. The assertion is clear for $a = 0$: any strategy for B restricts A's expected income to $1 + \epsilon$. Suppose the assertion is true for $a < a0$. If $a0$ is a limit ordinal, for any p and $\epsilon > 0$ there is a $a < a0$ with $u(a)(p) < u(a0, p)) + \epsilon$. By the induction hypothesis, B can, starting from p, restrict A to $u(a)(p) + \epsilon$, which does not exceed $u(a0(p)) + 2\epsilon$. The case $a0 = a + 1$ is more interesting. Let B play as follows, starting from p. Play in $g(u(a), p)$ to restrict A to $Tu(a)(p) + \epsilon = u(a0)(p) + \epsilon$. If H is ever hit strictly after p, say at p', then play in $G(f)$ from p' to restrict A to $u(a)(p') + \epsilon$.

To see that this strategy works, fix a strategy for A and put $x = u(a)(p')$ if p' is defined, i.e. if H is hit strictly after p, and $x = 0$ otherwise. Then $E(f|x) \leq x + \epsilon$. But x is A's payoff in $g(u(a), p)$, so $E(x) \leq u(a0) + \epsilon$. Thus $E(f) \leq u(a0) + 2\epsilon$, completing the induction and the proof of (3).

4. COMPLEXITY OF THE PROBLEM

Is there a more constructive solution to our problem – one not using transfinite iteration? The data is a set H of positions, and the problem is to calculate $v(H)$, the value of the game "hitting H infinitely often" (and also to calculate ϵ-optimal strategies, which we ignore in this discussion). The set Q of sets H is topologically the space of infinite sequences of 0s and 1s, and v, considered as a function on this space, turns out not to be Borel measurable. So something as complex as transfinite iteration is probably necessary.

To see that v is not Borel measurable, consider the simple case $I = \{0, 1\}$, $J = \{0\}$, so that we have a one-person game. We are given a set H of finite sequences of 0s and 1s, and must decide whether there is an infinite sequence of 0s and 1s with infinitely many initial segments in H. Let us call this the *H-problem.*

A problem that is no harder than the H-problem is the *S-problem*: given a set S of finite sequences of positive integers, is there an infinite sequence of positive integers with infinitely many initial segments in S? It is well-known that the class of sets S for which the answer is "yes" is an analytic non-Borel set. But every S-problem can be reduced to an H-problem by encoding: encode the positive integer n by the sequence of 0s and 1s of length $n+2$: 0 followed by n 1s followed by 0, and encode a finite sequence of positive integers by successive encoding of its elements, so that for instance the sequence (2,5,3) is encoded by the sequence of 16 0s and 1s 0110011111001110. An S-problem has a "yes" answer if and only if the encoded H-problem does, so that, since the set of S-problems with a "yes" answer is not Borel, neither is the set of H-problems with a "yes" answer.

REFERENCES

Blackwell, D. (1969), Infinite G_δ games with imperfect information, *Zastosowania Matematyki Applicationes Mathematicae*, Hugo Steinhaus Jubilee Volume X, 1969.

Karlin, S. (1950), Operator treatment of minimax principle, Contributions to the Theory of Games, *Annals of Mathematics Studies*, **24**, Princeton University Press.

Wolfe, P. (1955), The strict determinateness of certain infinite games, *Pacific Journal of Mathematics*, **5**, 891-897.

Smoothed Limit Theorems for Equilibrium Processes

Peter W. Glynn
and
Donald L. Iglehart
Stanford University

1. INTRODUCTION

Renewal ideas play a fundamental rule in applied probability. In particular, renewal methods are powerful tools for dealing with the asymptotic behavior of the regenerative processes that are typical of queueing and operations research applications. In this paper we survey the main results of the theory and develop some "smoothed" versions of the convergence theorems. These smoothed versions typically have much weaker associated regularity hypotheses than the classical version. In fact, these smoothed limit theorems are established under minimal conditions that are essentially necessary.

This paper is organized as follows. Section 2 describes the classical limit theory for solutions to renewal equations; Section 3 discusses the extensions to equilibrium processes. In Section 4, a smoothed version of the renewal theorem is described. It is applied to equilibrium processes in Section 5.

AMS 1980 Subject Classifications: Primary 60K05.
Key words and phrases: renewal theory, regerative processes

PROBABILITY, STATISTICS, AND MATHEMATICS
Papers in Honor of Samuel Karlin

Copyright © 1989 by Academic Press, Inc.
All rights of reproduction in any form reserved.
ISBN-0-12-058470-0

2. THE TWO BASIC FORMS OF THE RENEWAL THEOREM

In this section, we review the two basic forms of the renewal theorem. Let $\mathcal{F}$ be the family of probability distribution functions F corresponding to non-negative r.v.'s which are positive with positive probability (i.e. $F(0-) = 0$ and $F(0) < 1$). Given $F \in \mathcal{F}$ and a (measurable) function $b(t)$ which is a.e. finite-valued (a.e. with respect to Lebesgue measure), the **renewal equation** takes the form

$$a(t) = b(t) + \int_{[0,t]} a(t-u)F(du) \tag{2.1}$$

for a.e. $t \geq 0$. The function $a(\cdot)$ appearing in (2.1) is termed the **solution** to the renewal equation (when it exists) and the goal here is to study the behavior of $a(t)$ as t approaches infinity.

As is usual in the literature, we adopt the notation $F * a$ to denote the convolution $\int_{[0,t]} a(t-u)F(du)$. The renewal equation (2.1) can then be re-written symbolically in the form $a = b + F * a$.

Before proceeding to the statements of the two basic forms of the renewal theorem, we will first give conditions under which unique solutions to the renewal equation exist. Define $\mathcal{B}$ to be the class of Borel measurable functions $b : [0, \infty) \to I\!R$ such that $\sup\{|b(s)| : 0 \leq s \leq t\} < \infty$ for all $t \geq 0$. In other words, $\mathcal{B}$ corresponds to the family of real-valued Borel measurable functions on $[0, \infty)$ that are bounded on bounded intervals. For $F \in \mathcal{F}$, we let $U_F(t)$ be the function on $[0, \infty)$ defined by

$$U_F(t) = \sum_{n=0}^{\infty} F^{(n)}(t),$$

where $F^{(n)}$ is the n-fold convolution of F. The function $U_F(t)$ is called the **renewal function** associated with F.

The following well-known theorem establishes a uniqueness-existence result for solutions to the renewal equation. (For a proof, see p. 184–186 of Karlin and Taylor (1975).)

(2.2) THEOREM. If $F \in \mathcal{F}$ and $b \in \mathcal{B}$, then there exists a solution $\boldsymbol{a} \in \mathcal{B}$ to the renewal equation (2.1). The solution $\boldsymbol{a}$ is unique in the class $\mathcal{B}$ and is given by

$$a = U_F * b. \tag{2.3}$$

Furthermore, the solution a defined by (2.3) satisfies (2.1) at every $t \geq 0$.

In many applications, we can not assume that either a or b is in the class $\mathcal{B}$. For example, there may exist a non-empty Lebesgue set of measure zero on which $|b(t)|$ is infinite. To deal with this situation, we can often use the next result. (See Glynn (1988) for a proof.) Let $\mathcal{L}$ be the class of functions $b : [0, \infty) \rightarrow \mathbb{R}$ such that $\int_0^t |b(s)|ds < \infty$ for all $t \geq 0$.

(2.4) THEOREM. If $F \in \mathcal{F}$ and $b \in \mathcal{L}$, then there exists a solution $a \in \mathcal{L}$ to the renewal equation (2.1). The solution a is unique in the class $\mathcal{L}$ (i.e. if $a_1, a_2 \in \mathcal{L}$ satisfy (2.1), then $a_1 = a_2$ a.e.) and is given by $a = U_F * b$.

Typically, the solution $U_F * b$ to the renewal equation (2.1) can not be computed explicitly. However, the renewal theorem allows us to analyze the limiting behavior of the solution $(U_F * b)(t)$ as $t \rightarrow \infty$. Let $\lambda = (\int_{[0,\infty)} tF(dt))^{-1}$ and define λ to equal zero if F has infinite mean. Basically, the renewal theorem provides conditions under which

$$(U_F * b)(t) \rightarrow \lambda \int_0^\infty b(s)ds \tag{2.5}$$

as $t \rightarrow \infty$. The two versions of the renewal theorem give two different sets of conditions for establishing the validity of (2.5). The first version that we shall discuss is known as Smith's form of the renewal theorem; it imposes a fairly strong regularity hypothesis on F and, as a result, demands less structure on b.

(2.6) THEOREM. Suppose that $F \in \mathcal{F}$ and is spread-out (i.e. for some $n \geq 1$, $F^{(n)}$ has a non-trivial absolutely continuous component with respect to Lebesgue measure). If b is bounded, Lebesgue integrable on $[0, \infty)$, and $b(t) \rightarrow 0$ as $t \rightarrow \infty$, then $(U_F * b)(t) \rightarrow \lambda \int_0^\infty b(s)ds$.

For a proof, see Smith (1955, 1960). Looking at the conditions on b, the integrability hypothesis is natural in view of the fact that the integral $\int_0^\infty b(s)ds$ appears in the limit. As for the two other conditions, the following examples show that they are, in some sense, necessary.

(2.7) EXAMPLE. (necessity of boundedness): Suppose that b is non-negative, Lebesgue integrable on $[0,\infty)$, and $b(t) \to 0$ as $t \to \infty$, but that $\sup\{|b(s)| : 0 \le s \le 1\} = \infty$. We will show that there exists a spread-out $F \in \mathcal{F}$ such that $(U_F * b)(n) = \infty$ for $n \ge 1$.

There exists $\{t_n : n \ge 1\}$ such that $0 \le t_n \le 1$ and $b(t_n) \to \infty$. By picking a suitable subsequence, we may further assume that $b(t_n)\cdot 2^{-n} \to \infty$ as $n \to \infty$. Then, let $B(x) = \sum_{k=1}^{\infty} 2^{-k} I(x \ge 1 - t_k)$ and observe that

$$\int_{[0,1]} b(1-x)B(dx) = \sum_{k=1}^{\infty} 2^{-l} b(t_k) = \infty.$$

Now, let $F(x) = 2^{-1}(x \wedge 1) + \sum_{n=1}^{\infty} 2^{-n-1} B(x-n+1)$ for $x \ge 0$; note that F is spread-out. But

$$(U_F * b)(n) \ge \int_{[n-1,n]} b(n-x)F(dx) \ge 2^{-n-1} \int_{[0,1]} b(1-x)B(dx) = \infty.$$

(2.8) EXAMPLE. (necessity of $b(t) \to 0$ as $t \to \infty$): Suppose that b is bounded and Lebesgue integrable on $[0,\infty)$, but $b(t) \not\to 0$ as $t \to \infty$. We will show that there exists a spread-out $F \in \mathcal{F}$ such that $(U * b)(t) \not\to \lambda \int_0^\infty b(s)ds$. Choose $F(x) = 1 - \exp(-\lambda x)$ for $x \ge 0$. Then, $U_F(t) = 1 + \lambda t$ (as may be verified via Laplace transforms), so that

$$(U_F * b)(t) = b(t) + \lambda \int_0^t b(s)ds.$$

Since $\int_0^t b(s)ds \to \int_0^\infty b(s)ds$, it is evident that $b(t)$ must tend to zero, in order that $(U_F * b)(t)$ converge to the correct limit.

More recently, the following uniform version of Smith's theorem has been obtained by Arjas, Nummelin, and Tweedie (1978).

(2.9) THEOREM. Suppose that $F \in \mathcal{F}$ and is spread-out. If c is a non-negative bounded Lebesgue integrable function on $[0,\infty)$, and $c(t) \to 0$ as $t \to \infty$, then

$$\lim_{t\to\infty} \sup_{|b|\le c} |(U_F * b)(t) - \lambda \int_0^\infty b(s)ds| = 0.$$

Given an operator H, the notation $\sup_{|b| \leq c} H(b)$ means $\sup\{H(b) : b$ is a Borel measurable function on $[0, \infty)$ such that $|b(t)| \leq c(t)$ for each $t \geq 0\}$.

The second version of the renewal theorem is due to Feller. Rather than require F to be spread-out, it is demanded only that F be non-arithmetic (i.e. there exists no positive number η such that all the points of increase of F appears in the set $\{k\eta : k \geq 0\}$). On the other hand, the conditions on b are now strengthened somewhat. Let

$$\overline{\sigma}_n(b, h) = \sup\{b(s) : nh \leq s < (n+1)h\}$$
$$\underline{\sigma}_n(b, h) = \inf\{b(s) : nh \leq s < (n+1)h\}.$$

The function b is said to be **directly Rieman integrable** (d.R.i.) if $\sum_{n=0}^{\infty} \overline{\sigma}_n(|b|, h) < \infty$ for all $h > 0$ and if

$$\lim_{h \downarrow 0} h \cdot \sum_{n=0}^{\infty} (\overline{\sigma}_n(b, h) - \underline{\sigma}_n(b, h)) = 0.$$

Note that every directly Riemann integrable function b is necessarily bounded, Lebesgue integrable over $[0, \infty)$, and convergent to zero at infinity.

(2.10) THEOREM. Suppose that $F \in \mathcal{F}$ and is non-arithmetic. If b is d.R.i., then $(U_F * b)(t) \rightarrow \lambda \int_0^\infty b(s)ds$ as $t \rightarrow \infty$.

For a proof, see Feller (1970). With this strong hypothesis on b, it is well known that the non-arithmetic hypothesis on F is necessary and sufficient for $(U_F * b)(t)$ to converge to $\lambda \int_0^\infty b(s)ds$. By this, we mean that if $\lambda > 0$, then it is necessary and sufficient that F be non-arithmetic in order that $(U_F * b)(t) \rightarrow \lambda \int_0^\infty b(s)ds$ for all d.R.i. b. (For the necessity, note that if F is arithmetic with span η and $b(t) = I(0 \leq t < \eta/2)$, then $(U_F * b)(t) = U_F(t) - U_F(t - \eta/2)$, so that $(U_F * b)(n\eta) \rightarrow \lambda\eta$ by the discrete renewal theorem, whereas $(U_F * b)(n\eta + \frac{3}{4}\eta) \rightarrow 0$ as $n \rightarrow \infty$.)

3. LIMIT THEOREMS FOR EQUILIBRIUM PROCESSES

Our objective here is to discuss the implications for equilibrium processes of the renewal theorems stated in Section 2. An equilibrium process is a generalization of regenerative process that was introduced by Smith (1955).

Specifically, let $X = (X(t) : t \geq 0)$ be a (measurable) S-valued stochastic process. Given an increasing sequence of random times $0 \leq T(0) < T(1) < \ldots$, set $T(-1) = 0$ and let $N(t) = \max\{n \geq -1 : T(n) \leq t\}$. Put $\tilde{X}(t) = (X(t), t - T(N(t)))$. X is said to be an **equilibrium process** (with respect to $(T(n) : n \geq 0)$) if:

(3.1) i) $(\tilde{X}(T(n)+t) : t \geq 0) \stackrel{\mathcal{D}}{=} (\tilde{X}(T(0)+t) : t \geq 0)$ for $n \geq 0$ ($\stackrel{\mathcal{D}}{=}$ denotes equality in distribution).

ii) $T(n)$ is independent of $(\tilde{X}(T(n)+t) : t \geq 0)$ for $n \geq 0$.

X is a **non-delayed** equilibrium process if $T(0) = 0$ and is otherwise said to be **delayed**. Note that we do not demand that $(\tilde{X}(t) : t < T(n))$ be independent of $(\tilde{X}(t) : t \geq T(n))$, as would be the case for a regenerative process.

The principal application to equilibrium processes of renewal methods involves the study of the long-run behavior of the system. Specifically, let $\theta_t X = (X(t+s) : s \geq 0)$ and let f be a real-valued (suitably measurable) function. We will discuss the asymptotic behavior of $Ef(\theta_t X)$; of course, $Ef(X(t))$ is but a special case of this.

Let $\tau_i = T(i) - T(i-1)$ for $i \geq 0$. Set $a_f(t) = Ef(\theta_t X)$, $b_f(t) = E\{f(\theta_{T(0)+t}X); \tau_1 > t\}$, $c_f(t) = E\{f(\theta_t X); \tau_0 > t\}$, $G(t) = P\{\tau_0 \leq t\}$ and $F(t) = P\{\tau_1 \leq t\}$. Finally, put $V_i(f) = \sup\{f(\theta_t X) : T(i-1) \leq t < T(i)\}$ and $Y_i(f) = \int_{T(i-1)}^{T(i)} f(\theta_s X) ds$ for $i \geq 0$.

(3.2) PROPOSITION. i) Suppose that $EV_0(|f|) < \infty$ and $EV_1(|f|) < \infty$. Then, for all $t \geq 0$, $a_f(t)$ is finited-valued and satisfies, for all $t \geq 0$,

$$a_f(t) = c_f(t) + (G * U_F * b_f)(t). \tag{3.3}$$

ii) Suppose that $EY_0(|f|) < \infty$, $EY_1(|f|) < \infty$. Then, for a.e. $t \geq 0$ (a.e. with respect to Lebesgue measure), $a_f(t)$ is finite-valued. Furthermore, a_f satisfies (3.3) at a.e. $t \geq 0$.

PROOF. A standard argument shows that when f is bounded, (3.3) holds. (Theorem 2.2 can be applied directly in this case.) An easy approximation argument then shows that (3.3) holds for all non-negative f. To extend to the case where f is unbounded and of mixed sign, we use another approximation argument to show that (3.3) holds at each $t \geq 0$ for which $a_{|f|}(t) < \infty$. For i), note that $c_{|f|}(t) \leq EV_0(|f|)$, $b_{|f|}(t) \leq EV_1(|f|)$, and it follows that $a_{|f|}(t) \le EV_0(|f|) + U_F(t) \cdot EV_1(|f|) < \infty$. For ii), observe that $\int_0^\infty c_{|f|}(s)ds = EY_0(|f|)$, $\int_0^\infty b_{|f|}(s)ds = EY_1(|f|)$, and it follows that $\int_0^t a_{|f|}(s)ds \le EY_0(|f|) + U_F(t) \cdot EY_1(|f|) < \infty$ for all $t \ge 0$.

Note that the convolution $U_F * b_f$ appears in (3.3), thereby suggesting that the renewal theorem will play a central role in the analysis of the asymptotic behavior of $a_f(t)$. The following theorem is the translation of Theorem 2.9 to our current setting. (Observe that Fubini's theorem implies that $\int_0^\infty b_f(t)dt = EY_1(f)$.)

(3.4) THEOREM. Suppose that $F \in \mathcal{F}$ and is spread-out. If $E(V_0(|g|) + V_1(|g|) + Y_1(|g|)) < \infty$, then

$$\lim_{t\to\infty} \sup_{|f|\le|g|} |Ef(\theta_t X) - \lambda EY_1(f)| = 0.$$

PROOF. The proof is basically a consequence of Theorem 2.9. Note that $EV_0(|g|) < \infty$ implies that $c_f(t) \to 0$ uniformly in f satisfying $|f| \le |g|$. To handle $G * U_F * b_f$, we apply a slight extension of Theorem 2.9, stated as Theorem 1 of Arjas, Nummelin, and Tweedie (1978).

In terms of the hypotheses that enter Theorem 3.4, the moment conditions on $V_0(|g|)$ and $V_1(|g|)$ might, at first glance, seem unnecessary. They are basically imposed in order to guarantee that $c_{|g|}(t)$ and $b_{|g|}(t)$ be bounded functions that vanish at infinity. As pointed out in Examples 2.7 and 2.8, some condition of this kind is necessary.

Suppose that $E\tau_1 < \infty$ and let $\pi(\cdot)$ be the probability distribution on S defined by $\pi(\cdot) = \lambda E\{\int_{T(0)}^{T(1)} I(X(s) \in \cdot)ds\}$. An important corollary to Theorem 3.4 is that if F is spread-out, then $P\{X(t) \in \cdot\}$ converges to $\pi(\cdot)$ in total variation norm.

(3.5) COROLLARY. Suppose that $F \in \mathcal{F}$ and is spread-out. If $E\tau_1 < \infty$, then

$$||P\{X(t) \in \cdot\} - \pi(\cdot)|| \to 0 \text{ as } t \to \infty.$$

where $||\eta(\cdot)|| \triangleq \sup\{|\eta(A)|: A \text{ is Borel measurable}\}$.

Miller (1972) points out that, in some sense, the requirement that F be spread-out is necessary. Suppose that $F \in \mathcal{F}$ has finite mean and yet is not spread-out. For such an F, Miller showed that there exists a real-valued regenerative process X for which τ_1 has distribution F and yet $X(t)$ does not even converge in distribution as $t \to \infty$. Thus, in order to weaken the requirement on F to an assumption that F is non-arithmetic, it is necessary to include a hypothesis on X.

Suppose that S is a metric space. Let $D_S[0,\infty)$ be the Skorohod space of right-continuous functions $x : [0,\infty) \to S$ having left limits at every $t > 0$. The hypothesis on X involves assuming that X has paths in $D_S[0,\infty)$. The next result is basically Miller's extension of Feller's version of the renewal theorem to the equilibrium process setting.

(3.6) THEOREM. Suppose that $F \in \mathcal{F}$ is non-arithmetic and has finite mean. If S is a metric space and $X \in D_S[0,\infty)$, then $P\{X(t) \in \cdot\} \Rightarrow \pi(\cdot)$ as $t \to \infty$. ($\Rightarrow$ denotes weak convergence).

PROOF. This result was established by Miller (1972) in the special case that $S = \mathbb{R}$ (this argument is stated in terms of regenerative processes but easily extends to equilibrium processes). To extend to the general case, let $f : S \to \mathbb{R}$ be a continuous map. Then, $(f(X(t)) : t \geq 0)$ is an equilibrium process having paths in $D_{\mathbb{R}}[0,\infty)$. We can now apply Theorem 3.1 of Miller (1972) to conclude that $f(X(t)) \Rightarrow f(X(\infty))$ as $t \to \infty$, where $X(\infty)$ has distribution π. It follows from Theorem 5.2 ii) of Billingsley (1968) that $X(t) \Rightarrow X(\infty)$ as $t \to \infty$.

We note that Theorem 3.6, in contrast to Corollary 3.5, requires topological assumptions on S and path regularity of X. Furthermore, the total variation convergence of Corollary 3.5 is much stronger than the weak convergence of Theorem 3.6. In addition, note that Theorem 3.6 permits us only to assert that $Ef(X(t)) \to Ef(X(\infty))$ for all bounded continuous f; if f is either unbounded or discontinuous, Theorem 3.6 offers no conclusion, in contrast to Theorem 3.4. On the other hand, the non-arithmetic assumption on τ_1 is considerably weaker than the demand that τ_1 be spread-out, as appeared in Theorem 3.4 and Corollary 3.5. Thus, both the Smith and Feller versions described above are somewhat unsatisfactory.

4. A SMOOTHED VERSION OF THE RENEWAL THEOREM

In Sections 4 and 5, we state and prove the main results of this paper. We start in this section by describing a smoothed variant of the renewal theorem. This theorem will basically establish "smoothed convergence" of $(U_F * b)(t)$ to $\lambda \int_0^\infty b(s)ds$ under the weakest possible conditions, namely integrability of b over $[0, \infty)$ and non-arithmeticity of F.

Let K be an absolutely continuous distribution in $\mathcal{F}$ which has a Lebesgue density k with support contained in $[0, T](0 < T < \infty)$; we further require that k be continuous on $[0, T]$. We will use K to smooth out the solution $\boldsymbol{a} = U_F * b$ to the renewal equation. Specifically, we will consider the quantity $(K * \boldsymbol{a})(t) = \int_0^T k(s)\boldsymbol{a}(t-s)ds$ as $t \to \infty$. Thus, K is used to smooth the solution $\boldsymbol{a}$ in the region $[t-T, t]$. Since convolution with K ought to enhance the regularity of $\boldsymbol{a}$, we expect that some of the pathologies which we discussed in Section 2 will cease to be a problem. The next theorem is an illustration of this point (compare with Theorem 2.9).

(4.1) THEOREM. Suppose that $F \in \mathcal{F}$ is non-arithmetic. If c is a non-negative Lebesgue integrable function on $[0, \infty)$, then

$$\lim_{t \to \infty} \sup_{|b| \le c} |(K * U_F * b)(t) - \lambda \int_0^\infty b(s)ds| = 0.$$

PROOF. First we extend b and k to $(-\infty, \infty)$ by setting $b(t) = k(t) = 0$ for $t < 0$. Let $U_F'(t) = U_F(t) - \lambda t$ and $\varphi_b(t) = (K * b)(t)$. Observe that $K * U_F = U_F * K$ and $\int_0^\infty \varphi_b(s)ds = \int_0^\infty b(s)ds$ for b integrable. Hence,

$$\begin{aligned} &\sup_{|b| \le c} |(K * U_F * b)(t) - \lambda \int_0^\infty b(s)ds| \\ &= \sup_{|b| \le c} |(U_F * \varphi_b)(t) - \lambda \int_0^\infty \varphi_b(s)ds| \qquad (4.2) \\ &\le \sup_{|b| \le c} |(U_F' * \varphi_b)(t)| + \lambda \int_t^\infty \varphi_c(s)ds. \end{aligned}$$

Since φ_c is integrable over $[0, \infty)$, the second term tends to zero as $t \to \infty$. For the first term, we approximate φ_b by a suitably chosen piecewise constant function φ_b^h. Set $\varphi_b^h(t) = \varphi_b(h\lfloor t/h \rfloor)$ for $h > 0$ and write $U_F' * \varphi_b = U_F' * \varphi_b^h + U_F' * (\varphi_b - \varphi_b^h)$. Let $\ell = \lfloor t/h \rfloor$ and $r = t - h\lfloor t/h \rfloor$. It may be verified that

$$(4.3) \qquad \begin{aligned} &(U_F' * \varphi_b^h)(t) = \varphi_b(\ell h) \cdot U_F'(r) + \\ &\sum_{i=0}^{\ell-1} \varphi_b((\ell - i - 1)h) \cdot (U_F'(r + (i+1)h) - U_F'(r + ih)). \end{aligned}$$

It follows that

$$(4.4) \qquad \begin{aligned} &\sup_{|b| \le c} |(U_F' * \varphi_b^h)(t)| \le \varphi_c(h\lfloor t/h \rfloor)(U_F(h) + \lambda h) \\ &+ \sum_{j=0}^{\ell-1} \varphi_c(jh)|U_F(t - jh) - U_F(t - (j+1)h) - \lambda h|. \end{aligned}$$

For the first term on the right-hand side of (4.4), we note that $\varphi_c(t) = \int_0^t c(s)k(t-s)ds \to 0$ by the bounded convergence theorem (k is bounded because of continuity over its support). So the first term vanishes as $t \to \infty$. To deal with the second term, recall that $U_F(t - jh) - U_F(t - (j+1)h) \to \lambda h$ as $t \to \infty$ by Blackwell's theorem (Feller (1970), p. 360). Furthermore, these differences are bounded uniformly in t and j. Hence, the second term in (4.4) vanishes as $t \to \infty$, provided that we show $\sum_{j=0}^{\infty} \varphi_c(jh) < \infty$.

To prove this, let $M = \sup\{|k(s)| : 0 \le s \le T\}$ and let $\varepsilon(h) = \sup\{|k(t) - k(t-h)| : h \le t \le T\}$. Note that $\varepsilon(h) \downarrow 0$ as $h \downarrow 0$ by uniform continuity of k. For $|b| \le c$ and $0 < t' = h\lfloor t/h \rfloor < t$,

$$\begin{aligned} |\varphi_b(t) - \varphi_b(t')| &= |\int_{t-T}^{t} k(t-s)b(s)ds - \int_{t'-T}^{t'} k(t'-s)b(s)ds| \\ &\le \int_{t'}^{t} k(t-s)|b(s)|ds + \int_{t-T}^{t'} |k(t-s) - k(t'-s)||b(s)|ds \\ &+ \int_{t'-T}^{t-T} k(t'-s)|b(s)|ds \\ &\le M \int_{t-h}^{t} c(s)ds + \int_{h}^{T} |k(u) - k(u-t+t')|c(t-u)du \end{aligned}$$

(4.5)

$$
\begin{aligned}
&+\int_{t-t'}^{h} k(u)c(t-u)du + \int_{t-t'}^{h} k(u-t+t')c(t-u)du \\
&+M\int_{t-T-h}^{t-T} c(s)ds \\
&\leq M\left(\int_{t-h}^{t} c(s)ds + 2\int_{t-h}^{t+h} c(s)ds + \int_{t-T-h}^{t-T} c(s)ds\right) \\
&+\varepsilon(h)\int_{t-T}^{t} c(s)ds.
\end{aligned}
$$

Since $\int_{t-\alpha}^{t} c(s)ds$ is a Lebesgue integrable function of t over $[0,\infty)$ for any finite α, it is evident from (4.5) that

$$
\begin{aligned}
h\sum_{j=0}^{\infty}\varphi_c(jh) &= \int_0^{\infty}\varphi_c(h\lfloor t/h\rfloor)dt \\
&\leq \int_0^{\infty}\varphi_c(t)dt + \int_0^{\infty}|\varphi_c(t)-\varphi_c(h\lfloor t/h\rfloor)|dt < \infty.
\end{aligned}
$$

To complete the proof of the theorem, it is sufficient to prove that $\overline{\lim}_{t\to\infty}\sup_{|b|\leq c}|(U_F'*(\varphi_b-\varphi_b^h))(t)| \equiv \beta(h)$, with $\beta(h)\downarrow 0$ as $h\downarrow 0$. It is easily seen that (4.5) proves that $|\varphi_b(t)-\varphi_b^h(t)|\leq\psi_h(t)$, where

$$
\psi_h(kh) = M\left[3\int_{h(k-1)}^{h(k+2)} c(s)ds + \int_{h(k-1)-T}^{h(k+1)-T} c(s)ds\right] + \varepsilon(h)\int_{hk-T}^{h(k+1)} c(s)ds
$$

and $\psi_h(t)=\psi_h(h\lfloor t/h\rfloor)$. A routine calculation shows that $\psi_h(t)\to 0$ as $t\to\infty$ and that $\int_0^{\infty}\psi_h(t)dt\equiv\gamma(h)$, with $\gamma(h)\downarrow 0$ as $h\downarrow 0$. Now,

$$
\sup_{|b|\leq c}|(U_F'*(\varphi_b-\varphi_b^h))(t)| \leq \int_{[0,t]}(U_F(ds)+\lambda ds)\psi_h(t-s). \tag{4.6}
$$

Using the piecewise-constant structure of ψ_h, we can argue, as we did for (4.3), that

$$
\begin{aligned}
(U_F*\psi_h)(t) &\leq \psi_h(h\lfloor t/h\rfloor)U_F(h) \\
&+\sum_{j=0}^{\lfloor t/h\rfloor-1}\psi_h(jh)(U_F(t-jh)-U_F(t-(j-1)h)).
\end{aligned}
$$

Again, Blackwell's theorem asserts that $U_F(t-jh) - U_F(t-(j-1)h) \to \lambda h$ as $t \to \infty$. Using the aforementioned properties of ψ_h, we find that $\overline{\lim}_{t\to\infty}(U_F * \psi_h)(t) \le \lambda \int_0^\infty \psi_h(s)ds$. It follows from (4.6) that

$$\overline{\lim}_{t\to\infty} \sup_{|b|\le c} |(U_F' * (\varphi_b - \varphi_b^h))(t)| \le 2\lambda\gamma(h),$$

completing the proof.

The following "averaging" result is an immediate consequence of Theorem 4.1.

(4.7) COROLLARY. Suppose that $F \in \mathcal{F}$ is non-arithmetic and b is Lebesgue integrable on $[0,\infty)$. Then, for $h > 0$,

$$\frac{1}{h}\int_{t-h}^{t} (U_F * b)(s)ds \to \lambda \int_0^\infty b(s)ds \text{ as } t \to \infty.$$

A noteworthy feature of this result is that it holds for any positive h. Thus, if the solution $U_F * b$ is averaged in an arbitrarily small neighborhood of t, it behaves nicely.

5. SMOOTHED LIMIT THEOREMS FOR EQUILIBRIUM PROCESSES

In this section, we apply Theorem 4.1 to equilibrium processes. Our first result is obtained by combining Proposition 3.2 and Theorem 4.1.

(5.1) THEOREM. Suppose that $F \in \mathcal{F}$ is non-arithmetic and $E(V_0(|g|) + Y_1(|g|)) < \infty$. Assume G has a Lebesgue density with support contained in $[0,T](T < \infty)$. If the density is continuous on $[0,T]$, then

$$\lim_{t\to\infty} \sup_{|f|\le|g|} |Ef(\theta_t X) - \lambda EY_1(f)| = 0.$$

PROOF. This result follows immediately from Proposition 3.2 and Theorem 4.1, upon showing that $c_f(t)$ converges to zero uniformly in f. (Actually, Proposition 3.2 does not strictly apply here; the proof can be modified to show that (3.3) holds at all $t \ge 0$, however.) But $|c_f(t)| \le c_g(t) \le E\{V_0(|g|); \tau_0 > t\} \to 0$.

Total variation convergence is an easy corollary of Theorem 5.1.

(5.2) COROLLARY. Suppose that $F \in \mathcal{F}$ is non-arithmetic with finite mean. Assume G satisfies the conditions of Theorem 5.1. If π is defined as in Section 3, then

$$||P\{X(t)\varepsilon\cdot\} - \pi(\cdot)|| \to 0 \text{ as } t \to \infty.$$

The interesting point here is that the "delay cycle" can smooth the equilibrium process enough to obtain total variation convergence. Recall, from Section 3, that Miller's example shows that, without the smoothing of the delay cycle, we can not even expect convergence in distribution (without making further assumptions about the paths of X).

The results of Section 4 can also be used to obtain smoothed limit theorems that are more reminiscent of Corollary 4.7.

(5.3) THEOREM. Suppose that $F \in \mathcal{F}$ is non-arithmetic and $E(Y_0(|g|) + Y_1(|g|)) < \infty$. If K is as described in Section 4, then

$$\lim_{t\to\infty} \sup_{|f|\leq|g|} |\int_0^T k(s)Ef(\theta_{t-s}X)ds - \lambda EY_1(f)| = 0.$$

PROOF. To prove this result, we note that $\int_0^T k(s)Ef(\theta_{t-s}X)ds = (K*c_f)(t)+(G*K*U_f*b_f)(t)$ for $t \geq T$. Since $(K*U_F*b_f)(t) \to \lambda EY_1(f)$ uniformly in f (and can be bounded uniformly in both f and t), the bounded convergence theorem shows that $(G * K * U_F * b_f)(t) \to \lambda EY_1(f)$ uniformly in f. For the other term observe that $|(K*c_f)(t)| \leq (K*c_{|g|})(t) \leq M \int_{t-T}^t c_{|g|}(s)ds \to 0$. Here, M is a bound on k and we use the fact that $EY_0(|g|) < \infty$ to conclude that $c_{|g|}$ is integrable on $[0, \infty)$ (This implies that $\int_{t-T}^t c_{|g|}(s)ds \to 0$.)

Again, we can write down a total variation convergence corollary to this result; it is the equilibrium process analog of the averaging result given by Corollary 4.7.

(5.4) COROLLARY. Suppose that $F \in \mathcal{F}$ is non-arithmetic with finite mean. Let π be defined as in Section 3. Then for any $h > 0$,

$$||\frac{1}{h}\int_{t-h}^t P\{X(s)\varepsilon\cdot\}ds - \pi(\cdot)|| \to 0 \text{ as } t \to \infty.$$

A common feature of the smoothed limit theorems of this section is that convergence is proved under minimal hypotheses. In addition to its mathematical virtues, this should make hypothesis validation easier from an applications viewpoint. Furthermore, it is our opinion that these smoothed versions provide information that is, from a practical perspective, essentially equivalent to that obtained from an unsmoothed version.

REFERENCES

Arjas, E., Nummelin, E., and R. L. Tweedie (1974), Uniform limit theorems for non-singular renewal and Markov renewal processes, *Journal of Applied Probability*, **15**, 112–125.

Billingsley, P. (1964), *Convergence of Probability Measures*, John Wiley, New York.

Feller, W. (1970), *An Introduction to Probability Theory and Its Applications*, Vol. 2, John Wiley, New York.

Glynn, P. W. (1988), A low bias steady-state estimator for equilibrium processes, Technical Report, Department of Operations Research, Stanford University, Stanford, CA.

Karlin, S., and H. M. Taylor (1975), *A First Course in Stochastic Processes*, Academic Press, New York.

Miller, D. R. (1972), Existence of limits in regenerative processes, *Annals of Mathematical Statistics*, **43**, 1275–1242.

Smith, W. L. (1955), Regenerative stochastic processes, *Proceedings of the Royal Society London, Series A*, **232**, 6–31.

Smith, W. L. (1960), Remarks on the paper 'Regenerative stochastic processes,' *Proceedings of the Royal Society London, Series A*, **256**, 496–501.

Supercritical Branching Processes with Countably Many Types and the Size of Random Cantor Sets

Harry Kesten*
Cornell University

1. INTRODUCTION

Sam Karlin has made many fundamental contributions to branching processes. (See Athreya and Ney (1972) and its bibliography for descriptions of some of these papers.) Given his interest in the use of branching processes in population genetics, it is natural that most of these papers deal with branching processes of finitely many types. Here we wish to present a convergence theorem for a supercritical branching process with countably many types.

Branching processes with infinitely many types have, of course, been considered before. Moy (1967) considered countably many types, but it is

AMS 1980 Subject Classifications: Primary 60J80.

Key words and phrases: Branching processes, countably many types, random Cantor sets, Hausdorff dimension, projections.

*Research supported by the National Science Foundation through a grant to Cornell University.

PROBABILITY, STATISTICS,
AND MATHEMATICS
Papers in Honor of Samuel Karlin

Copyright © 1989 by Academic Press, Inc.
All rights of reproduction in any form reserved.
ISBN-0-12-058470-0

more common, when allowing infinitely many types, to consider a continuum of types. In the latter case the type of a particle is often identified with the position of the particle in some space; see Asmussen and Hering (1983). Practically always some compactness condition is imposed on the space of types, or, as in Moy (1967), it is assumed that the process can be approximated in some sense by a process with finitely many types, or with a compact set of types. This paper is no exception; the assumption (1.6) below is also of this nature, but we believe that our result is not covered by the existing literature.

The motivation for our theorem was an investigation by Dekking and Grimmet (1987) of projections of "random M-adic Cantor sets," which made quantitative some suggestions of Mandelbrot (1983), p. 218. The application to such Cantor sets will be described below. We first state our definitions and "pure" branching processes result.

$Z_t(j)$ will denote the number of particles of type j in the tth generation. Here, the time index t takes the values $0, 1, \cdots$, while the type index j takes its values in $\mathbb{N} = \{1, 2, \cdots\}$. The offspring numbers are represented by random vectors $X_t(i,j) = \{X_t(i,j)(k)\}_{k \in \mathbb{N}}$; $X_t(i,j)(k)$ denotes the number of children of type k of the ith individual of type j in the tth generation. Thus,

$$Z_{t+1}(k) = \sum_{j \in \mathbb{N}} \sum_{1 \le i \le Z_t(j)} X_t(i,j)(k).$$

What makes $\{Z_t\}$ a (Bienaymé-Galton-Watson) branching process are the following two assumptions which will be in force throughout.

(1.1) $\{X_t(i,j) : t \ge 0, \quad i, j \in \mathbb{N}\}$ are independent.

(1.2) For fixed j all random vectors $X_t(i,j)$, $t \ge 0$, $i \in \mathbb{N}$ have the same distribution.

Thus each particle produces offspring independently of all others, and the offspring distribution depends on the type of the particle only. We shall employ without further explanation the usual terminology for the family tree associated with the branching process, such as child(ren), parent, ancestor of a particle or individual. The reader is referred to Harris (1963), Sect. VI.2, or Jagers (1975), Sect. 1.2, for details. The mean offspring matrix M (which is assumed to be finite) has entries

$$M(j,k) = E\{X_t(i,j)(k)\},$$

which is independent of t and i. M^n denotes its nth matrix power.

To state the – unfortunately rather extensive – assumptions we need some more notation. $\{A_n\}$ will be a sequence of subsets of $\mathbb{N}$ and ϵ_n, $\gamma(n)$ will be certain strictly positive constants which satisfy the conditions listed below. I_A denotes the indicator function of the set A. For an arbitrary subset B of $\mathbb{N}$

$$Z_t(B) := \text{number of particles in the } t\text{th generation whose type belongs to } B,$$

$$Z^*_{T,t}(B) := \text{number of particles in the } t\text{th generation whose type belongs to } B, \text{ and whose ancestor in the } k\text{th generation had type in } A_k \text{ for } T \le k \le t (Z^*_{T,T}(j) = Z_T(j) I_{A_T}(j)).$$

$$M^*_t(j,k) = M(j,k) I_{A_t}(k) (= \text{restriction of } M \text{ to } \mathbb{N} \times A_t)$$

For a vector $v = \{v(j)\}_{j \in \mathbb{N}}$ we set

$$|v| = \sum_{j=1}^{\infty} |v(j)|.$$

We impose the following conditions:

$$\sum_{j=1}^{\infty} M^n(i,j) < \infty \quad \text{for all } n \ge 0, \quad i \in \mathbb{N}. \tag{1.3}$$

$$A_n \text{ increases to all of } \mathbb{N} \text{ and } 1 \in A_n \text{ for all } n. \tag{1.4}$$

$$\begin{gathered} \epsilon_n \downarrow 0, \gamma(i) \ge 1, \ \max\{\gamma(i) : i \in A_n\} \le e^{n\epsilon_n} \text{ and} \\ \#A_n := \text{ cardinality of } A_n \le e^{n\epsilon_n}. \end{gathered} \tag{1.5}$$

$$\begin{gathered} f_t := \sup_{\ell \ge 0} \frac{E|Z_\ell|}{E|Z_{\ell+t}|} \text{ satisfies} \\ f_t \sum_{j \notin A_t} E\, Z_t(j)\gamma(j) \le (t+1)^{-4}, \quad t \ge 0. \end{gathered} \tag{1.6}$$

$$\sup_n \frac{\Sigma_j M^n(i,j)}{\Sigma_j M^n(1,j)} \leq \gamma(i), \quad i \in \mathbb{N}. \tag{1.7}$$

For all $t \geq$ some t_0 and $\ell > t$, $i_1, i_2 \in A_t$

$$\frac{\sum_{j \in A_\ell} M^*_{t+1} \cdots M^*_\ell(i_1, j)}{\sum_{j \in A_\ell} M^*_{t+1} \cdots M^*_\ell(i_2, j)} \leq e^{t\epsilon_t}. \tag{1.8}$$

$$u(i) := \lim_{t \to \infty} \frac{\Sigma_j M^t(i,j)}{\Sigma_j M^t(1,j)} \text{ exists and } 0 \leq u(i) < \infty \text{ for all } i. \tag{1.9}$$

(1.10) ("second moment assumption") For all $j \in A_t$ and for all i

$$\sigma^2(i,j) := \mathrm{Var}\{X_1(1,i)(j)\} \leq e^{t\epsilon_t}.$$

(1.11) ("supercriticality") There exists a $\rho > 1$ such that a.e. on the set of non extinction of Z_t and on $Z_0(j) = \delta_{1,j}$ there exists a (random) S for which

$$\inf_{t \geq S} |Z^*_{S,t}|^{1/t} \geq \rho > 1.$$

Moreover $P\{Z_t \text{ does not become extinct}\} > 0$.

We shall comment on these conditions in Remark 2. Here, though, is our principal result.

THEOREM 1. *Under conditions (1.1)–(1.11) there exists a random number W such that for $Z_0(j) = \delta_{1,j}$ one has w.p.1*

$$\frac{|Z_t(\cdot) - WM^t(1,\cdot)|}{E|Z_t|} \to 0, \tag{1.12}$$

and

$$\frac{|Z_t|}{E|Z_t|} \to W. \tag{1.13}$$

Moreover,

$$W > 0 \text{ a.e. on the set of non-extinction of } Z. \tag{1.14}$$

REMARK 1. Note that under the initial condition $Z_0 = \delta_1$ (i.e., start with a single particle of type 1)

$$(1.15) \qquad E|Z_t| = \Sigma_j M^t(1,j).$$

(Compare Athreya and Ney (1972), equation V.2.2.) Thus (1.13) shows that the growth rate of $|Z_t|$ is deterministic, and specifies this rate. (1.12) shows that in an ℓ^1-sense even the direction of Z_t is asymptotically deterministic. The initial condition $Z_0 = \delta_1$ is chosen for simplicity only; a similar theorem holds for any finite $|Z_0|$.

REMARK 2. The idea of the proof is to truncate the types of Z_t, i.e., to replace Z_t by $Z^*_{T,t}$ for some large T, and to show that $Z^*_{T,t}$ grows more or less deterministically, by applying Chebyshev's inequality to $Z^*_{T,t+1} - Z_{T,t}M^*_t$ for $t \geq T$. The rather mild second moment assumption (1.10) is reasonable if one wants to apply Chebyshev. Some assumption of supercriticality is necessary for the theorem, as is well known from the behavior of single type branching processes (Athreya and Ney (1972), Ch. I). Unfortunately, we have to assume in (1.11) that the *truncated* process $Z^*_{S,t}$ is already supercritical. We must rule out that $|Z_t|$ grows exponentially merely by the presence of many particles of an extremely high type. In practice one verifies condition (1.11) by showing that Z_t dominates some finite type supercritical branching process. Condition (1.6) also serves to limit the presence of too many particles of a very large type, and is crucial for making the truncation possible; particles with type outside A_t at time t give a relatively small (expected) contribution to later generations. (1.7)–(1.9) are regularity conditions on the matrix M which mimic some properties which are automatic for finite positive matrices (by the Perron-Frobenius theorem). (1.6) and (1.8) turn out to be the most troublesome conditions in the applications mentioned below.

Despite the forbidding appearance of the conditions of Theorem 1, it can be applied in the following model from Dekking and Grimmet (1987). Let $\{U_t\}_{t\geq 0}$ be a branching process with finitely many types, chosen from a finite set $\mathcal{B}$, and assume that the offspring distribution is the *same for all particles, independent of their type*. To each particle of the tth generation we assign a label $(b_1, \cdots, b_t)$, where b_ℓ is the type of the ancestor in the ℓth generation of the given particle (b_t is the type of the particle itself). The labels can also be viewed as "words in the alphabet $\mathcal{B}$." One is interested in the asymptotic behavior of

$$N_t := \text{number of distinct words of length } t \text{ represented at least once in the } t\text{th generation.}$$

$\{N_t\}$ is *not* a branching process, but if one sets

$$Z_t(j) = \text{number of distinct words of length } t \text{ which are represented exactly } j \text{ times in the } t\text{th generation,}$$

then $\{Z_t\}$ is a branching process of countably many types as described above. Moreover $N_t = |Z_t|$ and (1.13) describes the growth of N_t in some cases.

Dekking and Grimmet had in mind a very specific interpretation of the labels in the U process. Let $M > 0$ and construct a random Cantor-like subset G of $[0,1]^d$ in the following way. Let $\mathcal{A} = \{0, \cdots, M-1\}^d$. For $a_1, \cdots, a_t \in \mathcal{A}$ (with $a_\ell = (a_\ell(1), \cdots, a_\ell(d))$ where each $a_j(i) \in \{0, 1, \cdots, M-1\}$) let $I_t(a_1, \cdots, a_t) = I_t^d(a_1, \cdots, a_t)$ be the cube

$$\prod_{i=1}^{d} \left[\sum_{\ell=1}^{t} a_\ell(i) M^{-\ell}, \; \sum_{\ell=1}^{t} a_\ell(i) M^{-\ell} + M^{-t} \right].$$

$I_t(a_1, \cdots, a_t)$ is a closed subcube of $I_{t-1}(a_1, \cdots, a_{t-1})$ (I_0 is the d-dimensional unit cube $[0,1]^d$). Further let μ be a probability distribution on the subsets of $\mathcal{A}$ and form random sets W_t of words of length t. W_0 consists of he empty word ony, and for each $(a_1, \cdots, a_t) \in W_t$ choose a subset $V_t(a_1, \cdots, a_{t+1})$ of it and take W_{t+1} the collection of all $(a_1, \cdots, a_{t+1})$ with $(a_1, \cdots, a_t) \in W_t$, $a_{t+1} \in V_t(a_1, \cdots, a_t)$. All sets V_t are chosen i.i.d. with distribution μ. Finally,

$$G_0 = I_0, \quad G_t = \bigcup_{(a_1, \cdots, a_t) \in W_t} I_t(a_1, \cdots, a_t),$$

$$G = \bigcap_t G_t.$$

In words, G_{t+1} is formed from G_t by subdividing each cube I_t in G_t into M^d subcubes of equal size, and by including in G_{t+1} a random subset of these subcubes. The subcubes of the various I_t to be included in G_{t+1} are

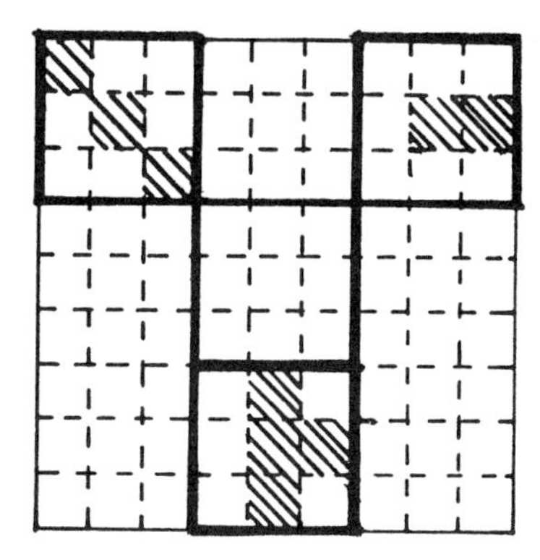

Fig. 1. Illustration of G_0 (the largest square), G_1 and G_2 for $d = 2$, $M = 3$. G_1 consists of the squares of area 3^{-2} with a boldly drawn perimeter. G_2 is the union of the hatched squares of area 9^{-2}.

chosen in an i.i.d. manner, governed by μ. An example of two steps of the construction is shown in Fig. 1. By construction $G_{t+1} \subset G_t$. G is called an M-adic Cantor set. The classical Cantor set is obtained by taking $d = 1$, $M = 3$, and $V_t(a_1, \cdots, a_t) = \{0, 2\}$ w.p.1 (i.e., μ is concentrated on one point: the subset $\{0, 2\}$ of $\mathcal{A} = \{0, 1, 2\}$).

Now consider the projection H_t, of G_t on the first $e \leq d$ coordinates. An e-dimensional cube $I_t^e(b_1, \cdots, b_t)$ with $b_\ell \in \mathcal{B} := \{0, 1, \cdots, M-1\}^e$ belongs to H_t if and only if there exist choices for $a_\ell(i)$, $e < i \leq d$, such that with $a_\ell = (b_\ell(d), \cdots, b_\ell(e), a_\ell(e+1), \cdots, a_\ell(d))$, $I_t^d(a_1, \cdots, a_t) \subset G_t$. Finally, view the cubes $I_t^d(a_1, \cdots, a_t)$ which occur in G_t as the particles of a branching process U_t. The type of "particle" $I_t(a_1, \cdots, a_t)$ is b_t, and its label is $(b_1, \cdots, b_t)$, where $b_\ell(i) = a_\ell(i)$, $1 \leq i \leq e$, $b_\ell \in \mathcal{B}$. Thus we obtain the types and labels by ignoring the last $(d - e)$ coordinates of all vectors a. As before

$$\begin{aligned} Z_t(j) :=\ & \text{number of particles whose label is represented} \\ & \text{exactly } j \text{ times in the } k\text{th generation} \\ =\ & \text{number of } (b_1, \cdots, b_t) \text{ for which } I^e(b_1, \cdots, b_t) \text{ is} \\ & \text{the projection of exactly } j \text{ cubes } I_t^d(a_1, \cdots, a_t) \text{ in } G_t. \end{aligned}$$

To state the result of Theorem 1 for $N_t := |Z_t|$ in this case define for $b \in \mathcal{B}$

$$m(b) = E\{\text{number of subcubes of type } b$$

$$\text{and size } M^{-t-1} \text{ in a given } I_t\}$$
$$(1.16) \qquad = \sum_{\substack{a \in \mathcal{A}^d \\ a(i)=b(i), 1 \le i \le e}} \mu(a \text{ belongs to } V_1).$$

Several cases have to be distinguished according to the behavior of the m's.

THEOREM 2. *Assume that for at least one $b \in \mathcal{B}$*

$$(1.17) \qquad m(b) > 0 \ \ and \ \ P\{\text{a particle has no children of type } b\} > 0.$$

If in addition to (1.17)

$$(1.18) \qquad \Gamma := \sum_b m(b) > 1$$

and

$$(1.19) \qquad \sum_b m(b) \log m(b) \le 0,$$

*then (1.12)–(1.14) hold. Moreover the asymptotic behavior of $E|Z_t|$ for large t is as follows:**

$$(1.20) \qquad E|Z_t| \asymp \Gamma^t \ \ if \ \sum_b m(b) \log m(b) < 0,$$

$$(1.21) \qquad E|Z_t| \asymp \frac{1}{\sqrt{t}} \Gamma^t \ \ if \ \sum_b m(b) \log m(b) = 0 \ \ and \\ 0 < m(b) < 1 \ \ for\ at\ least\ one\ b,$$

$$(1.22) \qquad E|Z_t| \asymp \frac{1}{t} \Gamma^t \ \ \text{if} \ \sum_b m(b) \log m(b) = 0 \ \ and \\ m(b) \ \ takes\ values\ 0\ and\ 1\ only.$$

---

* $a_t \asymp b_t$ means that $0 < C \le a_t/b_t \le C^{-1} < \infty$ for some constant C and all t.

THEOREM 3. *If*

$$\sum_b m(b) \log m(b) > 0 \ \ and \ \ \sum \log m(b) \leq 0. \tag{1.23}$$

then (1.12)-(1.14) holds. In this case the asymptotic behavior of $E|Z_t|$ *is as follows: If*

$$\sum_b \log m(b) < 0, \tag{1.24}$$

let θ *be the root in* $(0,1)$ *of the equation*

$$\sum_b \{m(b)\}^\theta \log m(b) = 0.$$

Under (1.24)

$$E|Z_t| \sim \frac{C}{t^{3/2}} \left[\sum_b \{m(b)\}^\theta \right]^t$$

for some constant $C > 0$. *If*

$$\sum_b \log m(b) = 0, \tag{1.25}$$

let λ *be the number of nonzero* $m(b)$. *Under (1.25) there exists a constant* $C > 0$ *such that*

$$E|Z_t| \sim \frac{C}{\sqrt{t}} \lambda^t. \tag{1.26}$$

THEOREM 4. *If*

$$\sum_b \log m(b) > 0 \ \ and \ (a \ fortiori) \ \sum m(b) \log m(b) > 0 \tag{1.27}$$

then for some constant $C > 0$ *and* λ *as in (1.26)*

$$E|Z_t| \sim C\lambda^t$$

and for some random variable W

$$\lambda^{-t}|Z_t| \to W \ \ w.p.1$$

$W > 0$ *almost everywhere on the set of non-extinction of* Z.

REMARK 3. It is easy to see that (1.11) can hold only if $E|Z_t|$ grows at least exponentially. Theorems 2–4 cover all cases in which $E|Z_t|$ grows exponentially. Theorem 1 does no apply when (1.27) holds. In particular the main contribution to $|Z_t|$ comes from $Z_t(j)$ with j exponentially large in t in this case (so that it is impossible to satisfy (1.5), (1.6)).

To find the asymptotic behavior of $E|Z_t|$ we rely heavily on Dekking (1987a and b). In these papers Dekking showed that $\lambda^{-t}E|Z_t|$ equals the probability of survival up till time t of a certain auxiliary branching process in random environment. Dekking and Grimmet (1987) identified the almost sure limit of $|Z_t|^{1/t}$ for the random Cantor sets, but clearly (1.12) and (1.13) give much more detailed information.

For the proof of Theorem 2 one chooses $A_n = \{1, \cdots, r_n\}$ with $r_n \sim \exp(2n^{1/2}\log n)$ and $\gamma(i) = u(i) = i$ in Theorem 1. For Theorem 3 one chooses $A_n = \{1, \cdots, r_n\}$ with $r_n \sim \exp(n^{3/4})$, $\gamma(i) = Ci^{\theta}(1 + \log i)$ ($\theta = 0$ under (1.25)), but $u(i)$ cannot be evaluated explicitly in this case.

REMARK 4. Trivially

$$M^{-et}|Z_t| = \text{ Lebesgue measure of } H_t.$$

Thus, Theorems 2–4 can be used to obtain information about the size of the projection H_t. Perhaps this will help in determining the Hausdorff measure with the correct measure function for $H := \cap H_t$. The box-counting dimension of H was determined in Dekking and Grimmett (1987), while the correct Hausdorff measure function for G itself (corresponding to $e = d$) was determined by Graf, Mauldin and Williams (1987). Very recently Falconer (1989) showed the equality of the box-counting dimension and the usual Hausdorff dimension. Other aspects of G for a special choice of μ have been investigated in Chayes, Chayes and Durrett (1988).

ACKNOWLEDGEMENT. The author is indebted to G. Grimmett for many helpful conversations concerning this paper.

2. PROOF OF THEOREM 1

We break the proof down into a sequence of lemmas, but for lack of space prove only the nontrivial ones. We do not prove Theorems 2–4 here.

We remind the reader of some standard moment formulae. Let $\mathcal{F}_t$ be the σ-field generated by $\{X_k(i,j) : k \le t, i, j \in I\!N\}$ (note that Z_k and $Z^*_{T,k}$ are $\mathcal{F}_t$-measurable for $k \le t$). As in Harris (1963), Sect. II.3, we have for $t > T$

$$E\{Z_t(j) \mid \mathcal{F}_T\} = Z_t M^{t-T}(j), \tag{2.1}$$

$$E\{Z^*_{T,t}(j) \mid \mathcal{F}_T\} = Z^*_{T,T} M^*_{T+1} \cdots M^*_t(j), \tag{2.2}$$

$$\sigma^2\{Z^*_{T,t}(j) \mid \mathcal{F}_{t-1}\} = \sum_i Z^*_{T,t-1}(i)\sigma^2(i,j), \quad j \in A_t \tag{2.3}$$

(see (1.10) for $\sigma^2(i,j)$).

(2.2), Markov's inequality and the Borel-Cantelli lemma yield the first lemma.

LEMMA 1. *W.p.1 there exist a (random) T_0 such that*

$$|Z^*_{T,t}| \le t^3 |Z^*_{T,T} M^*_{T+1} \cdots M^*_t| \text{ for all } t \ge T \ge T_0. \tag{2.4}$$

LEMMA 2.

$$\begin{aligned} &E\{\sup_{t \ge T}(E|Z_t|)^{-1}|Z_T M^{t-T} - Z^*_{T,T} M^*_{T+1} \cdots M^*_t|\} \\ &\qquad \le \sum_{t=T}^{\infty} \sum_{i \notin A_t} M^t(1,i)\gamma(i) f_t \le T^{-3}. \end{aligned} \tag{2.5}$$

PROOF: Set $D_{T,T}(j) = Z_T(j) I_{A^c_T}(j)$, and for $t > T$

(2.6) $D_{T,t}(B)$ = number of particles in tth generation with type in B, whose ancestor in the Tth generation had type *outside* A_T, but whose ancestor in the kth generation had type *in* A_k for $T < k \le t$.

Now $Z_t - Z^*_{T,t}$ counts the particles whose ancestor in some generation $k \in [T,t]$ had type outside A_k. Decomposition with respect to the last such k gives

$$Z_t - Z^*_{T,t} = \sum_{k=T}^{t} D_{k,t}. \tag{2.7}$$

Since $Z_t(j) \geq Z^*_{T,t}(j)$, this implies

$$\begin{aligned}
E\{|Z_t - Z^*_{T,t}| \mid \mathcal{F}_T\} &= |Z_T M^{t-T} - Z^*_{T,T} M^*_{T+1} \cdots M^*_t| \\
&= |Z_t M^{t-T}| - |Z^*_{T,T} M^*_{T+1} \cdots M^*_t| \\
&\leq \sum_{k=T}^{t} E\{|D_{k,t}| \mid \mathcal{F}_T\} \\
&= \sum_{k=T}^{t} \sum_{i \notin A_k} \sum_j (Z_T M^{k-T})(i) M^*_{k+1} \cdots M^*_t(i,j).
\end{aligned}$$

Consequently

$$\begin{aligned}
&\sup_{t \geq T} \{E|Z_t|\}^{-1} |Z_T M^{t-T} - Z^*_{T,T} M^*_{T+1} M^*_t| \\
&\qquad \leq \sup_{t \geq T} \sum_{k=T}^{t} \sum_{i \notin A_k} Z_T M^{k-T}(i) \frac{\Sigma_j M^{t-k}(i,j)}{E|Z_t|} \\
&\qquad \leq \sum_{k=T}^{\infty} \sum_{i \notin A_k} Z_T M^{k-T}(i) \gamma(i) f_k \quad \text{(by (1.7) and (1.6))}.
\end{aligned}$$

Taking expectations, replacing $E Z_T M^{k-T}$ by $E Z_k$ (see (2.1)), and using (1.6) now yields (2.5). ∎

LEMMA 3. *For every $\epsilon > 0$, a.e. on the set of non-extinction of $\{Z_t\}$, there exists a T_1 such that*

$$|Z^*_{T,T} M^*_{T+1} \cdots M^*_t| \geq (\rho - \epsilon)^t \quad \textit{for all } t \geq T \geq T_1. \tag{2.8}$$

PROOF: By (1.11)

$$P\{Z \text{ does not become extinct}\} = P\{|Z^*_{S,t}| \geq \rho^t \text{ for all } t \geq S\}.$$

Since $Z^*_{T,t} \geq Z^*_{S,t}$ for $T \geq S$ this gives for each fixed T

$$
\begin{aligned}
&P\{Z \text{ does not become extinct}\} - P\{S > T\} \\
&\quad \leq P\{|Z^*_{T,t}| \geq \rho^t \text{ for all } t \geq T\} \\
&\quad \leq E\{P|Z^*_{T,T}M^*_{T+1} \cdots M^*_t| \geq (\rho - \epsilon)^t \text{ for all } t \geq T \mid \mathcal{F}_T\}\} \\
&\qquad + \sum_{t=T}^{\infty} E\{P\{|Z^*_{T,T}M^*_{T+1} \cdots M^*_t| < (\rho-\epsilon)^t \text{ but } |Z^*_{T,t}| \geq \rho^t \mid \mathcal{F}_T\}\} \\
&\quad \leq P\{|Z^*_{T,T}M^*_{T+1} \cdots M^*_t| \geq (\rho-\epsilon)^t \text{ for all } t \geq T\} \\
&\qquad + \sum_{t=T}^{\infty} E\{\rho^{-t}(\rho-\epsilon)^t \mid \mathcal{F}_T\}\}
\end{aligned}
$$

(because $E\{|Z^*_{T,t}| \mid \mathcal{F}_T\} = |Z^*_{T,T}M^*_{T+1} \cdots M^*_t|$ (see (2.2)). Thus, as $T \to \infty$

$$
\begin{aligned}
&P\{|Z^*_{T,T}M^*_{T+1} \cdots M^*_t \geq (\rho-\epsilon)^t \text{ for all } t \geq T\} \\
&\qquad \geq P\{Z \text{ does not become extinct}\} + o_T(1).
\end{aligned} \tag{2.9}
$$

This implies the lemma since the event in the left hand side of (2.9) is contained in the event in the right hand side (provided $\rho - \epsilon > 1$). ∎

LEMMA 4. *For all* T

$$
\begin{aligned}
&P\{|Z^*_{T,t}(j) - Z^*_{T,t-1}M(j)| > t^2 e^{t\epsilon_t}|Z^*_{T,t-1}|^{1/2} \\
&\qquad \textit{for some } j \in A_t \mid \mathcal{F}_{t-1}\} \leq t^{-4}, \quad t > T.
\end{aligned} \tag{2.10}
$$

Also, for $D_{k,t}$ *as in (2.6),*

$$
\begin{aligned}
&P\{D_{k,t}(j) > D_{k,t-1}M(j) + t^2 e^{t\epsilon_t}|D_{k,t-1}|^{1/2} \\
&\qquad \textit{for some } j \in A_t \mid \mathcal{F}_{t-1}\} \leq t^{-4}, \quad t > k.
\end{aligned} \tag{2.11}
$$

PROOF: Use Chebyshev's inequality (and (2.2), (2.3), (1.10) and (1.5)). ∎

LEMMA 5. *Choose* $\rho - 2\epsilon > 1$. *Almost everywhere on the set of nonextinction there exists a* T_2 *such that for all* $t > T \geq T_2$

$$
Z_t(j) \leq t^3 e^{t\epsilon_t} M^t(1,j) = t^3 e^{t\epsilon_t} EZ_t(j) \quad \textit{for all } j \in A_t, \tag{2.12}
$$

$$\sup_{\ell \geq T}\{E|Z_\ell|\}^{-1}|Z_T M^{\ell-T} - Z^*_{T,T}M^*_{T+1}\cdots M^*_\ell| \leq T^{-1}, \tag{2.13}$$

$$\sum_{j \notin A_t} Z_t(j)\gamma(j)f_t \leq t^{-2}, \tag{2.14}$$

$$|Z^*_{T,t}(j) - Z^*_{T,t-1}M(j)| \leq t^2 e^{t\epsilon_t}|Z^*_{T,t-1}|^{1/2} \quad \textit{for all } j \in A_t, \tag{2.15}$$

$$\begin{aligned} D_{k,t}(j) \leq D_{k,t-1}M(j) + t^2 e^{t\epsilon_t}|D_{k,t-1}|^{1/2} \quad \textit{for all } j \in A_t,\\ T < k < t, \end{aligned} \tag{2.16}$$

$$|Z^*_{T,T}M^*_{T+1}\cdots M^*_t| \geq (\rho-\epsilon)^t, \quad \textit{(even true for } t = T\textit{)}, \tag{2.17}$$

and

$$|Z^*_{T,t-1}| \leq t^3|Z^*_{T,T}M^*_{T+1}\cdots M^*_{t-1}|. \tag{2.18}$$

Without loss of generality we further assume that T_2 is large enough so that for $c = \frac{1}{2}\Sigma_j M(1,j) > 0$

$$\begin{aligned} &T_2 > t_0 \text{ (see (1.8) for } t_0\text{)}, \sum_{j\in A_t} M^*_t(1,j) = \sum_{j \in A_t} M(1,j) \geq c \\ &\text{and } \sum_{k=t+1}^{\infty} k^{7/2}e^{4k\epsilon_{k-1}}(\rho-\epsilon)^{-(k-1)/2} \leq (\rho-2\epsilon)^{-t/2} \leq \frac{1}{2t}, \\ &t \geq T_2. \end{aligned} \tag{2.19}$$

The remainder of the proof is purely deterministic.

LEMMA 6. *For any sample point which satisfies (2.12)–(2.19) one has for* $t \geq T \geq$ *some* $T_3 \geq T_2$

$$|Z_t - Z^*_{T,T}M^*_{T+1}\cdots M^*_t| \leq 4T^{-1}\{E|Z_{t-1}| + E|Z_t|\} \tag{2.20}$$

and

$$|Z_t - Z^*_{T,T}M^{t-T}| \leq 5T^{-1}\{E|Z_{t-1}| + E|Z_t|\}. \tag{2.21}$$

PROOF: Let v_t be the vector with components

$$v_t(j) = \begin{cases} t^2 e^{t\epsilon_t} |Z^*_{T,t-1}|^{1/2} & \text{if } j \in A_t, \\ 0 & \text{if } j \notin A_t. \end{cases}$$

$\theta_t(j)$ will be a suitable number in $[-1, 1]$ and $\theta_k v_k$ will be the vector with components $\theta_k(j) v_k(j)$. Now (2.15) yields the following vector identity for $t > T \geq T_2$. (Note that the jth components for $j \notin A_t$ are all zero.)

$$\begin{aligned} Z^*_{T,t} &= Z^*_{T,t-1} M^*_t + \theta_t v_t = \cdots \\ &= Z^*_{T,T} M^*_{T+1} \cdots M^*_t + \sum_{k=T+1}^{t} \theta_k v_k M^*_{k+1} \cdots M^*_t. \end{aligned} \tag{2.22}$$

The error term in the right hand side is at most

$$\begin{aligned} &\sum_{k=T+1}^{t} |v_k M^*_{k+1} \cdots M^*_t| \\ &\quad = \sum_{k=T+1}^{t} k^2 e^{k\epsilon_k} |Z^*_{T,k-1}|^{1/2} \sum_{i \in A_k} \sum_{j \in A_t} M^*_{k+1} \cdots M^*_t(i,j) \\ &\quad \leq \sum_{k=T+1}^{t} k^{7/2} e^{k\epsilon_k} |Z^*_{T,T} M^*_{T+1} \cdots M^*_{k-1}|^{1/2} \\ &\qquad \cdot \sum_{i \in A_k} \sum_{j \in A_t} M^*_{k+1} \cdots M^*_t(i,j) \text{ by (2.18))} \\ &\quad \leq \sum_{k=T+1}^{t} k^{7/2} e^{k\epsilon_k} (\rho - \epsilon)^{-(k-1)/2} |Z^*_{T,T} M^*_{T+1} \cdots M^*_{k-1}| \\ &\qquad \cdot \sum_{i \in A_k} \sum_{j \in A_t} M^*_{k+1} \cdots M^*_t(i,j) \text{ (by (2.17)).} \end{aligned} \tag{2.23}$$

Next, by (1.8) and (2.19)

$$\begin{aligned} |Z^*_{T,T} M^*_{T+1} \cdots M^*_k| &= \sum_{i \in A_{k-1}} \sum_{j \in A_k} Z^*_{T,T} M^*_{T+1} \cdots M^*_{k-1}(i) M^*_k(i,j) \\ &\geq e^{-k\epsilon_{k-1}} \sum_{i \in A_{k-1}} Z^*_{T,T} M^*_{T+1} \cdots M^*_{k-1}(i) \sum_{j \in A_k} M^*_k(1,j) \\ &\geq c e^{-k\epsilon_{k-1}} |Z^*_{T,T} M^*_{T+1} \cdots M^*_{k-1}|. \end{aligned} \tag{2.24}$$

Also, by (1.8) and (1.5)

$$
\begin{aligned}
&|Z^*_{T,T}M^*_{T+1}\cdots M^*_k|\sum_{i\in A_k}\sum_{j\in A_t} M^*_{k+1}\cdots M^*_t(i,j)\\
&=\sum_{i,\ell\in A_k}\sum_{j\in A_t} Z^*_{T,T}M^*_{T+1}\cdots M^*_k(\ell)M^*_{k+1}\cdots M^*_t(i,j)\\
&\le e^{k\epsilon_k}\sum_{i,\ell\in A_k}\sum_{j\in A_t} Z^*_{T,T}M^*_{T+1}\cdots M^*_k(\ell)M^*_{k+1}\cdots M^*_t(\ell,j)\\
&\le e^{k\epsilon_k}(\#A_k)|Z^*_{T,T}M^*_{T+1}\cdots M^*_t|\le e^{2k\epsilon_k}|Z^*_{T,T}M^*_{T+1}\cdots M^*_t|.
\end{aligned}
\tag{2.25}
$$

Substitution of (2.24), (2.25) into (2.23) finally yields

$$
\sum_{k=T+1}^{t}|v_kM^*_{k+1}\cdots M^*_t|\le c^{-1}\sum_{k=T+1}^{\infty}k^{7/2}e^{4k\epsilon_{k-1}}(\rho-\epsilon)^{-(k-1)/2}
$$

$$
|Z^*_{T,T}M^*_{T+1}\cdots M^*_t|\le(\rho-2\epsilon)^{-T/2}|Z^*_{T,T}M^*_{T+1}\cdots M^*_t|,
$$

and hence, by virtue of (2.22)

$$
|Z^*_{T,t}-Z^*_{T,T}M^*_{T+1}\cdots M^*_t|\le(\rho-2\epsilon)^{-T/2}|Z^*_{T,T}M^*_{T+1}\cdots M^*_t| \tag{2.26}
$$

and a fortiori

$$
\sum_{j\in A_t}Z_t(j)\ge|Z^*_{T,t}|\ge[1-(\rho-2\epsilon)^{-T/2}]|Z^*_{T,T}M^*_{T+1}\cdots M^*_t|. \tag{2.27}
$$

Starting with (2.16) instead of (2.15), and taking into account that for $t>k$, $D_{k,t}(j)=0$, $j\notin A_t$, and $D_{k,t}(j)\le Z^*_{k+1,t}(j)$, we also obtain from (2.23)–(2.25) for $T_2\le k<t$

$$
|D_{k,t}|\le|D_{k,k+1}M^*_{k+1}\cdots M^*_t|+(\rho-2\epsilon)^{-k/2}|Z^*_{k+1,k+1}M^*_{k+2}\cdots M^*_t|. \tag{2.28}
$$

The second term in the right hand side of (2.28) is at most

$$
\begin{aligned}
&(\rho-2\epsilon)^{-k/2}|Z^*_{k+1,k+1}M^{t-k-1}|\\
&(\rho-2\epsilon)^{-k/2}\sum_{i\in A_{k+1}}(k+1)^3e^{(k+1)\epsilon_k}M^{k+1}(1,i)\sum_j M^{t-k-1}(i,j)\\
&(\text{by (2.12)})\ \le(\rho-2\epsilon)^{-k/2}(k+1)^3e^{(k+1)\epsilon_k}E|Z_t|.
\end{aligned}
\tag{2.29}
$$

To the first term in the right hand side of (2.28) we apply (2.16) once more. This yields

(2.30)

$$\begin{aligned}|D_{k,k+1}M^*_{k+2}\cdots M^*_t| &\le |D_{k,k}M^{t-k}| \\ &\quad + (k+1)^2 e^{(k+1)\epsilon_k}|D_{k,k}|^{1/2}\sum_{\ell\in A_{k+1}}\sum_{j\in A_t} M^*_{k+2}\cdots M^*_t(\ell,j) \\ &\le |D_{k,k}M^{t-k}| + (k+1)^2 e^{(k+1)\epsilon_k} \\ &\quad \cdot\left\{\sum_{i\notin A_k} Z_k(i)\right\}^{1/2}\sum_{\ell\in A_{k+1}}\sum_{j\in A_t} M^*_{k+2}\cdots M^*_t(\ell,j).\end{aligned}$$

Now, since $\gamma(i)\ge 1$ (cf. (1.5)) and $f_t \ge 1/E|Z_t|$ by definition,

$$|D_{k,k}| = \sum_{i\notin A_k} Z_k(i) \le \sum_{i\notin A_k} Z_k(i)\gamma(i) \le k^{-2}E|Z_k| \tag{2.31}$$

(cf. (2.14)), and also (by (1.6))

$$\sum_{i\notin A_k} EZ_k(i) \le k^{-4}E|Z_k| = k^{-4}\sum_{i\in A_k} EZ_k(i) + k^{-4}\sum_{i\notin A_k} EZ_k(i).$$

Finally, since $E|Z_k| \ge P\{S\le k\}\rho^k$ (cf. (1.11)), there exists a constant $C_0 > 0$ such that

$$E|Z_k| \ge C_0\rho^k.$$

Combining these estimates gives for large k

$$\begin{aligned}&\{\sum_{i\notin A_k} Z_k(i)\}^{1/2}\sum_{\ell\in A_{k+1}}\sum_{j\in A_t} M^*_{k+2}\cdots M^*_t(\ell,j) \\ &\quad\le k^{-1}C_0^{-1/2}\rho^{-k/2}E|Z_k|\sum_{\ell\in A_{k+1}}\sum_{j\in A_t} M^*_{k+2}\cdots M^*_t(\ell,j) \\ &\quad\le 2k^{-1}C_0^{-1/2}\rho^{-k/2}\sum_{i\in A_k} EZ_k(i)e^{2(k+1)\epsilon_k}\sum_{j\in A_t} M^{t-k-1}(i,j) \\ &\qquad\text{(by (1.8) and (1.5), compare (2.25))} \\ &\quad\le 2k^{-1}C_0^{-1/2}\rho^{-k/2}e^{2(k+1)\epsilon_k}E|Z_{t-1}|.\end{aligned}\tag{2.32}$$

A much simpler estimate gives

$$\begin{aligned}|D_{k,k}M^{t-k}| &\le \sum_{i\notin A_k} Z_k(i)\sum_j M^{t-k}(i,j) \\ &\le \sum_{i\notin A_k} Z_k(i)\gamma(i)f_k E|Z_t| \le k^{-2}E|Z_t|.\end{aligned}\tag{2.33}$$

(2.28)–(2.33) show that for large k and $t > k$

$$|D_{k,t}| \leq 2k^{-2}\{E|Z_{t-1}| + E|Z_t|\}. \tag{2.34}$$

This estimate remains valid for $t = k$ by (2.31).

(2.20) now follows from (2.26), (2.27), (2.12), (2.7) and (2.34). (2.21) follows from (2.20) and (2.13), since

$$Z^*_{T,T} M^*_{T+1} \cdots M^*_t(j) \leq Z^*_{T,T} M^{t-T}(j) \leq Z_T M^{t-T}(j).$$ ∎

It is now easy to obtain (1.12) and (1.13) from (2.21) and the next lemma, which is a simple consequence of (1.9) (and (1.7) and (1.6)).

LEMMA 7.

$$R = \lim_{n\to\infty} \frac{\Sigma_j M^{n+1}(1,j)}{\Sigma_j M^n(1,j)} = \sum_i M(1,i)u(i) \tag{2.35}$$

exists, and $0 < R < \infty$. *Consequently*

$$\lim_{n\to\infty} \frac{\Sigma_j M^n(i,j)}{\Sigma_j M^{n+k}(1,j)} = \frac{u(i)}{R^k}. \tag{2.36}$$

Lastly, for (1.14) we observe that by virtue of (2.5), for all large enough T and $t \geq T$

$$E|Z^*_{T,T} M^*_{T+1} \cdots M^*_t| = E \sum_{i\in A_T} \sum_j Z_T(i) M^*_{T+1} \cdots M^*_t(i,j) \geq \frac{1}{2} E|Z_t|.$$

In particular, there must exist an $i_t \in A_T$ such that

$$M^T(1,i_t) \sum_j M^*_{T+1} \cdots M^*_t(i_t,j) \geq [2(\#A_T)]^{-1} E|Z_t|.$$

By (1.5) and (1.8) this implies (if in addition $T > t_0$), for all $i \in A_T$

$$\begin{aligned}\sum_j M^*_{T+1} \cdots M^*_t(i,j) &\geq e^{-T\epsilon_T} \sum_j M^*_{T+1} \cdots M^*_t(i_t,j) \\ &\geq \frac{1}{2} e^{-2T\epsilon_T} [M^T(1,i_t)]^{-1} E|Z_t| \\ &\geq \frac{1}{2} e^{-2T\epsilon_T} [E|Z_T|]^{-1} E|Z_t| = C_T E|Z_t|\end{aligned}$$

for some $C_T > 0$. Therefore, if $T \geq t_0 + T_2$

$$W = \lim \frac{|Z_t|}{E|Z_t|} \geq \liminf \frac{|Z_{T,T} M_{T+1}^* \cdots M_t^*|}{E|Z_t|} \geq \sum_{i \in A_T} Z_T(i) C_T$$
$$= C_T |Z_{T,T}^*| \geq C_T (\rho - \epsilon)^T > 0 \text{ (on the set of nonextinction (by (2.17))}.$$

REFERENCES

Asmussen, S. and Hering, H. (1983), *Branching Processes*, Birkhäuser.

Athreya, K. B., and Ney, P. E. (1972), *Branching Processes*, Springer-Verlag.

Chayes, J. T., Chayes, L., and Durrett, R. (1988), Connectivity properties of Mandelbrot's percolation process, *Probability Theory and Related Fields,* **77**, 307–324.

Dekking, F. M. (1987a), Subcritical branching processes in a two state random environment and a percolation problem on trees, *Journal of Applied Probability,* **24**, 798–808.

Dekking, F. M. (1987b), On the survival probability of a branching process in a finite state i.i.d. environment, preprint.

Dekking, F. M., and Grimmett, G. R. (1987), Superbranching processes and projections of random Cantor sets, preprint.

Falconer, K. J. (1989), Projections of random Cantor sets, *Journal of Theoretical Probability,* **2**, 65–70.

Graf, S., Mauldin, R. D., and Williams, S. C. (1987), Exact Hausdorff dimension in random recursive constructions, Memoir #381, AMS; see also *Proceedings of the National Academy of Sciences,* **84**, (1987), 3959–3961.

Harris, T. E. (1963), *The Theory of Branching Processes*, Springer-Verlag.

Jagers, P. (1975), *Branching Processes with Biological Applications*, John Wiley & Sons, Inc., New York.

Mandelbrot, B. (1983), *The Fractal Geometry of Nature*, W. H. Freeman and Co.

Moy, S-T. C. (1967), Extensions of a limit theorem of Everett, Ulam and Harris on multiple branching processes to a branching process with countably many types, *Annals of Mathematical Statistics,* **38**, 992–999.

Maxima of Random Quadratic Forms on a Simplex

J. F. C. Kingman

University of Bristol and the Mathematical Sciences Research Institute, Berkeley

1. AN OLD PROBLEM IN POPULATION GENETICS

One of Karlin's many contributions to science has been to show how complex models of phenomena in populations genetics can be analyzed mathematically in order to clarify the questions biologists ask of their experiments (and of each other). In this way he and his collaborators have thrown light on factors whose effects only become apparent when they can be quantified.

This paper goes back some thirty years to one of the simpler problems, to a model which has been somewhat neglected because it seemed unable to predict observed behaviour without very arbitrary assumptions. I have recently argued (1988) that this neglect may not be entirely justified, for reasons which are mathematically far from trivial.

The model is that of selection at a single locus in a large, randomly mating, diploid population (Kingman (1961)). The dynamics are well-known: if the possible alleles at the locus are $A_1, A_2, \ldots, A_n$ then the vector

AMS 1980 Subject Classifications: Primary 92A10, Secondary 15A52.

Key words and phrases: Maxima, random quadratic forms, population genetics.

Copyright © 1989 by Academic Press, Inc.
All rights of reproduction in any form reserved.

ISBN-0-12-058470-0

$$p = (p_1, p_2, \dots, p_n) \tag{1.1}$$

of allele frequencies converges to a local maximum of the mean fitness function

$$V(p) = \sum_{i,j=1}^{n} f_{ij}\, p_i\, p_j. \tag{1.2}$$

Here f_{ij} denotes the fitness of the genotype $A_i A_j$, and satisfies

$$f_{ij} \geq 0, \quad f_{ij} = f_{ji}. \tag{1.3}$$

In recent years it has become clear that the number of alleles which can be produced by mutation at a locus may be very large. There is no need to assume that they are all present at some initial epoch; more probably they arise from time to time by mutation and the effects of selection and mutation alternate. A detailed discussion is given in Kingman (1988).

The vector p is of course restricted to the simplex Δ_n determined by the conditions

$$p_i \geq 0 \quad (i = 1, 2, \dots, n), \quad \sum_{j=1}^{n} p_j = 1, \tag{1.4}$$

and the maximum referred to is a restricted local maximum of $V(p)$ for $p \in \Delta_n$. If there is such a maximum in the interior of Δ_n (i.e., with $p_i > 0$ for all i), then barring degeneracies it is unique. If there is no internal maximum there may be several local maxima on the boundary of Δ_n, each with its own domain of attraction for the motion.

At first sight this model would seem to offer a plausible explanation for many genetic phenomena in which several alleles co-exist at a locus. For any p in Δ_n, it is easy to construct a fitness matrix (f_{ij}) satisfying (1.3) for which V has a strict global maximum at p. The problem comes when the scientist has to explain the particular form of fitness matrix needed to yield such a maximum.

In order that V have a maximum in the interior of Δ_n, the f_{ij} have to satisfy a number of algebraic inequalities which become increasingly implausible for larger values of n. It was for instance shown in Kingman

(1988) that, if Nature were to choose the f_{ij} independently (for $i \leq j$) and from the same uniform distribution, then the probability of an internal maximum is at most $1/n!$. This follows from a very crude argument and is undoubtedly a weak upper bound; Karlin (1981) asserts a bound of order $\exp(-\frac{1}{2}n^2 \log n)$.

This however is not the end of the story. If V does not have an internal maximum, the local maxima occur on the boundary of Δ_n, at points p for which some p_i are zero. Thus some alleles die out, but there may still be many alleles represented in the final composition. One of the results of Kingman (1988) is that the number of such alleles is at most $2.5n^{1/2}$, but this may still be large if n is large. Thus we arrive at the mathematical question: *what can be said about the local maxima on* Δ_n *of the quadratic form* (1.2), *where the* f_{ij} $(i \leq j)$ *are independent random variables uniformly distributed on* $(0,1)$*?* Of particular interest are asymptotic properties for large n.

2. THE PROBABILITY OF STABILITY

We address the problem stated at the end of the last section in the following way. Let I be a subset of $\{1, 2, \ldots, n\}$ of size r. The probability that V has a local maximum supported by I, i.e., with

$$\{i; p_i > 0\} = I, \tag{2.1}$$

clearly depends only on n and r and will be denoted by S_{nr}. Thus S_{nr} is the probability that V has a local maximum at some point p with

$$p_i > 0 \quad (i = 1, 2, \ldots, r), \quad p_i = 0 \quad (i = r+1, \ldots, n). \tag{2.2}$$

It is easy to see (Kingman (1961)) that this is equal to the probability that three conditions are satisfied:

feasibility — there exists p in the interior of Δ_r, and $v > 0$, such that

$$\sum_{j=1}^{r} f_{ij}\, p_j = v \quad (i = 1, 2, \ldots, r), \tag{2.3}$$

stability —

$$\sum_{i,j=1}^{r} f_{ij}\, u_i\, u_j \le 0 \tag{2.4}$$

whenever

$$\sum_{j=1}^{r} u_j = 0, \tag{2.5}$$

and **invulnerability** —

$$\sum_{j=1}^{r} f_{ij}\, p_j \le v \quad (i = r+1, \ldots, n). \tag{2.6}$$

This probability is to be calculated assuming that the f_{ij} $(i \le j)$ are independent and uniformly distributed on $(0,1)$, and that $f_{ij} = f_{ji}$. It can thus be expressed as an integral over the unit cube in $\frac{1}{2}n(n+1)$ dimensions, an integral which does not admit explicit evaluation but which (as we shall see) does have an asymptotic form for large n.

The vector p in (2.3) is itself important because it describes the genetic composition of the population in equilibrium. It would therefore be useful to complement S_{nr} by computing the probability $S_{nr}(A)$ that the three conditions hold with p in a given subset A of Δ_r.

It turns out that $S_{nr}(A)$ is much easier to handle than $S_{nr} = S_{nr}(\Delta_r)$ if A is *bounded* in the sense that, for some $\delta > 0$,

$$p_i \ge \delta \text{ for all } i \le r \text{ and all } p \in A. \tag{2.7}$$

It will be proved that, for each r, there is a constant s_r and a probability distribution ϕ_r on Δ_r such that, for each bounded A,

$$S_{nr}(A) \sim s_r n^{-\frac{1}{2}(r+1)} \phi_r(A) \tag{2.8}$$

as $n \to \infty$.

The methods of this paper do not however establish the plausible conjecture that

$$S_{nr} \sim s_r n^{-\frac{1}{2}(r+1)} \tag{2.9}$$

because of singularities on the boundary of Δ_r. This is mathematically unsatisfactory, but probably irrelevant biologically because an allele present in very small proportion will not be observed. Thus, even without (2.9), we can conclude from (2.8) that the probability that a particular r-set I supports a stable polymorphism is asymptotically $s_r n^{-\frac{1}{2}(r+1)}$. Since there are $\binom{n}{r}$ such sets, the expected number of stable polymorphisms of r alleles is asymptotically

$$s_r n^{\frac{1}{2}(r-1)}/r!. \tag{2.10}$$

As in Kingman (1988) it can be shown that this number has an asymptotic Poisson distribution.

Thus there are, for large enough n, many stable polymorphisms of r alleles. The distribution ϕ_r represents the joint distribution of the allele frequencies in a typical polymorphism arising in this way.

In the next section we establish an asymptotic formula for $S_{nr}(A)$, valid for all bounded A, which will suffice to prove (2.8) while also yielding expressions for s_r and ϕ_r.

3. THE ASYMPTOTIC FORM OF $S_{nr}(A)$

For any measurable $A \subset \Delta_n$, $S_{nr}(A)$ can be written as an $\frac{1}{2}n(n+1)$-fold integral with respect to the variables f_{ij} $(1 \le i \le j \le n)$, the region of integration being defined by (2.3)–(2.6) with $0 \le f_{ij} \le 1$ and $p \in A$. This can immediately be reduced to an $\frac{1}{2}r(r+1)$-fold integral, since only (2.6) involves f_{ij} with i or j greater than r. If, for $p \in \Delta_r$, we write

$$G(p, x) = \Pr\left\{\sum_{i=1}^{r} p_i U_i \le x\right\}, \tag{3.1}$$

where the U_i are independent random variables uniformly distributed on $(0,1)$, then

$$S_{nr}(A) = \int G(p, v)^{n-r}\, df, \tag{3.2}$$

where

$$df = \prod_{1 \le i \le j \le r} df_{ij} \tag{3.3}$$

and the region of integration is defined by

$$0 \leq f_{ij} \leq 1, \quad p \in A, \quad (2.3), \quad (2.4). \tag{3.4}$$

If with the same region of integration we write

$$P_r(z, A) = \int I\{G(p, v) > 1 - z\} \, df, \tag{3.5}$$

with $I\{.\}$ denoting indicator, (3.2) becomes

$$S_{nr}(A) = \int_0^1 (1 - z)^{n-r} \, dP_r(z, A). \tag{3.6}$$

To evaluate (3.5) we make a change of variables from f_{ij} $(1 \leq i \leq j \leq r)$ to

$$x_{ij} = 1 - f_{ij} \quad (i < j), \quad p_i \quad (1 \leq i \leq r - 1), \quad y = 1 - v. \tag{3.7}$$

Neglecting null sets, this is a bijection from (3.4) to the region determined by

$$0 \leq x_{ij} \leq 1, \quad 0 \leq y \leq 1, \quad p \in A \tag{3.8}$$

together with

$$0 \leq f_{ii} = 1 - p_i^{-1} \left(y - \sum_j x_{ij} \, p_j \right) \leq 1 \tag{3.9}$$

and the translation into the new variables of (2.4). It is convenient to write

$$\xi_i = \sum_{j \neq i} x_{ij} \, p_j, \tag{3.10}$$

so that (3.9) becomes

$$\xi_i \leq y \leq \xi_i + p_i \tag{3.11}$$

and (2.4) becomes

$$\sum_i p_i^{-1} (y - \xi_i) u_i^2 + \sum_{i \neq j} x_{ij} \, u_i \, u_j \geq 0, \tag{3.12}$$

Putting $u_i = p_i v_i$ in (3.12) and rearranging, the inequality takes the alternative form

$$\sum_{i,j=1}^{r} (y - x_{ij}) p_i p_j (v_i - v_j)^2 \geq 0, \tag{3.13}$$

and in this form the constraint (2.5) becomes irrelevant because only the differences of the v_i enter. In particular, taking all but one of the v_i equal to zero, we have

$$\xi_i \leq (1 - p_i) y, \tag{3.14}$$

showing that the left-hand inequality of (3.11) is redundant in the presence of (3.12).

The Jacobian of the transformation is easily calculated to be

$$J = \Pi^{-2} D(p, x, y), \tag{3.15}$$

where

$$\Pi = p_1 p_2 \ldots p_r \tag{3.16}$$

and

$$D(p, x, y) = - \begin{vmatrix} y - \xi_1 & x_{12} p_2 & \ldots & x_{1r} p_r & 1 \\ x_{21} p_1 & y - \xi_2 & \ldots & x_{2r} p_r & 1 \\ \cdots & \cdots & \cdots & \cdots & \cdots \\ x_{r1} p_1 & x_{r2} p_2 & \ldots & y - \xi_r & 1 \\ p_1 & p_2 & \ldots & p_r & 0 \end{vmatrix}, \tag{3.17}$$

the determinant being negative by (3.12). Hence, using the symmetry of the distribution (3.1),

$$P_r(z, A) = \iiint I\{G(p, y) < z\} \Pi^{-2} D(p, x, y)\, dx\, dy\, dp, \tag{3.18}$$

where

$$dx = \prod_{1 \leq i \leq j \leq r} dx_{ij}, \quad dp = \prod_{j=1}^{r-1} dp_j,$$

and the integral extends over the region defined by (3.8), (3.11) and (3.13).

Now suppose that A is bounded, so that

$$p_0 = \min(p_1, p_2, \ldots, p_r) \geq \delta > 0 \tag{3.19}$$

for all $p \in A$. We evaluate $P_r(z, A)$ for

$$z \leq \zeta = (\delta r)^r / r!. \tag{3.20}$$

It is easy to check that, for $y \leq p_0$,

$$G(p, y) = y^r / (r!\,\Pi), \tag{3.21}$$

so that $G(p, y) < z \leq \zeta$ implies that $y < \delta \leq p_0$ and (3.18) becomes

$$P_r(z, A) = \iiint I\{y^r < r!\,\Pi z\}\Pi^{-2} D(p, x, y)\, dx\, dy\, dp. \tag{3.22}$$

We have already noted that, in defining the range of this integral, the left-hand inequality in (3.11) is redundant. But now the right-hand inequality is also redundant, since $\xi_i \geq 0$ and $y < p_0 \leq p_i$, and so likewise is the inequality $y \leq 1$ in (3.8). Hence when (3.20) holds, the region of integration in (3.22) is adequately defined by

$$x_{ij} \geq 0, \quad y \geq 0, \quad p \in A, \quad (3.13). \tag{3.23}$$

For fixed p, these conditions are homogeneous in (x, y). Moreover, the function $D(p, x, y)$ is homogeneous of degree $(r - 1)$ in (x, y). Making the change of variable

$$x_{ij} = y x'_{ij}, \tag{3.24}$$

(3.22) becomes

$$P_r(z, A) = \iiint I\{y^r < r!\,\Pi z\}\Pi^{-2} D(p, yx', y) y^{\frac{1}{2}r(r-1)}\, dx'\, dy\, dp,$$

where the region of integration no longer contains explicit mention of $y > 0$. Since

$$D(p, yx', y) = y^{r-1} D(p, x', 1)$$

the integration with respect to y may be carried out to give

$$P_r(z, A) = \iint \Pi^{-2} D(p, x', 1) \frac{2}{r(r+1)} (r!\,\Pi z)^{\frac{1}{2}(r+1)}\, dx'\, dp.$$

Thus, for $z \leq \zeta$,

$$P_r(z, A) = z^{\frac{1}{2}(r+1)} \frac{2(r!)^{\frac{1}{2}(r+1)}}{r(r+1)} \int_A \Pi^{\frac{1}{2}(r-3)} K_r(p)\, dp, \tag{3.25}$$

where (dropping primes)

$$K_r(p) = \int D(p, x, 1)\, dx, \tag{3.26}$$

the integral extending over all $x_{ij} \geq 0$ such that

$$\sum_{i,j=1}^{r} (1 - x_{ij}) p_i\, p_j (v_i - v_j)^2 \geq 0 \tag{3.27}$$

for all v_i. Thus for small z, $P_r(z, A)$ is proportional to $z^{\frac{1}{2}(r+1)}$. Substituting into (3.6), we therefore have

$$S_{nr}(A) \sim n^{-\frac{1}{2}(r+1)} r^{-1} \Gamma(\frac{1}{2}r + \frac{1}{2})(r!)^{\frac{1}{2}(r+1)} \int_A \Pi^{\frac{1}{2}(r-3)} K_r(p)\, dp \tag{3.28}$$

as $n \to \infty$, for every bounded measurable $A \subset \Delta_r$.

4. THE FUNDAMENTAL CONE

If t_{ij} $(1 \leq i < j \leq r)$ are regarded as coordinates of $\frac{1}{2}r(r-1)$-dimensional space, and $t_{ij} = t_{ji}$, then the condition that

$$\sum_{i,j=1}^{r} t_{ij} (v_i - v_j)^2 \geq 0 \tag{4.1}$$

for all v_i defines a cone $C = C_r$ in that space, which merits the name *fundamental cone* in the present analysis. With the change of variable

$$t_{ij} = (1 - x_{ij}) p_i\, p_j, \tag{4.2}$$

(3.26) becomes

$$K_r(p) = \Pi^{1-r} \int_C D(p, x, 1) I\{t_{ij} \le p_i p_j (i < j)\}\, dt. \tag{4.3}$$

In order to prove (2.8) from (3.28), it suffices to prove that

$$k_r = \int_{\Delta_r} \Pi^{\frac{1}{2}(r-3)} K_r(p)\, dp < \infty, \tag{4.4}$$

since (2.8) then follows with

$$s_r = r^{-1}\Gamma(\frac{1}{2}r + \frac{1}{2})(r!)^{\frac{1}{2}(r+1)}\, k_r \tag{4.5}$$

and

$$\phi_r(A) = k_r^{-1} \int_A \Pi^{\frac{1}{2}(r-3)} K_r(p)\, dp. \tag{4.6}$$

To do this, first note that the elements of the determinant in (3.17) are non-negative, and the row sums are

$$y + 1, y + 1, \ldots, y + 1, 1,$$

so that

$$D(p, x, y) \le (y + 1)^r.$$

By homogeneity

$$D(p, x', 1) \le y^{-r+1}(1 + y)^r,$$

and taking $y = r - 1$,

$$D(p, x, 1) \le er. \tag{4.7}$$

Now substitute (4.7) into (4.3), and (4.3) into (4.4), to give

$$k_r \le er \int_{\Delta_r} \int_C \Pi^{-\frac{1}{2}(r+1)} I\{t_{ij} \le p_i p_j \quad (i < j)\}\, dt\, dp. \tag{4.8}$$

If C' denotes the subset of C on which no t_{ij} is zero, then $C - C'$ is null and C' is a disjoint union

$$C' = \bigcup_G C(G), \tag{4.9}$$

where G ranges over subsets of $\{(i,j); 1 \le i < j \le r\}$ and $C(G)$ consists of those points of C for which $t_{ij} > 0$ when $(i,j) \in G$ and $t_{ij} < 0$ otherwise. To prove (4.4) it is thus sufficient to show that, for all G,

$$\int_{\Delta_r} \int_{C(G)} \Pi^{-\frac{1}{2}(r+1)} I\{t_{ij} \le p_i\, p_j \quad (i<j)\}\, dt\, dp < \infty. \tag{4.10}$$

In an obvious way, each G can be identified with an undirected graph on the vertices $1, 2, \ldots, r$. Taking $v_i = I\{i \in B\}$ in (4.1) we see that

$$\sum_{\substack{i \in B \\ j \notin B}} t_{ij} \ge 0, \tag{4.11}$$

so that $t_{ij} > 0$ for some $i \in B$, $j \notin B$. Since this holds for every proper subset B of $\{1, 2, \ldots, r\}$, $C(G)$ is non-empty only if the graph G is connected.

Taking $B = \{i\}$ in (4.11) we have

$$\sum_{j \ne i} t_{ij} \ge 0 \tag{4.12}$$

for all i. Hence, in the integral (4.10),

$$\begin{aligned} \sum_j (-t_{ij}) I\{(i,j) \notin E\} &\le \sum_j t_{ij}\, I\{(i,j) \in E\} \\ &\le \sum_j p_i\, p_j \le p_i, \end{aligned}$$

where E is the set of edges of G. Hence for $(i,j) \notin E$,

$$-t_{ij} \le p_i,$$

and by symmetry

$$-t_{ij} \le \min(p_i, p_j).$$

Thus in (4.10) each variable is restricted either to the interval

$$[-\min(p_i, p_j), 0)$$

or to the shorter interval

$$(0, p_i p_j],$$

and the integral cannot exceed

$$\int_{\Delta_r} \Pi^{-\frac{1}{2}(r+1)} \prod_{i<j} \min(p_i, p_j)\, dp,$$

which can easily be shown to be convergent. This completes the proof of (4.4), and thus of (2.8).

5. THE SHAPE OF A TYPICAL POLYMORPHISM

The distribution ϕ_r is the joint distribution of the allele frequencies in a typical stable polymorphism of r alleles. When $r = 2$ the calculations are easily carried out, to give

$$s_2 = (\frac{1}{2}\pi)^{3/2} \tag{5.1}$$

and

$$\phi_2(A) = \pi^{-1} \int_A (p_1 p_2)^{-1/2}\, dp_1, \tag{5.2}$$

the arc-sine law or symmetric Dirichlet distribution with parameter $\frac{1}{2}$.

It is interesting, and not a little disturbing, to find the Dirichlet distribution here, since it also occurs frequently in neutral models of polymorphisms (Kingman (1980)). This suggests that it might be impossible to distinguish, by statistical analysis of a number of polymorphisms, whether single-locus selection is the underlying mechanism. Fortunately, as we shall see, this is a phenomenon peculiar to the case $r = 2$.

It is also worth noting that the distribution (5.2) is quite different from that obtained by Lewontin, Ginsburg, and Tuljapurkar (1978), who also started with diploid fitnesses chosen independently from a uniform distribution. But they considered a two-allele system with no possibility of mutation, and conditioned on stability. In other words, they computed

$$S_{22}(A)/S_{22}$$

rather than

$$\phi_2(A) = \lim_{n\to\infty} S_{n2}(A)/S_{n2}.$$

When $r = 3$ the computations are more complex but still feasible. From (3.17),

(5.3)
$$D(p,x,1) = p_1(1-x_{12})(1-x_{13}) + p_2(1-x_{12})(1-x_{23}) + p_3(1-x_{13})(1-x_{23}),$$

so that (4.3) becomes

(5.4)
$$K_3(p) = \Pi^{-3} \iiint_{C_3} (t_{12}t_{13} + t_{12}t_{23} + t_{13}t_{23}) I\{t_{ij} \leq p_i\, p_j\}\, dt_{12}\, dt_{13}\, dt_{23}.$$

As in the last section, this integral is a sum of integrals over $C(G)$, where G runs over the connected graphs on $\{1,2,3\}$. There are four of these:

$$\begin{aligned} G_0 &= \{(1,2),(1,3),(2.3)\}, \\ G_1 &= \{(1,2),(1,3)\}, \\ G_2 &= \{(1,2),(2,3)\}, \\ G_3 &= \{(1,3),(2,3)\}. \end{aligned}$$

It is clear from (4.1) that

$$C(G_0) = \{t_{ij}; t_{ij} > 0 \quad (i < j)\},$$

from which the integral over $C(G_0)$ is at once computed to be $\frac{1}{4}\Pi^3$. Similarly $C(G_1)$ is determined by the inequalities

$$t_{12} > 0, \quad t_{13} > 0, \quad 0 < -t_{23} \leq t_{12}t_{13}/(t_{12} + t_{13}).$$

The contribution from $C(G_1)$ then turns out to be

$$\frac{1}{2} p_1^5\, p_2^3\, p_3^3 J(p_2, p_3),$$

where

$$J(a,b) = \int_0^1 \int_0^1 \frac{u^2 v^2}{au + bv}\, du\, dv, \tag{5.5}$$

which has the explicit, if unrevealing, expression

(5.6)
$$J(a,b) = \frac{a^2}{5b^3} \log\left(\frac{a+b}{a}\right) + \frac{b^2}{5a^3} \log\left(\frac{a+b}{b}\right) - \frac{a}{5b^2} + \frac{1}{10b} + \frac{1}{10a} - \frac{b}{5a^2}.$$

Dealing similarly with G_2 and G_3 we arrive finally at

$$K_3(p) = \frac{1}{4} + \frac{1}{2}p_1^2 J(p_2, p_3) + \frac{1}{2}p_2^2 J(p_1, p_3) + \frac{1}{2}p_3^2 J(p_2, p_3), \tag{5.7}$$

whence

$$k_3 = \frac{\pi^2}{30} - \frac{1}{8} = 0.204, \quad s_3 = 12k_3 = 2.45. \tag{5.8}$$

A useful approximation to (5.7) follows from the observation that $(a + b)J(a, b)$ is a continuous function of a/b on $[0, \infty]$, whose exact upper and lower bounds are

$$J(0, 1) = \frac{1}{6}$$
$$\text{and } 2J(1, 1) = \frac{2}{5}(2 \log 2 - 1) = \frac{1}{6.472}.$$

Substituting

$$J(a, b) \doteq \frac{1}{6(a + b)} \tag{5.9}$$

into (5.7) gives

$$K_3(p) \doteq \frac{1}{12}\left(\frac{1}{1 - p_1} + \frac{1}{1 - p_2} + \frac{1}{1 - p_3} - 1\right),$$

and thus, on normalising,

$$\phi_3(A) \doteq \frac{2}{5}\int_A \left(\frac{1}{1 - p_1} + \frac{1}{1 - p_2} + \frac{1}{1 - p_3} - 1\right) dp_1\, dp_2. \tag{5.10}$$

The error in the approximation (5.10) can be shown to be less than 0.02 for all A.

Thus ϕ_3 is qualitatively quite different from any Dirichlet distribution. The marginal density of a particular allele frequency is, under the approximation,

$$\frac{2}{5}(p - 2 \log p),$$

so that the frequency spectrum is

$$\frac{6}{5}(p - 2 \log p), \tag{5.11}$$

as opposed to the well-known expression

$$\theta p^{-1}(1-p)^{\theta-1}$$

(where θ is a mutation rate) from the neutral theory.

6. A LOWER BOUND FOR GENERAL r

The reader will have noted that, when $r = 3$, $\Pi D(p, x, 1)$ is a function only of the variables t_{ij} defined by (4.2). This is a general phenomenon. If (4.2) is substituted into (3.17) we obtain

$$\Pi D(p,x,1) = -\begin{vmatrix} t_1 + p_1^2 & p_1 p_2 - t_{12} & \dots & p_1 p_r - t_{1r} & p_1 \\ p_2 p_1 - t_{21} & t_2 + p_2^2 & \dots & p_2 p_r - t_{2r} & p_2 \\ \dots & \dots & \dots & \dots & \dots \\ p_r p_1 - t_{r1} & p_r p_2 - t_{r2} & \dots & t_r + p_r^2 & p_r \\ p_1 & p_2 & \dots & p_r & 0 \end{vmatrix}$$

$$= -\begin{vmatrix} t_1 & -t_{12} & \dots & -t_{1r} & p_1 \\ -t_{21} & t_2 & \dots & -t_{2r} & p_2 \\ \dots & \dots & \dots & \dots & \dots \\ -t_{r1} & -t_{r2} & \dots & t_r & p_r \\ p_1 & p_2 & \dots & p_r & 0 \end{vmatrix} \tag{6.1}$$

where

$$t_i = \sum_{j \neq i} t_{ij} = p_i(1 - p_i - \xi_i). \tag{6.2}$$

In (6.1) add the first $(r-1)$ columns to the rth and expand by the $(r+1, r)$ element to give

$$\Pi D = \begin{vmatrix} t_1 & -t_{12} & \dots & -t_{1,r-1} & p_1 \\ -t_{21} & t_2 & \dots & -t_{2,r-1} & p_2 \\ \dots & \dots & \dots & \dots & \dots \\ -t_{r1} & -t_{r2} & \dots & t_{r,r-1} & p_r \end{vmatrix}.$$

Add the first $(r-1)$ rows to the last, and expand by the (r, r) element to give

$$\Pi D(p,x,1) = \begin{vmatrix} t_1 & -t_{12} & \dots & -t_{1,r-1} \\ -t_{21} & t_2 & \dots & -t_{2,r-1} \\ \dots & \dots & \dots & \dots \\ -t_{r-1,1} & -t_{r-1,2} & \dots & t_{r-1} \end{vmatrix}. \tag{6.3}$$

This determinant has the well-known expression

$$\Pi D(p, x, 1) = \sum \prod_{(i,j) \in T} t_{ij}, \tag{6.4}$$

where the sum extends over all trees T spanning the vertices $1, 2, \ldots, r$. Thus (4.3) becomes

$$K_r(p) = \Pi^{-r} \int_C \sum \left\{ \prod_{(i,j) \in T} t_{ij} \right\} I\{t_{ij} \le p_i p_j \quad (i < j)\} \, dt. \tag{6.5}$$

Although this expression defies explicit evaluation for general r, it does yield a crude lower bound for K_r, and hence for k_r and s_r. From (4.1), C contains all t with $t_{ij} \ge 0$ for all $i < j$, and so

$$\begin{aligned} K_r(p) &\ge \Pi^{-r} \int \sum \left\{ \prod_{(i,j) \in T} t_{ij} \right\} I\{0 \le t_{ij} \le p_i p_j \quad (i < j)\} \, dt \\ &= \Pi^{-r} \sum \prod_{(i,j) \in T} \{\frac{1}{2}(p_i p_j)^2\} \prod_{(i,j) \notin T} (p_i p_j) \\ &= \Pi^{-1} 2^{1-r} \sum \prod_{(i,j) \in T} p_i p_j. \end{aligned}$$

This last sum may be evaluated using the special case of (6.4) when $t_{ij} = p_i p_j$, to give finally

$$K_r(p) \ge 2^{1-r} \tag{6.6}$$

(cf (5.7)). From this inequality, (4.4) and (4.5) imply that

$$s_r \ge \frac{r-1}{r} \frac{(r!)^{\frac{1}{2}(r+1)}}{2^r} \frac{\Gamma(\frac{1}{2}r - \frac{1}{2})^{r+1}}{\Gamma(\frac{1}{2}r^2 - \frac{1}{2}r)}. \tag{6.7}$$

For large n, the expected number of stable r-allele polymorphisms is asymptotically

$$s_r n^{\frac{1}{2}(r-1)} / r!,$$

and Stirling's formula applied to (6.7) shows that there is an absolute constant $c > 0$ such that

$$\frac{s_r}{r!} \ge c r^{-\frac{1}{4}} (\frac{1}{2} \pi^3 r)^{\frac{1}{4} r} \tag{6.8}$$

In particular,

$$s_r n^{\frac{1}{2}(r-1)}/r! \to \infty \tag{6.9}$$

as $n, r \to \infty$ with $r \leq \log n$.

This suggests, but does not prove, that

$$S_{nr} \binom{n}{r} \to \infty \tag{6.10}$$

under the same conditions. The best I can do is to demonstrate (6.10) under the stronger condition

$$r = O\big((\log n)^{1-\epsilon}\big), \tag{6.11}$$

for any $\epsilon > 0$. This is achieved by using (3.18) to obtain a lower bound for $P_r(z, \Delta_r)$, by restricting the integral to values of x, y and p with

$$x_{ij} < y < p_i \quad (i < j). \tag{6.12}$$

Using (3.21) and (6.4) this eventually yields the inequality

$$\begin{aligned} S_{nr} \geq\ & r^{-1}(r!)^{\frac{1}{2}(r+1)}2^{1-r} \\ & \int_{\Delta_r}\int_0^1 I\{p_0^r > r!\,\Pi z\}\Pi^{\frac{1}{2}(r-3)}z^{\frac{1}{2}(r-1)}(1-z)^{n-r}\,dz\,dp \end{aligned} \tag{6.13}$$

from which (6.10) follows when (6.11) holds.

All this is very crude, and should certainly be improved. The inequality (6.12) means for instance that we are only considering equilibria with

$$f_{ij} > v, \tag{6.14}$$

polymorphisms in which every heterozygote is fitter than the population as a whole. This is a very strong form of heterosis, yet our analysis shows that, for a typical array of fitnesses for a large number n of alleles, there are many such equilibria with r alleles represented, for all values of r up to a limit of the form (6.11). The $(r \times r)$ submatrix of fitnesses would look highly unnatural if one could not observe the fitnesses associated with the extinct alleles. Thus we should not be too eager to rule out the possiblity that fairly large polymorphisms can be maintained by single-locus selection.

REFERENCES

Karlin, S. (1981), Some natural viability systems for a multiallelic locus: a theoretical study, *Genetics*, **97**, 457–473.

Kingman, J. F. C. (1961), A mathematical problem in population genetics, *Proceedings of the Cambridge Philosophy Society*, **57**, 574–582.

Kingman, J. F. C. (1980), *Mathematics of Genetic Diversity*, Society for Industrial and Applied Mathematics.

Kingman, J. F. C. (1988), Typical polymorphisms maintained by selection at a single locus, *Journal of Applied Probability*, to appear.

Lewontin, R. C., Ginzburg, L. R., and Tuljapurkar, S. D. (1978), Heterosis as an explanation for large amounts of genic polymorphism, *Genetics*, **88**, 149–170.

Mandel, S. P. H. and Scheuer, P. A. G. (1959), An inequality in population genetics, *Heredity*, **13**, 519–524.

Total Positivity and Renewal Theory

Thomas M. Liggett
University of California, Los Angeles

Suppose that $f(k)$ is a probability density on the positive integers, and let $u(k)$ be the corresponding renewal sequence. Kaluza (1928) and de Bruijn and Erdös (1953) proved several results which relate convexity properties of f to convexity properties of u. We first note that these convexity properties can be formulated in terms of the total positivity of certain orders of the functions $f(i+j+1)$ and $u(i+j)$. This observation permits us to prove an infinite collection of implications which contain the Kaluza and de Bruijn and Erdös results as special cases. In our second result, we show how the imposition of a mild total positivity assumption on $f(k)$ permits one to give a straightforward proof of the fact that $u(n) - u(n+1)$ is asymptotic to a constant multiple of the tail probabilities of f. Continuous time versions of these results are discussed briefly. This work was motivated by a problem in interacting particle systems.

AMS 1980 Subject Classifications: Primary 60K05, Secondary 15A48.
Key words and phrases: renewal theory, total positivity, Kaluza sequences, moment sequences.
Research supported in part by NSF Grant DMS 86-01800.

PROBABILITY, STATISTICS,
AND MATHEMATICS
Papers in Honor of Samuel Karlin

Copyright © 1989 by Academic Press, Inc.
All rights of reproduction in any form reserved.
ISBN-0-12-058470-0

1. INTRODUCTION

Samuel Karlin has made fundamental contributions to the two fields of total positivity and renewal theory. In this paper, we will investigate some connections between these two areas. We begin by recalling some basic definitions. The equation

$$u(n) = \sum_{k=1}^{n} f(k)u(n-k), \quad n \geq 1, \quad u(0) = 1 \tag{1.1}$$

gives a one-to-one mapping between sequences $\{f(k), k \geq 1\}$ and $\{u(k), k \geq 0\}$ of real numbers with $u(0) = 1$. When $f(k)$ is a probability density on the positive integers, (1.1) is known as the renewal equation. In this case, we can let $\{X_i, i \geq 1\}$ be a sequence of independent and identically distributed random variables with distribution given by $P(X_i = k) = f(k)$, and set $S_n = X_1 + \cdots + X_n$ and $S_0 = 0$. Then $u(n)$ has the following probabilistic interpretation:

$$u(n) = \sum_{m=0}^{\infty} P(S_m = n) \quad \text{for} \quad n \geq 0. \tag{1.2}$$

The basic renewal theorem states under a mild aperiodicity assumption that

$$u(\infty) = \lim_{n\to\infty} u(n) = \left[\sum_{k=1}^{\infty} kf(k)\right]^{-1}. \tag{1.3}$$

A matrix $M = (m_{i,j})$ is said to be totally positive of order $r \geq 1$ (TP_r) if for every $k \leq r$, every k by k submatrix of M has a determinant which is nonnegative. If these determinants are all strictly positive, the matrix M is said to be strictly totally positive of order r (STP_r). Given a function $g(n)$ on the nonnegative integers, one can consider total positivity properties of the matrix with entries $m_{i,j} = g(i+j)$. Samuel Karlin has been the primary developer of the extensive theory of total positivity, and has demonstrated its usefulness in differential equations, probability theory, and many other parts of mathematics – see his book Karlin (1968), for example.

Karlin (1964) discusses applications of total positivity to Markov chains, and hence indirectly to renewal theory, insofar as renewal theory occurs in their study. There have been a number of investigations of relations between monotonicity and renewal theory – see Brown (1980) and the references there, for example. Little if anything, however, has been done on direct connections between total positivity and renewal theory. We hope to fill this gap to some extent in this paper.

About twenty years ago, when I began to work with Sam Karlin as a graduate student at Stanford, he was heavily involved in total positivity, and encouraged me to work in that area. I read his book, which was then in galley form, but ended up working in pure probability theory instead. I did find one of the Karlin-McGregor (1959) theorems very useful in one paper which I wrote shortly after my thesis – Liggett (1970). It is the rather striking assertion that for a one dimensional continuous time stochastic process, continuity of paths is equivalent to the total positivity of the transition probabilities of the process in the spatial variables. With that exception, this is my first paper related to total positivity. I hope that it partially makes up for my not having followed some of Sam's advice twenty years ago.

The starting point for our first result is the following collection of four facts, which at first glance seem to be unrelated, concerning sequences $f(k)$ and $u(k)$ which are connected by (1.1):

(a) If $f(k) \geq 0$ for all $k \geq 1$, then $u(k) \geq 0$ for all $k \geq 0$.

(b) If $f(k-1)f(k+1) \geq f^2(k)$ for all $k \geq 2$ then $u(k-1)u(k+1) \geq u^2(k)$ for all $k \geq 1$.

(c) If $u(k-1)u(k+1) \geq u^2(k)$ for all $k \geq 1$, then $f(k) \geq 0$ for all $k \geq 1$.

(d) $f(k) = \int_0^\infty x^k d\mu$ for some measure μ and all $k \geq 1$ if and only if $u(k) = \int_0^\infty x^k d\nu$ for some measure ν and all $k \geq 0$.

The first fact is obvious, the second was proved by de Bruijn and Erdös (1953) and the last two were proved by Kaluza (1928) – see also Horn (1970) and Shanbhag (1977).

In order to interpret these facts in terms of total positivity, note that a function $g(k)$ on the nonnegative integers is nonnegative if and only if $g(i+j)$ is TP_1, and that it satisfies $g(k-1)g(k+1) \geq g^2(k)$ for all $k \geq 1$ if and only if $g(i+j)$ is TP_2. Furthermore, g is a moment sequence if and only if $g(i+j)$ is TP_r for all $r \geq 1$ (see Section 7 of Chapter 2 of Karlin (1968) for a continuous version of this result, or obtain the discrete version directly by combining Theorem 1.3 of Shohat and Tamarkin (1943) with the comments on page 18 of Karlin (1968)). Therefore the four facts above are special cases of the following result. Its proof is based on a set of identities which relate determinants based on u to determinants based on f. It will be carried out in the next section.

THEOREM 1. *Suppose that the sequences $f(k)$ and $u(k)$ are connected by (1.1), and let r be a positive integer.*

(a) *If $f(i+j+1)$ is TP_r, then $u(i+j)$ is TP_r.*
(b) *If $u(i+j)$ is TP_{r+1}, then $f(i+j+1)$ is TP_r.*

There are some connections between Theorem 1 and some of the results proved in Karlin (1964). In that paper, Karlin proves that if X_n is a discrete time Markov chain on $\{0, 1, \ldots\}$ whose transition probabilities are TP_r in the spatial variables, then $u(n+m) = P^0(X_{n+m} = 0)$ is TP_r in $n, m \geq 0$ (Theorem 2.7) and $f(n+m) = P^0(\tau_0 = n+m)$ is TP_r in $n \geq 0$ and $m \geq 1$, where τ_0 is the hitting time of 0 (Proposition 10.4). By our Theorem 1, the second of these results implies the first.

Over the years, many theorems have been proved which assert that under some assumptions, and in one sense or another, $u(n) - u(n+1)$ behaves like the tail

$$F(n) = \sum_{k=n}^{\infty} f(k)$$

of the probability density f. Karlin (1955) proved an early result in this direction: If r is a positive integer and

$$\sum_{n=1}^{\infty} n^{r+1} f(n) < \infty,$$

then

$$\sum_{n=1}^{\infty} n^{r-1} |u(n) - u(\infty)| < \infty.$$

Other results along these lines are given by Stone (1965) and by Grübel (1982, 1983) – see also the references in the latter papers. Using results of Chover, Ney and Wainger (1973), Embrechts and Omey (1984) proved that

$$\lim_{n\to\infty} \sum_{k=1}^{n-1} \frac{F(k)F(n-k)}{F(n)} = 2\sum_{k=1}^{\infty} kf(k) < \infty$$

implies that

$$\lim_{n\to\infty} \frac{u(n)-u(n+1)}{F(n+2)} = \left[\sum_{k=1}^{\infty} kf(k)\right]^{-2}.$$

In our second theorem, we show that by imposing a mild total positivity assumption on F, one can give a relatively simple proof of a similar result. Our proof is essentially the same as one of the usual proofs of the ordinary renewal theorem (1.3), but it yields the more refined statement about convergence of ratios.

THEOREM 2. *Suppose that $f(k)$ is a probability density on the positive integers. Assume that $F(i+j+1)$ is* TP_2 *and that $F(2)>0$. Then*

$$0 \le u(n) - u(n+1) \le F(n+2). \tag{1.4}$$

Now put

$$\rho = \lim_{n\to\infty} \frac{F(n+1)}{F(n)} \le 1, \tag{1.5}$$

where the limit exists and is positive by monotonicity since $F(i+j+1)$ is TP_2*, and $\rho \le 1$ since $F(n)$ is decreasing. Then*

$$\lim_{n\to\infty} \frac{u(n)-u(n+1)}{F(n+2)} = 0 \quad \textit{if} \quad \sum_{n=1}^{\infty} F(n)\rho^{-n} = \infty \tag{1.6}$$

and

$$\lim_{n\to\infty} \frac{u(n)-u(n+1)}{F(n+2)} = \left[\sum_{k=0}^{\infty} F(k+1)\rho^{-k}\right]^{-2} > 0 \quad \textit{if} \quad \sum_{n=1}^{\infty} \frac{F^2(n)}{F(2n)} < \infty. \tag{1.7}$$

REMARKS. (a) The property that $F(i+j+1)$ is TP_2 is also known as DFR (decreasing failure rate). It is slightly weaker than the property that $f(i+j+1)$ is TP_2.

(b) It would be interesting to determine the rate of convergence to zero in (1.6).

EXAMPLES. (a) Fix $0 < \rho \leq 1$, and define $f(k)$ by

$$F(n+1) = \frac{(2n)!}{n!(n+1)!}\left(\frac{\rho}{4}\right)^n \quad \text{for} \quad n \geq 0.$$

Using identity (1.20) in Chapter VI of Liggett (1985), it is not hard to check that the corresponding renewal sequence is determined by

$$u(n) - u(n+1) = \frac{\rho}{4}F(n+1),$$

so that the limit of $[u(n) - u(n+1)]/F(n+2)$ is $1/4$ in this case. This is a case in which (1.7) holds, since $F(n)$ is asymptotic to a constant multiple of $n^{-3/2}\rho^n$.

(b) Consider the distribution of one half of the time until the first return to the origin for the simple symmetric random walk on the integers. It has density

$$f(n) = \frac{1}{2n-1}\binom{2n-1}{n}2^{-2n+1}.$$

The corresponding renewal sequence is of course given by

$$u(n) = \binom{2n}{n}2^{-2n}.$$

Both $u(n) - u(n+1)$ and $f(n)$ are asymptotic to a constant multiple of $n^{-3/2}$, so this provides an example of (1.6).

Continuous time analogues of Theorems 1 and 2 will be discussed briefly in Section 4. We will conclude this section by describing two applications of Theorems 1 and 2 which played important roles in my recent solution in Liggett (1989) of a problem involving exponential rates of convergence for certain interacting particle systems which are known as nearest particle systems. (See Chapter VII of Liggett (1985) for more about these processes.) The first makes precise the statement that if $f(n)$ has exponential tails, then $u(n)$ converges to its limit exponentially rapidly. For our application to interacting particle systems, it is not sufficient to say that $u(n) - u(\infty)$ is exponentially small – the following stronger statement in terms of ratios is needed.

COROLLARY. *Suppose that f is a probability density on the positive integers which satisfies $f(1) < 1$ and is not a geometric density. If either*

(a) $f(i+j+1)$ *is* TP_3, *or*

(b) $F(i+j+1)$ *is* TP_2 *and* $\displaystyle\sum_{n=1}^{\infty}\frac{F^2(n)}{F(2n)} < \infty$, *then*

$$\limsup_{n\to\infty}\frac{u(n)-u(\infty)}{u(n)-u(n+1)}\leq\frac{1}{1-\rho}, \tag{1.8}$$

where ρ *is defined in* (1.5).

REMARK. In Liggett (1989), we used the obvious fact that (1.8) holds under the stronger assumption that $f(k)$ (and hence by Kaluza's result, also $u(k)$) is a moment sequence. We can now replace the moment sequence assumption in Theorem 1.5 of that paper by the assumption of the above corollary, at the expense of changing the constants somewhat.

PROOF. We can assume that $\rho < 1$, since otherwise (1.8) is trivial. Suppose first that (a) holds. By Theorem 1, $u(i+j)$ is also TP_3. Therefore, after performing two column operations, we see that for $n \geq 0$ and $k \geq 1$, we have

$$\begin{vmatrix} u(n)-u(n+1) & u(n+1)-u(n+2) & u(n+2) \\ u(n+1)-u(n+2) & u(n+2)-u(n+3) & u(n+3) \\ u(n+k)-u(n+k+1) & u(n+k+1)-u(n+k+2) & u(n+k+2) \end{vmatrix} \geq 0.$$

Letting k tend to ∞, it follows that $\Delta(n) = u(n) - u(n+1)$ has the property that $\Delta(i+j)$ is TP_2. Therefore, either $\Delta(n) > 0$ for all $n \geq 0$ or $\Delta(n) = 0$ for all $n \geq 1$. We have excluded the geometric case, so the former case holds. It then follows by monotonicity that

$$\lim_{n\to\infty}\frac{\Delta(n+1)}{\Delta(n)}=\gamma$$

exists. By (1.4), $\gamma \leq \rho$. Now write

$$\frac{u(n)-u(\infty)}{u(n)-u(n+1)}=\sum_{k=0}^{\infty}\frac{\Delta(n+k)}{\Delta(n)}$$

and use the dominated convergence theorem to get (1.8). Now assume that (b) holds. Then the limit of $[u(n)-u(n+1)]/F(n+2)$ exists and is strictly positive by Theorem 2, so that (1.8) will follow from the inequality

$$\sum_{k=n}^{\infty}F(k)\leq\frac{F(n)}{1-\rho}\quad \text{for}\quad n\geq 1.$$

But this is a consequence of $F(k+1) \leq \rho\, F(k)$, which in turn follows from the monotonicity in the convergence in (1.5).

Our second application of total positivity in renewal theory provides a representation of a stationary renewal process in terms of a sequence of independent and identically distributed random variables. Suppose that $u(0) = 1$, $u(i+j)$ is TP_2, and $0 < u(\infty) < \infty$. By Theorem 1, the corresponding $f(k)$ is a probability density with finite mean. Define a probability measure on the nonnegative integers by $\pi(0) = u(1)$ and

$$\pi(n) = \frac{u(n+1)}{u(n)} - \frac{u(n)}{u(n-1)} \text{for} \quad n \geq 1,$$

which is nonnegative by the total positivity assumption. Let $\{X_i, -\infty < i < \infty\}$ be independent random variables with $P(X_i = n) = \pi(n)$. Now let $\{\eta_i, -\infty < i < \infty\}$ be Bernoulli random variables defined in terms of the X_i's by

$$\eta_i = 1 \quad \text{if and only if} \quad X_{i+k} \leq k \quad \text{for all} \quad k \geq 0.$$

In Theorem 4.6 of Liggett (1989), we proved that $\{\eta_i, -\infty < i < \infty\}$ has the distribution of the set of renewal times of a stationary renewal process whose interarrival times have density $f(k)$. This construction of a stationary renewal process has a number of applications which are discussed in that paper. Of course, for any stationary renewal process, the interarrival times form an i.i.d. sequence. However, the renewal process cannot be expressed directly as a function of this sequence because of the difficulty in initializing the process. Our construction is therefore often quite useful, even though it can only be used if $u(i+j)$ is TP_2.

ACKNOWLEDGEMENT. I wish to thank R. Grübel for bringing several recent papers on renewal theory to my attention, and in particular for pointing out the connection between Theorem 2 and Theorem 3.2 in Embrechts and Omey (1984).

2. THE TOTAL POSITIVITY CONNECTION

This section is devoted to the proof of Theorem 1. The main work is in finding and checking a set of identities which relate determinants based on f

with determinants based on u. They are given in the following proposition. We will use the following notation for determinants of matrices constructed from a sequence $g(\cdot)$:

$$g(i_0, \ldots, i_k) = \begin{vmatrix} g(i_0) & g(i_0+1)\cdots g(i_0+k) \\ g(i_1) & g(i_1+1)\cdots g(i_1+k) \\ \cdot & \cdot \quad\quad \cdot \\ \cdot & \cdot \quad\quad \cdot \\ g(i_k) & g(i_k+1)\cdots g(i_k+k) \end{vmatrix}.$$

PROPOSITION 2.1. *Suppose that the sequences f and u are related by* (1.1). *The following four identities are valid for $n \geq 1$ and $k \geq 1$:*

$$(a) f(n+1, n+2, \ldots, n+k)u(n, n+1, \ldots, n+k) = \sum_{j=1}^{n} f(j, n+1, n+2, \ldots, n+k)u(n-j, n, n+1, \ldots, n+k-1).$$

$$(b) f(n+1, n+2, \ldots, n+k)u(n, n+1, \ldots, n+k-1) = \sum_{j=1}^{n} f(j, n+1, n+2, \ldots, n+k-1)u(n-j, n, n+1, \ldots, n+k-1).$$

$$(c) \qquad u(1, 2, \ldots, k) = f(1, 2, \ldots, k).$$

$$(d) \qquad u(0, 1, \ldots, k) = f(2, 3, \ldots, k+1).$$

REMARK. Note that identities (a) and (b) bear a resemblance to (1.1). Special cases of the identities in Proposition 2.1 have appeared earlier. Identities (c) and (d) are given in equation (32) in Kaluza (1928). Identity (b) for $k = 1$ is equation (15) in Kaluza (1928). Identity (a) for $k = 1$ is equation (7) in de Bruijn and Erdös (1953).

PROOF. First we show that (c) and (d) follow from (a) and (b). In doing so, we may assume that $f(1, 2, \ldots, k) \neq 0$ and $f(2, 3, \ldots, k+1) \neq 0$ for all $k \geq 1$, since the sequences satisfying these inequalities are dense in the collection of all sequences. Identities (a) and (b) with $n = 1$ give respectively

$$f(2,\dots,k+1)u(1,\dots,k+1) = f(1,\dots,k+1)u(0,\dots,k)$$

and

$$f(2,\dots,k+1)u(1,\dots,k) = f(1,\dots,k)u(0,\dots,k)$$

for $k \geq 1$. Therefore

$$\frac{u(1,\dots,k+1)}{f(1,\dots,k+1)} = \frac{u(0,\dots,k)}{f(2,\dots,k+1)} = \frac{u(1,\dots,k)}{f(1,\dots,k)}$$

for $k \geq 1$. It follows that the expression above is independent of k, and is hence equal to one since $u(1) = f(1)$. This gives both (c) and (d). Turning now to the proof of (a), let $f_i = f(n+1,\dots,n+i,n+i+2,\dots,n+k+1)$ and $u_l = u(n,\dots,n+l-1,n+l+1,\dots,n+k)$ for $0 \leq i,\ l \leq k$, and expand the determinants on the right side of (a) along their top rows to get

$$\mathrm{RHS}(a) = \sum_{j=1}^{n}\sum_{i=0}^{k}\sum_{l=0}^{k} f(j+i)u(n-j+l)(-1)^{i+l} f_i u_i.$$

Now carry out the sum on j using (1.1) to get

$$\begin{aligned}
\text{(2.1)}\qquad \mathrm{RHS}(a) &= \sum_{i,l=0}^{k} f_i u_l (-1)^{i+l}\Bigg[u(n+i+l) - \sum_{j=1}^{i} f(j)u(n+i+l-j) \\
&\qquad\qquad - \sum_{j=i+n+1}^{i+n+l} f(j)u(n+i+l-j)\Bigg] \\
&= \sum_{i,l=0}^{k} f_i u_l (-1)^{i+l}\Bigg[u(n+i+l) - \sum_{j=1}^{i} f(j)u(n+i+l-j) \\
&\qquad\qquad - \sum_{j=0}^{l-1} f(n+i+l-j)u(j)\Bigg].
\end{aligned}$$

Next we examine separately the sums corresponding to each of the three terms in brackets:

$$\text{(2.2)}\quad \sum_{l=0}^{k} u_l(-1)^l u(n+i+l-j) = u(n+i-j,n,n+1,\dots,n+k-1) = 0$$

for $0 \le i-j \le k-1$, where the first equality comes from expanding the determinant on the right along its top row, and the second comes from the fact that the determinant of a matrix with two equal rows is zero. Similarly,

$$(2.3)\quad \sum_{i=0}^{k} f_i(-1)^i f(n+i+l-j) = f(n+l-j, n, n+1, \ldots, n+k-1) = 0$$

for $0 \le l-j \le k-1$. Finally,

$$(2.4)\quad \sum_{l=0}^{k} u_l(-1)^l u(n+k+l) = u(n+k, n, n+1, \ldots, n+k-1) = (-1)^k u(n, \ldots, n+k).$$

Using (2.2), (2.3), and (2.4) in (2.1) gives (a). The proof of (b) is similar. This time, let $f_i = f(n+1, \ldots, n+i, n+i+2, \ldots, n+k)$ and $u_l = u(n, \ldots, n+l-1, n+l+1, \ldots, n+k)$, and compute

$$\mathrm{RHS}(b) = \sum_{j=1}^{n} \sum_{i=0}^{k-1} \sum_{l=0}^{k} f(j+i)u(n-j+l)(-1)^{i+l} f_i u_l.$$

Using (1.1) again gives

$$(2.5)\quad \mathrm{RHS}(b) = \sum_{i=0}^{k-1} \sum_{l=0}^{k} (-1)^{i+l} f_i u_l \times \left[u(n+i+l) - \sum_{j=1}^{i} f(j)u(n+i+l-j) - \sum_{j=0}^{l-1} f(n+i+l-j)u(j)\right].$$

To prove (b), use (2.2) and the following identity in (2.5):

$$\sum_{i=0}^{k-1} f_i(-1)^i f(n+i+l-j) = f(n+l-j, n+1, \ldots, n+k-1) = \begin{cases} 0 & \text{if } 1 \le l-j \le k-1 \\ (-1)^{k+1} f(n+1, \ldots, n+k) & \text{if } l-j = k. \end{cases}$$

This completes the proof of the proposition. ∎

The following result is a special case of Theorem 3.3 of Chapter 2 of Karlin (1968).

PROPOSITION 2.2. *The sequence* $g(\cdot)$ *has the property that* $g(i+j)$ *is* STP_r *for* $0 \le i,\ j \le N$ *if and only if* $g(n,\ n+1, \dots, n+k) > 0$ *for all* $n \ge 0$ *and all* $k \ge 0$ *with* $k < r$ *and* $n + 2k \le 2N$.

PROOF OF THEOREM 1. We will prove the slightly modified version of the theorem in which TP is replaced by STP in both the hypotheses and the conclusions. To then remove the S's, it is necessary to approximate totally positive sequences by strictly totally positive sequences. The simplest way to do this is to add a small strictly totally positive sequence to the given sequence. See Chapters 2 and 3 of Karlin (1968) for details. To prove part (a) of the theorem, assume that r is a positive integer and that $f(i+j+1)$ is STP_r for $i,\ j \ge 0$. By Proposition 2.2, in order to show that $u(i+j)$ is STP_r for $i,\ j \ge 0$, it is enough to show that

$$u(n, \dots, n+k) > 0 \tag{2.6}$$

for all $n \ge 0$ and all $0 \le k < r$. This is true for $n = 0$ and $n = 1$ by parts (c) and (d) of Proposition 2.1. We proceed now by induction on n. Suppose that (2.6) is true for $n < N$ and all $0 \le k < r$. Put $n = N$ in part (a) of Proposition 2.1. Then all of the determinants which appear in that identity except $u(N, \dots, N+k)$ are strictly positive by the total positivity assumption on $f(\cdot)$, the induction assumption on $u(\cdot)$, and Proposition 2.2. Therefore, $u(N, \dots, N+k) > 0$ as well. The proof of part (b) of Theorem 1 is similar, using part (b) of Proposition 2.1 in place of part (a) for the induction step. ∎

3. THE CONVERGENCE THEOREM

Here we prove Theorem 2. We will assume throughout that $f(\cdot)$ is a probability density on the positive integers whose tail probabilities $F(\cdot)$

have the property that $F(i+j+1)$ is TP_2. The key step is to find a useful recurrence relation satisfied by

$$v(n) = \frac{u(n) - u(n+1)}{F(n+2)}. \tag{3.1}$$

The first thing one might think of doing is to take differences in (1.1), and then divide by $F(n+2)$. This does lead to a recursion for $v(n)$, but one which is not particularly useful, and which does not take sufficient advantage of the total positivity assumption in Theorem 2.

A better recursion is obtained by coupling together two copies of the renewal process corresponding to the density $f(\cdot)$. To carry out this coupling, let

$$p(k) = \frac{f(k)}{F(k)} \quad \text{for} \quad k \geq 1 \tag{3.2}$$

be the conditional probability of having a renewal at time k given that there has been no renewal prior to that time. Note that

$$\prod_{k=1}^{n} [1 - p(k)] = F(n+1) \tag{3.3}$$

and that the total positivity assumption on $F(\cdot)$ is equivalent to the statement that $p(k)$ is a decreasing function of k. Construct two sequences $\{\eta(n), n \geq 1\}$ and $\{\zeta(n), n \geq 0\}$ of Bernoulli random variables with values 0 and 1 in the following way: $\eta(1) = \zeta(0) = 1$, $P\{\zeta(1) = 1\} = f(1)$, $\zeta(n) \leq \eta(n)$ for all $n \geq 1$, and conditional on $\eta(k)$ and $\zeta(k)$ for $k < n$,

$$\begin{aligned} \eta(n) = \zeta(n) = 1 &\text{ with probability } p(n - \max\{j < n : \zeta(j) = 1\}), \\ \eta(n) = \zeta(n) = 0 &\text{ with probability } 1 - p(n - \max\{j < n : \eta(j) = 1\}), \text{ and} \\ \eta(n) = 1, \zeta(n) = 0 &\text{ with probability } p(n - \max\{j < n : \eta(j) = 1\}) \\ &\qquad - p(n - \max\{j < n : \zeta(j) = 1\}). \end{aligned}$$

This construction is possible because of the monotonicity of $p(k)$. It is easy to check that the marginal distribution of the sequence $\{\eta(n),\ n \geq 1\}$ is that of a sequence of renewals with first renewal at time one and inter-renewal density $f(\cdot)$. Similarly, the marginal distribution of the sequence $\{\zeta(n),\ n \geq 0\}$ is that of a sequence of renewals with first renewal at time zero and inter-renewal density $f(\cdot)$. Therefore, we can use a decomposition according to the time of the last discrepancy between η and ζ to perform the following calculation for $n \geq 1$, keeping in mind that if there is a discrepancy at m, then $\zeta(k) = 0$ for $k \leq m$:

$$\begin{aligned}
u(n) - u(n+1) &= P[\eta(n+1) = 1] - P[\zeta(n+1) = 1] \\
&= P[\eta(n+1) = 1, \zeta(n+1) = 0] \\
&= \sum_{k=1}^{n} P[\eta(n+1) = 1,\ \zeta(n+1) = 0,\ \eta(k) = 1,\ \zeta(k) = 0,\ \eta(j) = 0 \\
&\qquad\qquad\qquad\qquad\qquad\qquad \text{for all } \ k < j \leq n] \\
&= \sum_{k=1}^{n} P[\eta(k) = 1,\ \zeta(k) = 0] \left\{ \prod_{j=1}^{n-k} [1 - p(j)] \right\} [p(n-k+1) - p(n+1)] \\
&= \sum_{k=1}^{n} [u(k-1) - u(k)] F(n-k+1) \left[\frac{F(n+2)}{F(n+1)} - \frac{F(n-k+2)}{F(n-k+1)} \right],
\end{aligned}$$

where the last equality follows from (3.3). Rewriting this in terms of $v(\cdot)$ using (3.1), and then making a change of variable in the summation gives

$$\begin{aligned}
v(n) &= \sum_{k=1}^{n} v(k-1) F(k+1) \left[\frac{F(n-k+1)}{F(n+1)} - \frac{F(n-k+2)}{F(n+2)} \right] \\
&= \sum_{k=1}^{n} g_{n-k}(k) v(n-k),
\end{aligned} \tag{3.4}$$

where

$$g_m(k) = F(m+2) \left[\frac{F(k)}{F(m+k+1)} - \frac{F(k+1)}{F(m+k+2)} \right] \tag{3.5}$$

for $m \geq 0$ and $k \geq 1$. Note that $g_m(k) \geq 0$ by the total positivity assumption, and is a sub probability density:

$$\sum_{k=1}^{\infty} g_m(k) = 1 - F(m+2)\rho^{-m-1} < 1 \tag{3.6}$$

by (1.5). Thus since $v(0) = 1$, (3.4) exhibits $v(n)$ as the renewal sequence corresponding to inter-renewal times which are independent, but whose distributions are time dependent and defective.

Note that if $F(n)$ is replaced by $F(n)s^{n-1}$ for some $0 < s < 1$, then $g_m(k)$, and hence $v(n)$, remains unchanged. This explains why the statement of Theorem 2 does not depend on the "exponential part" of the density $f(n)$. We can now proceed to the proof of Theorem 2.

PROOF OF THEOREM 2. Since $g_m(k)$ is a sub probability density and $v(0) = 1$, (3.4) implies that

$$0 \le v(n) \le 1 \tag{3.7}$$

for all $n \ge 0$. This gives (1.4). Now note that

$$g(k) = \lim_{m \to \infty} g_m(k) = F(k)\rho^{-k+1} - F(k+1)\rho^{-k} \tag{3.8}$$

for $k \ge 1$ by (1.5), and that $g(\cdot)$ sums to one if and only if $\lim_k F(k)\rho^{-k} = 0$. By (3.6), (3.8) and Scheffé's Theorem,

$$\lim_{m \to \infty} \sum_{k=1}^{\infty} |g_m(k) - g(k)| = 0. \tag{3.9}$$

Define a sequence of Bernoulli random variables $\{\eta(n), n \ge 0\}$ by $\eta(0) = 1$ and

$$P[\eta(k+1) = \cdots = \eta(n-1) = 0, \eta(n) = 1 \mid \eta(0), \eta(1), \ldots, \eta(k-1), \eta(k) = 1] = g_k(n-k).$$

Using (3.4) and induction, it follows that $v(n) = P[\eta(n) = 1]$. Then, considering the distances between N and the nearest indices smaller and larger than N at which η is one, we have

$$\begin{aligned} P[\eta(m) = 1 \text{ for some } m > N] &= \sum_{k=0}^{N} \sum_{l=1}^{\infty} v(N-k) g_{N-k}(k+l) \\ &= \sum_{k=0}^{N} v(N-k) F(N-k+2) \left[\frac{F(k+1)}{F(N+2)} - \rho^{-N+k-1} \right] \end{aligned} \tag{3.10}$$

by (3.5) and (1.5). Now consider a subsequence of the N's along which $v(N+n)$ converges for each $-\infty < n < \infty$, say to $w(n)$. Every sequence has such a subsequence by (3.7). Then we can replace n by $N+n$ in (3.4) and pass to the limit along that subsequence using (3.9) to obtain

$$w(n) = \sum_{k=1}^{\infty} g(k)w(n-k) \quad \text{for} \quad -\infty < n < \infty. \tag{3.11}$$

If the sum of $g(\cdot)$ is strictly less than one, then $w(n) = 0$ for all n, so we can conclude that $\lim_n v(n) = 0$. Therefore, we may assume from now on that $g(\cdot)$ sums to one, and hence that $\lim_k F(k)\rho^{-k} = 0$. In this case, (3.11) says that $w(\cdot)$ is a (bounded) harmonic function for a random walk, and hence is a constant C by the Choquet-Deny Theorem. Now pass to the limit in (3.10) along our subsequence using Fatou's Lemma to conclude that

$$C\sum_{k=0}^{\infty} F(k+1)\rho^{-k} \le 1.$$

If the sum above is infinite, it follows that $C=0$, and hence that $\lim_n v(n) = 0$. This proves (1.6). Proceeding to the proof of (1.7), we may assume that

$$\sum_{n=1}^{\infty} \frac{F^2(n)}{F(2n)} < \infty. \tag{3.12}$$

The TP_2 assumption on the tail probabilities implies that

$$\frac{F(k+1)F(N-k+2)}{F(N+2)} \le \frac{F^2(k)}{F(2k)}$$

if $2k \le N+2$. Therefore (3.12) and the dominated convergence theorem imply that

$$\lim_{N\to\infty} \sum_{k=0}^{N/2} v(N-k)F(N-k+2)\left[\frac{F(k+1)}{F(N+2)} - \rho^{-N+k-1}\right] = C\sum_{k=0}^{\infty} F(k+1)\rho^{-k}.$$

After making the change of variable $l = N-k$, the same argument applied to the other half of the terms in the sum on the right of (3.10) gives

$$\lim_{N\to\infty} \sum_{l=0}^{N/2} v(l)F(l+2)\left[\frac{F(N-l+1)}{F(N+2)} - \rho^{-l-1}\right] = 0.$$

So, we can pass to the limit along the subsequence of N's in (3.10) to conclude that

$$P[\eta(m) = 1 \text{ for infinitely many } m] = C\sum_{k=0}^{\infty} F(k+1)\rho^{-k}. \tag{3.13}$$

This determines the value of C, so it follows that the limit of $v(n)$ exists along the full sequence, and

$$\lim_{n\to\infty} v(n) = \frac{P[n(m) = 1 \text{ for infinitely many } m]}{\sum_{k=0}^{\infty} F(k+1)\rho^{-k}}. \tag{3.14}$$

It remains to show that

$$P[\eta(m) = 1 \text{ for infinitely many } m] = \left[\sum_{k=0}^{\infty} F(k+1)\rho^{-k}\right]^{-1}. \tag{3.15}$$

To do so, let

$$z(n) = P[\eta(m) = 1 \text{ for some } m \geq n]. \tag{3.16}$$

The idea is to show that $z(n)$ is the renewal sequence associated with the density $g(k)$. Once this is done, we can let n tend to ∞ in (3.16) using the ordinary renewal theorem to obtain (3.15). To see that $z(n)$ satisfies the renewal equation for g, write for $n \geq 1$:

$$\begin{aligned}
&\sum_{k=1}^{n} g(k)z(n-k) - z(n)\\
&\quad= \sum_{k=1}^{n} F(k)\rho^{-k+1}[z(n-k) - z(n-k+1)] - F(n+1)\rho^{-n}\\
&\quad= \sum_{k=1}^{n} F(k)\rho^{-k+1}P[\eta(n-k) = 1,\ \eta(j) = 0 \text{ for all } j > n-k]\\
&\qquad - F(n+2)\rho^{-n}\\
&\quad= \sum_{k=1}^{n} F(k)\rho^{-k+1}v(n-k)F(n-k+2)\rho^{-(n-k+1)} - F(n+1)\rho^{-n}\\
&\quad= \rho^{-n}\left\{\sum_{k=1}^{n} F(k)[u(n-k) - u(n-k+1)] - F(n+1)\right\}\\
&\quad= \rho^{-n}\left\{\sum_{k=1}^{n} u(n-k)f(k) - u(n)\right\} = 0.
\end{aligned}$$

In the above computation, we have summed by parts in the first and next to last steps, and have used the renewal equation (1.1) in the last step. The third equality follows from the construction of the sequence $\eta(n)$ and (3.6), and the fourth comes from the definition of $v(n)$ in (3.1). ∎

4. CONTINUOUS TIME

In this section, we will give a brief description of the continuous time analogues of the results which were presented earlier for discrete time. Let $f(t)$ be a continuous probability density on $[0, \infty)$, and let $u(t)$ be the corresponding renewal density, i.e., the sum for $k \geq 1$ of the k-fold convolutions of f with itself. Then $u(t)$ satisfies the renewal equation

$$u(t) = f(t) + \int_0^t f(s)u(t-s)\,ds, \quad t \geq 0. \tag{4.1}$$

The analogue of Theorem 1 is

THEOREM 3. *Let r be a positive integer.*

(a) *If* $f(s+t)$ *is* TP_r, *then* $u(s+t)$ *is* TP_r.
(b) *If* $u(s+t)$ *is* TP_{r+1}, *then* $f(s+t)$ *is* TP_r.

The proof of Theorem 3 is most easily carried out by discretizing $f(\cdot)$, applying Theorem 1 to the discretization, and then passing to the limit. There are continuous time analogues of the identities in Proposition 2.1, but they are not particularly useful for proving Theorem 3. Still, it is probably worth stating some of these analogues. To do so, we need some notation. If $g(\cdot)$ is a function on $[0, \infty)$ which has k continuous derivatives, let

$$g_{k+1}(t_0, \ldots, t_k) = \begin{vmatrix} g(t_0) & g'(t_0) \cdots g^{(k)}(t_0) \\ g(t_1) & g'(t_1) \cdots g^{(k)}(t_1) \\ \cdot & \cdot \quad\quad \cdot \\ \cdot & \cdot \quad\quad \cdot \\ g(t_k) & g'(t_k) \cdots g^{(k)}(t_k) \end{vmatrix}$$

for $0 \leq t_0 < t_1 < \cdots < t_k$. When there are coincidences among two or more of the t_i's, successive rows are differentiated, so that for example

$$g_3(s,t,t) = \begin{vmatrix} g(s) & g'(s) & g''(s) \\ g(t) & g'(t) & g''(t) \\ g'(t) & g''(t) & g'''(t) \end{vmatrix}.$$

The continuous time analogues of identities (a) and (c) of Proposition 2.1 then become for $k \geq 1$

$$f_k(t,\ldots,t)u_{k+1}(t,\ldots,t) = f_{k+1}(t,\ldots,t)u_k(t,\ldots,t) + \int_0^t f_{k+1}(s,t,\ldots,t)u_{k+1}(t-s,t,\ldots,t)\,ds$$

and

$$u_k(0,\ldots,0) = f_k(0,\ldots,0)$$

respectively. The proof of the first of these is the same as the proof in discrete time – one simply expands the determinants in the integral along the first row, and then uses (4.1). To prove the second identity, simply put $t = 0$ in the first, and then use induction on k. The analogues of (b) and (d) are more complicated, so they will be omitted.

In order to state the continuous time version of Theorem 2, let

$$F(t) = \int_t^\infty f(s)\,ds$$

be the tail probabilities corresponding to the density $f(\cdot)$. Note that $F(s+t)$ is TP_2 in $s, t \geq 0$ if and only if $f(t)/F(t)$ is decreasing in t, so that in this case, we may define

$$\lambda = \lim_{t\to\infty} \frac{f(t)}{F(t)} \geq 0.$$

THEOREM 4. *Suppose that f is a probability density on $[0,\infty)$ such that the tail probabilities satisfy $F(s+t)$ is TP_2 in $s, t \geq 0$. Then the renewal density $u(t)$ is decreasing on $[0,\infty)$, so that we may define a measure ν on $[0,\infty)$ by*

$$d\nu(t) = -\frac{1}{F(t)}\,du(t).$$

For $T \geq 0$, let ν_T be the measure on $[-T, \infty)$ which is obtained by translating ν to the left by a distance T. Then ν_T converges vaguely as T tends to infinity to the zero measure if

$$\int_0^\infty F(t)e^{\lambda t}\, dt = \infty,$$

and converges vaguely to a strictly positive multiple of Lebesgue measure if

$$\int_0^\infty \frac{F^2(t)}{F(2t)}\, dt < \infty.$$

The proof of Theorem 4 is similar to the proof of Theorem 2 which is given in Section 3. We will simply write down the continuous time analogues of the more important formulas used there. The continuous time versions of (3.4) and (3.5) are

$$d\nu(t) = -d\frac{f(t)}{F(t)} + \int_0^t g_s(t-s)\, d\nu(s)\, dt$$

where the first term on the right is the (finite) measure obtained by differentiating the increasing function $-f(t)/F(t)$, and

$$g_s(t) = -F(s)\frac{d}{dt}\frac{F(t)}{F(t+s)}$$

respectively. Note that the TP_2 assumption on $F(\cdot)$ implies again that g_s is a sub probability density on $[0, \infty)$, and that

$$\lim_{s\to\infty} g_s(t) = f(t)e^{\lambda t} - \lambda F(t)e^{\lambda t}.$$

Furthermore,

$$\int_0^T g_s(t)\, dt = 1 - \frac{F(s)F(T)}{F(T+s)},$$

which is increasing in s for each T, so that the measures with densities g_s are stochastically decreasing in s. This can be used to show that $\nu(T+[a,b])$ is uniformly bounded in T for fixed $a < b$. Therefore the proof can be carried out as before.

REFERENCES

Brown, M. (1980), Bounds, inequalities, and monotonicity properties for some specialized renewal processes, *Annals of Probability*, **8**, 227-240.

Chover, J., Ney, P. and Wainger, S. (1973), Functions of probability measures, *Journal of Analyse Mathematics*, **26**, 255-302.

de Bruijn, N. G. and Erdös, P. (1953), On a recursion formula and on some Tauberian theorems, *Journal of Research of the Nattional Bureau of Standards*, **50**, 161-164.

Embrechts, P. and Omey, E. (1984), Functions of power series, *Yokohama Mathematics Journal*, **32**, 77-88.

Grübel, R. (1982), Eine restgliedabschätzung in der erneuerungstheorie, *Arkiv der Mathematik*, **39**, 187-192.

Grübel, R. (1983), Functions of discrete probability measures: rates of convergence in the renewal theorem, *Zeitschrift für Wahrscheinlichkeitstheorie und Verwandte Gebiete* **64**, 341-357.

Horn, R. A. (1970), On moment sequences and renewal sequences, *Journal of Mathematical Analysis and Application*, **31**, 130-135.

Kaluza, T. (1928), Ueber die koeffizienten reziproker funktionen, *Mathematische Zeitschrift*, **28**, 161-170.

Karlin, S. (1955), On the renewal equation, *Pacific Journal of Mathematics*, **5**, 229-257.

Karlin, S. (1964), Total positivity, absorption probabilities and applications, *Transactions of the American Mathematical Society*, **111**, 33-107.

Karlin, S. (1968), *Total Positivity*, Stanford University Press, Stanford.

Karlin, S. and McGregor, J. (1959), Coincidence probabilities, *Pacific Journal of Mathematics*, **9**, 1141-1164.

Liggett, T. M. (1970), On convergent diffusions: the densities and the conditioned processes, *Indiana University Mathematics Journal*, **20**, 265-279.

Liggett, T. M. (1985), *Interacting Particle Systems*, Springer-Verlag, New York.

Liggett, T. M. (1989), Exponential L_2 convergence of attractive reversible nearest particle systems, *Annals of Probability*, to appear.

Shanbhag, D. N. (1977), On renewal sequences, *Bulletin of the London Mathematical Society*, **9**, 79-80.

Shohat, J. A. and Tamarkin, J. D. (1943), *The problem of Moments*, American Mathematical Society, Providence, Rhode Island.

Stone, C. (1965), On characteristic functions and renewal theory, *Transactions of the American Mathematical Society*, **120**, 327-342.

Some Remarks on Nonnegative Polynomials on Polyhedra

Charles A. Micchelli
IBM, Yorktown Heights

Allan Pinkus
Technion

1. INTRODUCTION

A classical theorem of Lukács, see Szegö (1939), p. 4, states that an algebraic polynomial $p(t)$ of degree at most n, nonnegative on the interval $[a, b]$, has a representation of the form

$$p(t) = \begin{cases} (t-a)(b-t)R^2(t) + Q^2(t), & n = 2m \\ (t-a)S^2(t) + (b-t)T^2(t), & n = 2m-1 \end{cases} \tag{1.1}$$

where R, S and T are real polynomials of degree $\leq m-1$, and Q is a polynomial of degree $\leq m$. This result and its corresponding versions on $[0, \infty)$ and $(-\infty, \infty)$ have spawned numerous generalizations and applications. They have been used, for example, in the theory of orthogonal polynomials, in the solution of moment problems, and in the study of certain extremal problems for polynomials. One recent use was given in Edelman, Micchelli (1987), where they studied the problem of interpolation by monotone piecewise polynomials.

AMS 1980 Subject Calssification: Primary 26C99, Secondary 41A63.
Key words and phrases: Polynomials, polyhedra, degree raising, Bernstein-Bézier form.

Copyright © 1989 by Academic Press, Inc.
All rights of reproduction in any form reserved.
ISBN-0-12-058470-0

A refinement of Lukács' Theorem was given in Karlin, Shapley (1953), where it was proven that an algebraic polynomial $p(t)$ of degree at most n, nonnegative on $[a,b]$, with at most $n-1$ zeros on $[a,b]$ counting multiplicities, admits a unique representation of the form

$$p(t) = \alpha(t-a)(b-t)\prod_{j=1}^{m-1}(t-t_{2j})^2 + \beta\prod_{j=1}^{m}(t-t_{2j-1})^2, \tag{1.2a}$$

for $n = 2m$, and

$$p(t) = \alpha(t-a)\prod_{j=1}^{m-1}(t-t_{2j})^2 + \beta(b-t)\prod_{j=1}^{m-1}(t-t_{2j-1})^2, \tag{1.2b}$$

for $n = 2m-1$, where $\alpha, \beta > 0$ and $a \le t_1 \le \cdots \le t_{n-1} \le b$. In Karlin (1963), see also Karlin, Studden (1966), this result is generalized to Chebyshev systems. The role of this result in the geometry of moment spaces is well documented in Karlin, Studden (1966). In this regard (1.2) not only identifies the extreme rays of the cone of nonnegative polynomials of degree n on $[a,b]$, but also states that any other element of this cone is a positive combination of at most two such rays.

Attempts to generalize Lukács' Theorem in its various forms to polynomials of more than one variable quickly lead to negative results. For instance, in one dimension, if $p(t)$ is nonnegative for all real t, it is a sum of two squares. However, Hilbert (1888) showed that there exists a real polynomial of degree six in two variables which is nonnegative on $\mathbb{R}^2$, but which is not a finite sum of squares. His argument did not lead to an explicit polynomial. Motzkin (1967) was the first to provide such an example. Subsequently, several authors have provided other examples cf. Robinson (1973), Choi (1975), Schmüdgen (1979) and Berg, Christensen, Jensen (1979). It is interesting to note that Hilbert (1893) later showed that nonnegative polynomials are a sum of squares of rational functions.

Our initial starting point was the modest problem of a representation theorem, similar to (1.2), for *quadratic* polynomials of two variables on a *triangle*. Keeping (1.2) in mind for $n = 2$, we believed that any quadratic polynomial $p(x)$ nonnegative on the triangle $\sigma = [v^1, v^2, v^3]$, with vertices $v^1, v^2, v^3 \in \mathbb{R}^2$ could be represented in the form

$$p(x) = \alpha\lambda_1(x)\lambda_2(x) + \beta\lambda_1(x)\lambda_3(x) + \gamma\lambda_2(x)\lambda_3(x) + q(x) \tag{1.3}$$

where the $(\lambda_1(x), \lambda_2(x), \lambda_3(x))$ are the barycentric coordinates of x relative to σ, the numbers α, β, and γ are nonnegative, and q is a bivariate quadratic polynomial nonnegative on all of $\mathbb{R}^2$, and vanishing at some point in σ. To somewhat explain this analogue to (1.2), each of the first three terms vanish on two sides of the triangle and are nonnegative thereon, and therefore correspond in a sense to the factor $(t-a)(b-t)$ of (1.2a). Our intuition was correct and we will prove this result in the course of our discussion.

Nevertheless, even representations of quadratic polynomials in more variables on a simplex becomes a murky problem indeed. We were only able to prove an analogue of (1.1), and not (1.2), for nonnegative quadratic polynomials on simplices in $\mathbb{R}^3$, while an analogue of both (1.1) and (1.2) on simplices in $\mathbb{R}^4$ is in general false. The reason, as we shall later see, is that this question relates to the notion of *copositive matrices* introduced by Motzkin (1952), (1965). Unfortunately the extreme rays of $n \times n$ copositive matrices have yet to be classified and they are not, for $n \geq 5$, what is demanded by a representation of the form (1.3) for simplices in $\mathbb{R}^{n-1}$. More explanation will be provided shortly.

Returning for the moment to univariate polynomials to obtain some guidance and motivation, we remark that there is yet another characterization theorem for *strictly* positive polynomials on $[0,1]$. It is given in terms of a nonnegative sum of *Bernstein* polynomials of some degree, generally higher than the degree of the polynomial they represent, cf. Pólya, Szegö (1976), p. 78, and Karlin, Studden (1966), p. 126. We will generalize this theorem to any number of variables by using the concept of *degree raising*. This idea provides us with a simple iterative test for determining when a polynomial is strictly positive on a simplex, and a rather simple iterative test for strict copositivity of a matrix. Tests for copositivity and strict copositivity have been the subject of numerous papers, cf. Motzkin (1952), (1965), Garsia (1964), Cottle, Habetler, Lemke (1970), Martin (1981), Hadeler (1983), and Väliaho (1986), (1988). The above test is linear in the matrix elements as opposed to various determinantal criteria of other authors.

We will now discuss these ideas and related ones in detail. Our first task is the precise formulation of the problem which we will address.

2. NONNEGATIVE POLYNOMIALS: PROJECTIVE COORDINATES

We let $\Pi_n(\mathbb{R}^d)$ denote the class of real algebraic polynomials of degree at most n on $\mathbb{R}^d$, i.e., $p \in \Pi_n(\mathbb{R}^d)$ if

$$p(x) = \sum_{|\alpha| \le n} a_\alpha x^\alpha \left(= \sum_{|\alpha| \le n} a_\alpha x_1^{\alpha_1} \cdots x_d^{\alpha_d} \right)$$

where $\alpha = (\alpha_1, \ldots, \alpha_d) \in \mathbb{N}_+^d$, $|\alpha| = \alpha_1 + \cdots + \alpha_d$, and $a_\alpha \in \mathbb{R}$. We consider polynomials $p \in \Pi_n(\mathbb{R}^d)$ nonnegative over a convex polyhedron Ω in $\mathbb{R}^d$. The polyhedron is described by a finite set of linear inequalities. Namely,

$$\Omega = \{x : \Lambda(x) \ge 0,\ x \in \mathbb{R}^d\}$$

where the mapping $\Lambda : \mathbb{R}^d \to \mathbb{R}^m$, $\Lambda(x) = (\lambda_1(x), \ldots, \lambda_m(x))$ has affine coordinates given by

$$\lambda_i(x) := \delta^i \cdot x + \mu_i, \quad i = 1, \ldots, m,$$

with $\{\delta^1, \ldots, \delta^m\} \subseteq \mathbb{R}^d$, $\mu = (\mu_1, \ldots, \mu_m) \in \mathbb{R}^m$, and $\delta \cdot x = \sum_{j=1}^d \delta_j x_j$. Since we are concerned with nonnegative polynomials over Ω, we can and will always assume, without loss of generality, that $0 \in \Omega$. This simply means that $\mu \ge 0$. We also suppose that $\mu \ne 0$. This can always be assumed by adding redundant inequalities in the description of Ω.

We first list several concrete examples of polyhedra Ω to motivate and illustrate our subsequent remarks.

EXAMPLE 2.1 (Simplex). Let $\sigma := [v^1, \ldots, v^{d+1}]$ denote the convex hull of $v^1, \ldots, v^{d+1}$ in $\mathbb{R}^d$, where we assume that the $v^1, \ldots, v^{d+1}$ do not all lie on a hyperplane in $\mathbb{R}^d$. In other words, this is an arbitrary simplex in $\mathbb{R}^d$. We define $\Lambda(x) = (\lambda_1(x), \ldots, \lambda_{d+1}(x))$ as the *barycentric coordinates* of x relative to σ. The $\{\lambda_i(x)\}_{i=1}^{d+1}$ are determined by the equations

$$x = \sum_{j=1}^{d+1} \lambda_j(x) v^j$$

$$1 = \sum_{j=1}^{d+1} \lambda_j(x) .$$

Then $\Omega = \sigma$ and the map $c_\sigma(\lambda) := \sum_{j=1}^{d+1} \lambda_j v^j$ takes the standard simplex $\{\lambda : \lambda \geq 0, \sum_{j=1}^{d+1} \lambda_j = 1\}$ onto σ.

EXAMPLE 2.2 (First Quadrant in $\mathbb{R}^2$). Let $x = (x_1, x_2)$ and $\lambda_1(x) = x_1$, $\lambda_2(x) = x_2$, $\lambda_3(x) = 1$. (Note the redundant inequality $\lambda_3(x) \geq 0$.)

EXAMPLE 2.3 (Vertical Slab in $\mathbb{R}^2$). $\lambda_1(x) = x_1$, $\lambda_2(x) = x_2$ and $\lambda_3(x) = 1 - x_1$.

EXAMPLE 2.4 (Oblique Slab in $\mathbb{R}^2$). $\lambda_1(x) = x_1$, $\lambda_2(x) = x_2$ and $\lambda_3(x) = 1 - x_1 + x_2$.

EXAMPLE 2.5 (Square in $\mathbb{R}^2$). $\lambda_1(x) = x_1$, $\lambda_2(x) = x_2$, $\lambda_3(x) = 1 - x_1$ and $\lambda_4(x) = 1 - x_2$.

EXAMPLE 2.6 (Right-Half Plane in $\mathbb{R}^2$). $\lambda_1(x) = x_1$ and $\lambda_2(x) = 1$.

Each of these examples offer certain peculiarities in delineating a representation for quadratic polynomials on their respective domains. We will deal with each in detail in Section 3.

We now consider the problem of when a polynomial p is nonnegative on Ω. We will first introduce projective coordinates so that later, in the case when p is a quadratic polynomial, we can relate the nonnegativity of p on Ω to the notion of conditionally positive semi-definite matrices. To this end we introduce $y = (x, \tau) \in \mathbb{R}^{d+1}$ and define $L(y) = \Delta x + \tau\mu$, where Δ is the $m \times d$ matrix whose columns are $\delta^1, \ldots, \delta^m$. We assume that the linear map $L : \mathbb{R}^{d+1} \to \mathbb{R}^m$ has a trivial null space, i.e., the $m \times (d+1)$ matrix $[\Delta, \mu]$ is of rank $d+1$. Examples 2.1–2.5 have this property. Example 2.6 does not. Under this condition, there exist integers $1 \leq i_1 < \cdots < i_{d+1} \leq m$ such that the map $\widetilde{L} : \mathbb{R}^{d+1} \to \mathbb{R}^{d+1}$ given by

$$\widetilde{L}(y) = (\delta^{i_1} \cdot x + \tau\mu_{i_1}, \ldots, \delta^{i_{d+1}} \cdot x + \tau\mu_{i_{d+1}})$$

is invertible on $\mathbb{R}^{d+1}$. By reordering, we may assume that $i_j = j$, $j = 1, \ldots, d+1$. Thus there exist vectors $v^1, \ldots, v^{d+1} \in \mathbb{R}^{d+1}$ and constants $w_1, \ldots, w_{d+1}$ such that

$$\begin{aligned} x &= \sum_{i=1}^{d+1} v^i \lambda_i(x) \\ 1 &= \sum_{i=1}^{d+1} w_i \lambda_i(x). \end{aligned} \tag{2.1}$$

We use the tilde notation $\widetilde{\Lambda}(x) = (\lambda_1(x), \ldots, \lambda_{d+1}(x)) = \widetilde{\Delta} x + \widetilde{\mu}$, where $\widetilde{\Delta}$ is the matrix obtained from the first $d+1$ rows of Δ, and $\widetilde{\mu}$ is the vector of the first $d+1$ elements of μ. Note the following correspondance with projective coordinates.

$$L(y) = \begin{cases} \tau\Lambda(\tau^{-1}x), & y = (x,\tau),\ \tau \neq 0 \\ \Delta x, & \tau = 0 \end{cases}$$

and similarly for the tilde maps.

From the above we can associate with every polynomial $p \in \Pi_n(\mathbb{R}^d)$ a unique polynomial $\widetilde{p}$ of degree n on $\mathbb{R}^{d+1}$ which is homogeneous of degree n in its variables $(\lambda_1, \ldots, \lambda_{d+1})$, such that

$$\widetilde{p}(\widetilde{\Lambda}(x)) = p(x), \quad x \in \mathbb{R}^d .$$

The existence of $\widetilde{p}$ is easily established. For a given monomial $x_1^{\alpha_1} \cdots x_d^{\alpha_d}$ with $|\alpha| = \alpha_1 + \cdots + \alpha_d \leq n$, we can use (2.1) to rewrite it as

$$x_1^{\alpha_1} \cdots x_d^{\alpha_d} = \prod_{j=1}^{d} \Big(\sum_{i=1}^{d+1} v_j^i \lambda_i(x) \Big)^{\alpha_j} \Big(\sum_{i=1}^{d+1} w_i \lambda_i(x) \Big)^{n-|\alpha|}$$

which is homogeneous of degree n in the λ_i. The uniqueness of $\widetilde{p}$ is obvious.

Any given p has, of course, different representations from different choices of the tilde map. We formalize this observation in the following:

PROPOSITION 2.1. *Given any invertible linear map $\widetilde{L} : \mathbb{R}^d \times \mathbb{R} \to \mathbb{R}^{d+1}$, there is a one-to-one correspondance between $p \in \Pi_n(\mathbb{R}^d)$ and homogeneous $\widetilde{p} \in \Pi_n(\mathbb{R}^{d+1})$ given by the equation*

$$p(x) = \widetilde{p}(\widetilde{L}(x,1)), \quad x \in \mathbb{R}^d .$$

Going back to the polyhedron Ω, we see that fixing a choice of the tilde map leads to a cone in $\mathbb{R}^{d+1}$, namely

$$\widetilde{\Omega}_+ := \{\nu : \nu = \widetilde{L}(x,\tau),\ \tau \geq 0,\ x \in \mathbb{R}^d,\ L(x,\tau) \geq 0\} .$$

The nonnegativity of p on Ω transforms into the nonnegativity of $\widetilde{p}$ on $\widetilde{\Omega}_+$. For clearly if $\widetilde{p}(\nu) \geq 0$ for all $\nu \in \widetilde{\Omega}_+$ then $p(x) = \widetilde{p}(\widetilde{L}(x,1)) \geq 0$ for all

$x \in \Omega$ since $\nu = \widetilde{L}(x,1) \in \widetilde{\Omega}_+$. Conversely, assume $p(x) \geq 0$ for all $x \in \Omega$. Let $\nu_0 \in \widetilde{\Omega}_+$ be such that $\nu_0 = \widetilde{L}(x_0, \tau_0)$, $\tau_0 \geq 0$, $L(x_0, \tau_0) \geq 0$. If $\tau_0 > 0$, then $\tau_0^{-1}\nu_0 = \widetilde{L}(\tau_0^{-1}x_0, 1)$ and $\tau_0^{-1}x_0 \in \Omega$. Therefore

$$0 \leq p(\tau_0^{-1}x_0) = \widetilde{p}(\tau_0^{-1}\nu_0).$$

By the homogeneity of $\widetilde{p}$, we get

$$0 \leq \widetilde{p}(\tau_0^{-1}\nu_0) = \tau_0^{-n}\widetilde{p}(\nu_0)$$

and thus $\widetilde{p}(\nu_0) \geq 0$. If $\tau_0 = 0$, i.e., $\nu_0 = \widetilde{L}(x_0, 0)$, we let $\nu_\varepsilon := \widetilde{L}(\varepsilon x_0, 1)$. Since by assumption $0 \in \Omega$, we have $\nu_\varepsilon \in \widetilde{\Omega}_+$. Therefore

$$0 \leq \lim_{\varepsilon \to 0+} \varepsilon^{-n} p(\varepsilon x_0) = \lim_{\varepsilon \to 0+} \varepsilon^{-n} \widetilde{p}(\nu_\varepsilon) = \widetilde{p}(\nu_0).$$

We can improve the above observation by noting that under an additional hypothesis on Ω, the polynomial $\widetilde{p}$ is actually nonnegative on the larger cone

$$\widetilde{\Omega} = \{\nu : \nu = \widetilde{L}(x,\tau),\ L(x,\tau) \geq 0,\ (x,\tau) \in I\!R^{d+1}\}$$

which is determined by one less inequality (i.e., $\tau \geq 0$ is deleted), and as a further consequence *strict positivity* of p on Ω is then equivalent to strict positivity of $\widetilde{p}$ on $\widetilde{\Omega}$.

PROPOSITION 2.2. *With the above assumptions, set*

$$C = \{x : \Delta x \geq 0,\ x \in I\!R^d\}.$$

Assume $C = \{0\}$. Then $\widetilde{\Omega}_+ = \widetilde{\Omega}$ and futhermore, whenever p is a polynomial, then $p(x) > 0$, $x \in \Omega$ if and only if $\widetilde{p}(\nu) > 0$, $\nu \in \widetilde{\Omega}\backslash\{0\}$.

REMARK 2.1. Since $0 \in \Omega$ we have $\mu \geq 0$ and therefore $C \subseteq \Omega$. Furthermore if Ω is bounded it is easily seen that $C = \{0\}$. Therefore this condition holds for Examples 2.1 and 2.5, but does not hold for the other examples. However, as may be seen from Example 2.3, it is not necessary that $C = \{0\}$ in order that $\widetilde{\Omega}_+ = \widetilde{\Omega}$.

PROOF. Let $(x,\tau) \in I\!R^{d+1}$ satisfy $L(x,\tau) \geq 0$. Then $\Delta x \geq -\tau\mu$ and so, if $\tau \leq 0$ we conclude (since $\mu \geq 0$) that $x \in C$. Therefore $x = 0$, and since

$\mu \neq 0$, we also get $\tau = 0$. We have established not only that $\widetilde{\Omega}_+ = \widetilde{\Omega}$, but also that the only $\nu = \widetilde{L}(x, 0) \in \widetilde{\Omega}$ is $\nu = 0$. Hence the remaining claim of the proposition is also immediate. ∎

REMARK 2.2. It is easy to see that the condition $C = \{0\}$ is equivalent to the fact that the cone spanned by the $\delta^1, \ldots, \delta^m$ is all of $\mathbb{R}^d$. We also remark that since we have shown that the inequalities $L(x, \tau) \geq 0$ imply $\tau \geq 0$ whenever $C = \{0\}$, it follows that there exist, in this case, nonnegative γ_i, $i = 1, \ldots, m$, for which

$$1 = \sum_{i=1}^{m} \gamma_i \lambda_i(x), \quad x \in \mathbb{R}^d .$$

We now investigate in some detail the special cases alluded to earlier. Most of our remarks pertain to Example 2.1. Here Ω is a given simplex σ, $m = d+1$ so that the tilde and un-tilde maps are the same, and $\widetilde{\Omega} = \mathbb{R}_+^{d+1}$. Furthermore, in this case Proposition 2.1 merely gives us what is commonly referred to as the Bernstein-Bézier representation of a polynomial $p \in \Pi_n(\mathbb{R}^d)$ in barycentric coordinates relative to σ. We adopt the usual notation and write

$$p(x) = \sum_{|\alpha|=n} b_\alpha B_\alpha^n(\lambda) \tag{2.2}$$

where $\lambda^\alpha = \lambda_1^{\alpha_1} \cdots \lambda_{d+1}^{\alpha_{d+1}}$, and $B_\alpha^n(\lambda) = \binom{n}{\alpha} \lambda^\alpha$.

From Proposition 2.2 we have that $p(x)$ is nonnegative (positive) on σ if and only if the Bernstein-Bézier polynomial is nonnegative (positive) on $\mathbb{R}^{d+1}$ ($\mathbb{R}^{d+1} \backslash \{0\}$). Recall that an $m \times m$ symmetric matrix A is *copositive* if $(Ax, x) \geq 0$ for all $x \in \mathbb{R}_+^m$. It is *strictly copositive* if $(Ax, x) > 0$ for all $x \in \mathbb{R}_+^m \backslash \{0\}$. As a corollary we have:

COROLLARY 2.3. *Let p be a quadratic polynomial on $\mathbb{R}^d$. Then p is nonnegative (positive) on a simplex σ if and only if in the Bernstein-Bézier form (2.2), the $(d+1) \times (d+1)$ matrix $A = (a_{ij})$ defined by $a_{ij} = b_{e^i + e^j}$ where $e^i \in \mathbb{R}^{d+1}$, $(e^i)_k = \delta_{ik}$, is copositive (strictly copositive).*

Using a result of Hadeler (1983) characterizing 3×3 copositive matrices, we immediately obtain a condition for a bivariate quadratic polynomial to be nonnegative over a triangle σ, in terms of the coefficients of its Bernstein-Bézier representation of degree 2.

COROLLARY 2.4. *Let p be a bivariate quadratic polynomial. Then $p(x) \geq 0$, $x \in \sigma$, σ a given triangle if and only if when*

$$p(x) = \sum_{i,j=1}^{3} b_{ij}\lambda_i(x)\lambda_j(x) \tag{2.3}$$

where $\lambda = (\lambda_1, \lambda_2, \lambda_3)$ are the barycentric coordinates of x, we have
(i) $b_{ii} \geq 0, \quad i = 1, 2, 3$
(ii) $b_{ij} \geq -\sqrt{b_{ii}b_{jj}}$, *all* $i \neq j$
and at least one of the following
(iii) $b_{12}\sqrt{b_{33}} + b_{23}\sqrt{b_{11}} + b_{13}\sqrt{b_{22}} + \sqrt{b_{11}b_{22}b_{33}} \geq 0$
(iv) $\det B \geq 0$
holds. Furthermore, p is positive on σ if and only if (i) and (ii) hold with strict inequality, and either (iii) holds or (iv) holds with strict inequality.

REMARK 2.3. In (2.3) it is to be assumed that $b_{ij} = b_{ji}$ for all i and j.

REMARK 2.4. This result was recently discussed by Nadler (1988) from another point of view.

Another characterization of 3×3 copositive matrices leads us to an analogue of (1.1). To explain, we first observe that any symmetric matrix A which is a sum of a positive semi-definite matrix plus a matrix with nonnegative elements is necessarily copositive. More importantly, it is known that all copositive matrices of order ≤ 4 can be so represented, see Diananda (1962). However there are 5×5 copositive matrices which cannot be decomposed in this form. This leads us to the following representation of bivariate and trivariate quadratic polynomials, nonnegative on a simplex.

COROLLARY 2.5. *Let p be a quadratic polynomial of either two or three variables. Then $p(x)$ is nonnegative on the simplex σ if and only if $p(x)$ has the form*

$$p(x) = \sum_{1 \leq i < j \leq d+1} \alpha_{ij}\lambda_i(x)\lambda_j(x) + q(x), \quad x \in I\!R^d, \tag{2.4}$$

$d = 2,3$, where $\lambda = (\lambda_1, \ldots, \lambda_{d+1})$ are barycentric coordinates of x relative to σ, $\alpha_{ij} \geq 0$, $1 \leq i < j \leq d+1$, and q is nonnegative for all $x \in \mathbb{R}^d$.

REMARK 2.5. It is known that the homogeneous quadratic

$$(A\lambda, \lambda) := (\lambda_1+\lambda_2+\lambda_3+\lambda_4+\lambda_5)^2 - 4\lambda_1\lambda_2 - 4\lambda_2\lambda_3 - 4\lambda_3\lambda_4 - 4\lambda_4\lambda_5 - 4\lambda_5\lambda_1$$

is nonnnegative on $\mathbb{R}^5_+$, but the matrix A is not the sum of a positive semi-definite and nonnegative matrix, cf. Hall (1986). The corresponding quadratic polynomial $p(x) := (A\lambda(x), \lambda(x))$, $x \in \sigma$, σ a simplex, cannot be written in the form (2.4).

The representation (2.4) is an analogue of (1.1) in that we only know that q is nonnegative on $\mathbb{R}^d$. In the next section we show that in the case $d = 2$, we can choose q so that it vanishes at some point in σ. Firstly, however, we give an alternative characterization of *positive* polynomials on σ, of any degree in any number of variables.

To this end, we need to review the process of degree raising, an idea well-known in the methodology associated with modelling curves and surfaces by the Bernstein-Bézier technique. Any $p \in \Pi_n(\mathbb{R}^d)$ has a representation in Bernstein-Bézier form

$$p(x) = \sum_{|\alpha|=m} b_\alpha^m B_\alpha^m(\lambda), \tag{2.5}$$

$b_\alpha^m = b_\alpha^m(p)$, for all $m \geq n$. Furthermore, there is a simple iterative relationship between successive representations as witnessed by the formula

$$b_\alpha^{m+1} = \frac{1}{m+1} \sum_{j=1}^{d+1} \alpha_j b_{\alpha-e^j}^m, \quad |\alpha| = m+1.$$

At each stage, the new representation (2.5) is formed by a simple convex combination of the previous coefficients. Hence

$$\max_{|\alpha|=m+1} |b_\alpha^{m+1}| \leq \max_{|\alpha|=m} |b_\alpha^m|, \quad m \geq n.$$

We combine this with the fact that there exists a positive constant c, depending on σ and n, such that for all $p \in \Pi_n(\mathbb{R}^d)$

$$c \max_{|\alpha|=n} |b_\alpha^n(p)| \leq \max_{x \in \sigma} |p(x)|$$

to obtain

$$c \max_{|\alpha|=m} |b_\alpha^m(p)| \leq \max_{x \in \sigma} |p(x)|, \quad m \geq n\,.$$

Using the Bernstein operator

$$B_m(f(\sigma))(x) := \sum_{|\beta|=m} f\Big(c_\sigma\Big(\frac{\beta}{m}\Big)\Big) B_\beta^m(\lambda)$$

where, as before, $x = c_\sigma(\lambda) = \sum_{j=1}^{d+1} \lambda_j v^j$, we get

$$c \max_{|\alpha|=m} \Big| b_\alpha^m - p\Big(c_\sigma\Big(\frac{\alpha}{m}\Big)\Big)\Big| \leq \max_{x \in \sigma} |p(x) - B_m(p(\sigma))(x)|$$

since for every m, $B_m(p(\sigma))$ is in fact a polynomial of degree n, cf. Dahmen, Micchelli (1988). It is well-known that there exists a constant K depending on p and σ such that the error in approximating p by its Bernstein operator value is bounded by $\frac{K}{m}$. We therefore finally obtain

$$\max_{|\alpha|=m} \Big| b_\alpha^m - p\Big(c_\sigma\Big(\frac{\alpha}{m}\Big)\Big)\Big| \leq \frac{K}{cm}\,. \tag{2.6}$$

This demonstrates the well-known fact that in degree raising the coefficients converge uniformly to values of the polynomial. As a consequence we obtain:

THEOREM 2.6. *Let $p \in \Pi_n(\mathbb{R}^d)$. Then $p(x) > 0$, all $x \in \sigma$, σ some simplex, if and only if there exists an $m \geq n$ such that*

$$p(x) = \sum_{|\alpha|=m} b_\alpha^m B_\alpha^m(\lambda)$$

with $b_\alpha^m > 0$, all α, where $\lambda = (\lambda_1, \ldots, \lambda_{d+1})$ are the barycentric coordinates of x relative to σ.

REMARK 2.6. For polynomials of one variable, the above result says that $p(x) = \sum_{i=0}^{n} a_i x^i$ is positive on $[0,1]$ if and only if there exists an $m \geq n$ such that in the representation

$$p(x) = \sum_{j=0}^{m} b_j \binom{m}{j} x^j (1-x)^{m-j} \tag{2.7}$$

we have $b_j > 0$ for $j = 0, 1, \ldots, m$, see Pólya, Szegö (1976), p. 78, Karlin, Studden (1966), p. 126. This result continues to fascinate. See the recent paper Erdélyi, Szabados (1988) where a study is made of the least m needed in (2.7).

Somewhat more can be said in the case where $d = 1$. Namely, if $p(x) > 0$ on the open interval $(0, 1)$, then we can always divide $p(x)$ by $x^r(1-x)^t$, where r and t are the multiplicities of the zero of p at zero and one, respectively, to obtain a polynomial strictly positive on $[0, 1]$. Thus (2.7) persists (with some zero coefficients) when p is positive on $(0, 1)$. In two or more dimensions we cannot relax the hypothesis of strict positivity. For example, if p is nonnegative on all of $\mathbb{R}^2$ and vanishes only at a point $x^0 \in \partial\sigma$, which is not a vertex (say for definiteness on the face $\lambda_1(x) = 0$) then $\lambda_1(x^0) = 0$ and $\lambda_2(x^0), \ldots, \lambda_{d+1}(x^0) > 0$. Thus if

$$p(x) = \sum_{|\alpha|=m} b_\alpha^m B_\alpha^m(\lambda) \tag{2.8}$$

and $b_\alpha^m \geq 0$, it follows that $b_\alpha^m = 0$ if $\alpha_1 = 0$. But then the polynomial on the right of equation (2.8) vanishes identically on the face $\lambda_1(x) = 0$. Thus for such a polynomial we never have $b_\alpha^m \geq 0$ for all $|\alpha| = m$. The above Theorem 2.6 and Corollary 2.3 also give us a test for *strict* copositivity of a matrix.

COROLLARY 2.7. *Let $B = (b_{ij})$ be a $(d+1) \times (d+1)$ symmetric matrix. Set*

$$b_{e^i+e^j}^2 := b_{ij}, \quad i, j = 1, \ldots, d+1,$$

and

$$b_\alpha^{m+1} = \frac{1}{m+1} \sum_{i=1}^{d+1} \alpha_i b_{\alpha-e^i}^m, \quad m \geq 2,$$

where e^i is the ith unit vector in $\mathbb{R}^{d+1}$ and $\alpha = (\alpha_1, \ldots, \alpha_{d+1}) \in \mathbb{N}^{d+1}$, $|\alpha| = \alpha_1 + \cdots + \alpha_{d+1} = m+1$. Then B is strictly copositive if and only if there exists an $m \geq 2$ such that $b_\alpha^m > 0$ for all $|\alpha| = m$.

As the previous example shows, this test may fail if the matrix B is only copositive. However from (2.6), we see that the above B is copositive if and only if

$$\liminf_{m\to\infty} \min_{|\alpha|=m} b_\alpha^m \geq 0\,.$$

Our final remark of this section pertains to even degree polynomials strictly positive on bounded polyhedra. For this purpose we require the following result.

PROPOSITION 2.8. *Let Γ be an $m \times (d+1)$ matrix of rank $d+1$, and p a homogeneous polynomial of degree $2k$ on $\mathbb{R}^{d+1}$ satisfying $p(x) > 0$ if $\Gamma x \geq 0$, $x \neq 0$. There then exists a homogeneous polynomial q of degree $2k$ on $\mathbb{R}^m$ such that $q(\Gamma x) = p(x)$, $x \in \mathbb{R}^{d+1}$, and $q(y) > 0$ if $y \in \mathbb{R}^m_+ \backslash \{0\}$.*

PROOF. Let $Q : \mathbb{R}^m \to \mathbb{R}^{d+1}$ be the orthogonal projection of $\mathbb{R}^m$ onto the columns of Γ. That is,

$$Qy = \Gamma(\Gamma^T \Gamma)^{-1}\Gamma^T y\,.$$

For any $t \in \mathbb{R}$, set

$$q_t(y) = t[(y, y) - (Qy, y)]^k + p((\Gamma^T \Gamma)^{-1}\Gamma^T y)\,.$$

Then q_t is a homogeneous polynomial of degree $2k$ on $\mathbb{R}^m$. Furthermore, for all $x \in \mathbb{R}^{d+1}$

$$q_t(\Gamma x) = t[(\Gamma x, \Gamma x) - (\Gamma x, \Gamma x)]^k + p(x) = p(x)\,.$$

We claim that there exists a $t_0 > 0$ for which $q_{t_0}(y) > 0$ if $y \in \mathbb{R}^m_+ \backslash \{0\}$. (Note that if q_{t_0} is nonnegative for t_0, then it is nonnegative for all $t \geq t_0$.) Assume this is false. There then exists a sequence $y_n \in \mathbb{R}^m_+ \backslash \{0\}$, $(y_n, y_n) = 1$, such that $q_n(y_n) \leq 0$. We choose a limit point y_0 of the sequence $\{y_n\}$. Then $(y_0, y_0) = 1$ and $y_0 \in \mathbb{R}^m_+ \backslash \{0\}$. Futhermore we have

$$[(y_n, y_n) - (Qy_n, y_n)]^k \leq -\frac{1}{n} p((\Gamma^T \Gamma)^{-1}\Gamma^T y_n)$$

for each n. Let $n \uparrow \infty$. Then

$$[(y_0, y_0) - (Qy_0, y_0)]^k \leq 0\,.$$

Since Q is an orthogonal projection, $(y, y) - (Qy, y) \geq 0$ for all $y \in \mathbb{R}^m$. Thus

$$(y_0, y_0) = (Qy_0, y_0)\,.$$

But this implies, since Q is an orthogonal projection, that there exists an $x_0 \in \mathbb{R}^{d+1} \backslash \{0\}$ such that $y_0 = \Gamma x_0$. Thus $\Gamma x_0 \geq 0$, $x_0 \neq 0$, implying by assumption that $p(x_0) > 0$. However

$$p((\Gamma^T \Gamma)^{-1} \Gamma^T y_n) \leq -n[(y_n, y_n) - (Qy_n, y_n)]^k \leq 0$$

for all n, and therefore $p((\Gamma^T \Gamma)^{-1} \Gamma^T y_0) \leq 0$. But $p((\Gamma^T \Gamma)^{-1} \Gamma^T y_0) = p(x_0)$. This contradiction proves the proposition. ∎

REMARK 2.7. This proposition extends a result of Martin, Jacobson (1981) (Theorem 4.2, p. 239) which corresponds to the case $k = 1$. The proof is patterned after that special case.

From this above result we obtain using Propositions 2.1, 2.2 and 2.8:

COROLLARY 2.9. *Let p be a polynomial of even degree on $\mathbb{R}^d$ which is strictly positive on a bounded polyhedron $\Omega = \{x : \Lambda(x) \geq 0, x \in \mathbb{R}^d\}$, where $\Lambda(x) = \Delta x + \mu$, and $[\Delta, \mu]$ is a matrix of rank $d+1$. There then exists a homogeneous polynomial q on $\mathbb{R}^m$, of the same degree as p, satisfying $q(\Lambda(x)) = p(x)$, all $x \in \mathbb{R}^d$, and $q(y) > 0$ for all $y \in \mathbb{R}^m_+ \backslash \{0\}$.*

3. NONNEGATIVE BIVARIATE QUADRATIC POLYNOMIALS

We begin this section with a proof of the theorem mentioned in the introduction.

THEOREM 3.1. *Let $p(x)$, $x \in \mathbb{R}^2$ be any quadratic polynomial nonnegative on a triangle σ with vertices v^1, v^2, v^3. Let $(\lambda_1(x), \lambda_2(x), \lambda_3(x))$ denote the barycentric coordinates of x relative to σ. Then there are nonnegative constants α, β, γ, and a quadratic polynomial q which is nonnegative on all of $\mathbb{R}^2$ and vanishes at some point $y \in \sigma \backslash \{v^1, v^2, v^3\}$ such that*

$$p(x) = \alpha\lambda_1(x)\lambda_2(x) + \beta\lambda_1(x)\lambda_3(x) + \gamma\lambda_2(x)\lambda_3(x) + q(x). \tag{3.1}$$

We base the proof of this result on the following proposition.

PROPOSITION 3.2. *Let C be a positive semi-definite 3×3 (real) matrix. There then exist 3×3 matrices A and N such that $C = A + N$, where*
(i) $N = (n_{ij})_{i,j=1}^{3}$ satisfies $n_{ij} \geq 0$, $n_{ii} = 0$, $i, j = 1, 2, 3$.
(ii) A is a positive semi-definite matrix with a vector λ^0 in its null space, where $\lambda^0 \in \mathbb{R}_+^3$ and at least two components of λ^0 are strictly positive.

PROOF. We divide the proof into various cases, depending on the sign patterns of the elements of C. We first observe that we may assume that the diagonal elements of C are all positive. For if not, then after permutation of rows and columns, if necessary, we may assume that C has the form

$$C = \begin{pmatrix} 0 & 0 & 0 \\ 0 & b & f \\ 0 & f & c \end{pmatrix}$$

with $b, c \geq 0$ and $bc \geq f^2$. We then set

$$A = \begin{pmatrix} 0 & 0 & 0 \\ 0 & b & -\sqrt{bc} \\ 0 & -\sqrt{bc} & c \end{pmatrix}, \quad N = \begin{pmatrix} 0 & 0 & 0 \\ 0 & 0 & f + \sqrt{bc} \\ 0 & f + \sqrt{bc} & 0 \end{pmatrix}$$

and note that $\lambda^0 = (0, \sqrt{c}, \sqrt{b})$ is an eigenvector of A. If $b = 0$, then $\lambda^0 = (1, 1, 0)$ is also an eigenvector of A. The similar result holds if $c = 0$.

We therefore assume that

$$C = \begin{pmatrix} a & d & e \\ d & b & f \\ e & f & c \end{pmatrix}$$

with $a, b, c > 0$.

We consider the various possible sign patterns of the off-diagonal elements of C. Since we can always permute rows and columns, we may assume that d and e have the same sign (weakly). Thus we need only consider four cases.

CASE 1. $d, e, f \geq 0$. In this case we set

$$A = \begin{pmatrix} a & 0 & 0 \\ 0 & b & -\sqrt{bc} \\ 0 & -\sqrt{bc} & c \end{pmatrix}, \quad N = \begin{pmatrix} 0 & d & e \\ d & 0 & f + \sqrt{bc} \\ e & f + \sqrt{bc} & 0 \end{pmatrix}.$$

$\lambda^0 = (0, \sqrt{c}, \sqrt{b})$ is an eigenvector of A.

CASE 2. $d, e \geq 0$ and $f \leq 0$. We decompose C exactly as in Case 1.

CASE 3. $d, e, f \leq 0$. We assume from Case 2 that $d, e < 0$. We claim that there exists an $\alpha_0 \geq 1$ such that the matrix

$$A = \begin{pmatrix} a & \alpha_0 d & \alpha_0 e \\ \alpha_0 d & b & \alpha_0 f \\ \alpha_0 e & \alpha_0 f & c \end{pmatrix}$$

is positive semi-definite and singular. To see this note that $\det A$ is a cubic or quadratic polynomial in α_0 with leading coefficient negative, and is nonnegative for $\alpha_0 = 1$. Thus there exists an $\alpha_0 \geq 1$ for which

$$\det A = abc + 2\alpha_0^3 def - \alpha_0^2(af^2 + be^2 + cd^2) = 0. \tag{3.2}$$

Furthermore, it necessarily follows from (3.2) that $ab \geq \alpha_0^2 d^2$, $ac \geq \alpha_0^2 e^2$ and $bc \geq \alpha_0^2 f^2$. Thus A is positive semi-definite. In fact if $ab = \alpha_0^2 d^2$, then from (3.2),

$$2\alpha_0^3 def - \alpha_0^2(af^2 + be^2) = 0\,.$$

Since all terms are nonpositive, this implies that $be^2 = 0$ contradicting our previous hypothesis. Thus $ab > \alpha_0^2 d^2$, and similarly $ac > \alpha_0^2 e^2$.

We set $N = C - A$. Since $\alpha_0 \geq 1$ and $d, e, f \leq 0$, N has the desired form. The vector

$$\lambda^0 = (\alpha_0^2 df - \alpha_0 eb, -\alpha_0 af + \alpha_0^2 ed, ab - \alpha_0^2 d^2)$$

is an eigenvector of A. Since $a, b, c > 0$, $d, e, < 0$, $f \leq 0$ and $ab > \alpha_0^2 d^2$, it follows that all three components of λ^0 are strictly positive.

CASE 4. $d, e \leq 0$ and $f \geq 0$. If $def = 0$, then we are in one of the previous cases. As such we assume that $d, e < 0$ and $f > 0$.

If $abc - be^2 - cd^2 \geq 0$, set

$$\widetilde{C} = \begin{pmatrix} a & d & e \\ d & b & 0 \\ e & 0 & c \end{pmatrix}, \quad \widetilde{N} = \begin{pmatrix} 0 & 0 & 0 \\ 0 & 0 & f \\ 0 & f & 0 \end{pmatrix}.$$

Then $\widetilde{C}$ is positive semi-definite and we have reduced our problem to case 3.

As such we assume that

$$abc - be^2 - cd^2 < 0\,. \tag{3.3}$$

Since $\det C = abc + 2def - (af^2 + be^2 + cd^2)$, and (3.3) holds, there exists an $\alpha \in (0,1]$ such that

$$C_\alpha = \begin{pmatrix} a & d & e \\ d & b & \alpha f \\ e & \alpha f & c \end{pmatrix}$$

is singular and positive semi-definite. Since $f \geq 0$, $C - C_\alpha$ has nonnegative entries. As such it suffices to work with C_α.

If $a(\alpha f)^2 \geq cd^2$, set

$$A = \begin{pmatrix} a & d & -\sqrt{ac} \\ d & b & -\sqrt{\frac{c}{a}}d \\ -\sqrt{ac} & -\sqrt{\frac{c}{a}}d & c \end{pmatrix},$$

and

$$N = \begin{pmatrix} 0 & 0 & e + \sqrt{ac} \\ 0 & 0 & \alpha f + \sqrt{\frac{c}{a}}d \\ e + \sqrt{ac} & \alpha f + \sqrt{\frac{c}{a}}d & 0 \end{pmatrix}.$$

By assumption $\alpha f + \sqrt{\frac{c}{a}}d \geq 0$, and since C_α is positive semi-definite $e + \sqrt{ac} \geq 0$. The vector $\lambda^0 = (\sqrt{c}, 0, \sqrt{a})$ is an eigenvector of A. If $a(\alpha f)^2 \geq be^2$, a similar analysis holds.

The final possibility we must consider is that $a(\alpha f)^2 < cd^2$ and $a(\alpha f)^2 < be^2$, while C_α is positive semi-definite and singular. In this case we set $A = C_\alpha$. The vector

$$\lambda^0 = (bc - (\alpha f)^2, e(\alpha f) - cd, d(\alpha f) - be)$$

is an eigenvector of A. The first component is nonnegative since C_α is positive semi-definite. Now

$$c(-d) = \sqrt{c}\sqrt{cd^2} > \sqrt{c}\sqrt{a(\alpha f)^2} = \sqrt{ac}(\alpha f) \geq -e(\alpha f)$$

and similarly

$$b(-e) = \sqrt{b}\sqrt{be^2} > \sqrt{b}\sqrt{a(\alpha f)^2} = \sqrt{ab}(\alpha f) \geq -d(\alpha f).$$

Thus the second and third components of λ^0 are strictly positive. This proves the proposition. ∎

PROOF OF THEOREM 3.1. Assume p is a quadratic polynomial nonnegative over a triangle σ with vertices v^1, v^2, v^3. From Corollary 2.3 (see also Corollary 2.5), p may be written in the form

$$p(x) = (B\lambda(x), \lambda(x))$$

where $\lambda(x) = (\lambda_1(x), \lambda_2(x), \lambda_3(x))$ are the barycentric coordinates of x relative to σ, and B is a 3×3 copositive matrix. From Diananda (1962), B can be decomposed as $C + \widetilde{N}$ where C is positive semi-definite and $\widetilde{N}$ is a matrix with nonnegative entries. Since adding positive terms to the diagonal entries of a positive semi-definite matrix maintains the positive semi-definiteness of the matrix, we may assume that the diagonal entries of $\widetilde{N}$ are all zero. We now decompose C as in Proposition 3.2 to obtain

$$B = A + N$$

where A and N are as in the statement of Proposition 3.2. Thus

$$p(x) = (A\lambda(x), \lambda(x)) + (N\lambda(x), \lambda(x)).$$

¿From the definition of N,

$$(N\lambda(x), \lambda(x)) = \alpha\lambda_1(x)\lambda_2(x) + \beta\lambda_1(x)\lambda_3(x) + \gamma\lambda_2(x)\lambda_3(x)$$

for some $\alpha, \beta, \gamma \geq 0$. In addition,

$$q(x) = (A\lambda(x), \lambda(x))$$

is a quadratic polynomial nonnegative on all of $\mathbb{R}^2$, and vanishes at some point $y \in \sigma \backslash \{v^1, v^2, v^3\}$ corresponding to $\lambda(y) = \lambda^0$, where λ^0 has been normalized so that the sum of its coefficients is one. ∎

We now present representation theorems for each of Examples 2.2–2.6.

THEOREM 3.3 (First Quadrant). *Let p be a nonnegative quadratic polynomial on $\mathbb{R}^2_+$. Then there exist nonnegative constants α, β, γ, and a quadratic polynomial q which is nonnegative on all of $\mathbb{R}^2$ such that*

$$p(x) = \alpha x_1 + \beta x_2 + \gamma x_1 x_2 + q(x), \quad x = (x_1, x_2).$$

PROOF. We write $p(x)$ in the form

$$p(x) = (B\Lambda(x), \Lambda(x))$$

where $\Lambda(x) = (x_1, x_2, 1)$ and B is a 3×3 copositive matrix. Here we have made use of Proposition 2.1 and the remarks thereafter. We now decompose B as a positive semi-definite matrix A plus a matrix N with nonnegative entries and zero diagonal entries. ∎

THEOREM 3.4 (Vertical Slab). *Let p be a nonnegative quadratic polynomial on the vertical slab of Example 2.3. Then there exist nonnegative constants α, β, γ, and a quadratic polynomial q which is nonnegative on all of $\mathbb{R}^2$ and which vanishes at some point y in the slab, but not at $(0,0)$ or $(1,0)$, such that*

$$p(x) = \alpha x_1 x_2 + \beta x_1(1 - x_1) + \gamma x_2(1 - x_1) + q(x).$$

PROOF. First we write p in homogeneous coordinates as

$$p(x) = (B\Lambda(x), \Lambda(x))$$

where $\Lambda(x) = (x_1, x_2, 1 - x_1)$, and $x = (x_1, x_2)$. B is a 3×3 matrix and has the property that $(Bz, z) \geq 0$ whenever $z = (x_1, x_2, \tau - x_1)$ with $x_1 \geq 0$, $x_2 \geq 0$, $\tau - x_1 \geq 0$ and $\tau \geq 0$. Even though the hypothesis of Proposition 2.2 is not satisfied, the inequality $\tau \geq 0$ is still redundant. Hence B is copositive. We now proceed as in the proof of Theorem 3.1. The polynomial q vanishes at some point y as claimed since the eigenvector λ^0 of A can be normalized so that its first and third components sum to one. ∎

THEOREM 3.5 (Oblique Slab). *Let p be a nonnegative quadratic polynomial on the oblique slab of Example 2.4. Then there exist nonnegative constants $\alpha, \beta, \gamma, \delta, \varepsilon, \mu$, and a quadratic polynomial q which is nonnegative on all of $\mathbb{R}^2$ such that*

$$\begin{aligned} p(x) = \alpha x_1 + &\beta x_2 + \gamma(1 - x_1 + x_2) + \delta x_1 x_2 + \varepsilon x_1(1 - x_1 + x_2) \\ &+ \mu x_2(1 - x_1 + x_2) + q(x)\,. \end{aligned}$$

For a proof of this result, we first appeal to the following theorem.

THEOREM (Martin, Jacobson (1981)). *Let Q be an $n \times n$ symmetric matrix and A an $m \times n$ matrix with $m \leq 4$. Suppose $(Qx, x) \geq 0$ whenever $Ax \geq 0$, and there exists an $\tilde{x} \in \mathbb{R}^n$ with $A\tilde{x} > 0$. Then there exists a copositive $m \times m$ matrix C, and a positive semi-definite $n \times n$ matrix S such that*

$$Q = A^T C A + S \,. \tag{3.4}$$

REMARK 3.1. Necessary and sufficient conditions for the existence of a decomposition of the type (3.4) is given in Martin, Powell, Jacobson (1981).

PROOF OF THEOREM 3.5. By Proposition 2.1 we write

$$p(x) = (Q\Lambda(x), \Lambda(x))$$

where $\Lambda(x) = (x_1, x_2, 1 - x_1 + x_2)$, $x = (x_1, x_2)$ and Q is a 3×3 matrix. We know that $(Qz, z) \geq 0$ whenever $z = (x_1, x_2, \tau - x_1 + x_2)$ with $x_1 \geq 0$, $x_2 \geq 0$, $\tau - x_1 + x_2 \geq 0$ and $\tau \geq 0$. The last condition is not redundant here. Thus $(Qz, z) \geq 0$ if $Az \geq 0$ where

$$A = \begin{pmatrix} 1 & 0 & 0 \\ 0 & 1 & 0 \\ 0 & 0 & 1 \\ 1 & -1 & 1 \end{pmatrix} .$$

We use the above stated theorem to express p in the form

$$p(x) = (CA\Lambda(x), A\Lambda(x)) + (S\Lambda(x), \Lambda(x))$$

where C is a 4×4 copositive matrix and S is a 3×3 positive semi-definite matrix. We use the result of Diananda (1962) to decompose C as $B + N$ where B is positive semi-definite, and N has nonnegative entries with zero diagonals, plus the fact that $A\Lambda(x) = (x_1, x_2, 1 - x_1 + x_2, 1)^T$ to obtain the statement of the theorem. ∎

The representation theorem for the square is proved in a similar fashion. We omit the details.

THEOREM 3.6 (Square). *Let p be a nonnegative quadratic polynomial on the square of Example 2.5. Then there exist nonnegative constants $\alpha, \beta, \gamma, \delta, \varepsilon, \mu$, and a quadratic polynomial q which is nonnegative on all of $\mathbb{R}^2$ such that*

$$p(x) = \alpha x_1 x_2 + \beta x_1(1 - x_1) + \gamma x_1(1 - x_2) + \delta x_2(1 - x_1) + \varepsilon x_2(1 - x_2) + \mu(1 - x_1)(1 - x_2) + q(x).$$

REMARK 3.2. In Theorems 3.3 and 3.5 we cannot choose the q appearing therein to vanish at some point in the respective polyhedra. This may be seen by representing the polynomial $p(x) = 1$ thereon. In these cases we necessarily have $q(x) = 1$. It may be that the q appearing in Theorem 3.6 and the q of three variables in Corollary 2.5 can be chosen to vanish in their respective polyhedra. This fact would seem to require us to prove an analogue of Proposition 3.2 for 4×4 matrices. The existence of such an analogue remains an open question.

Our final characterization theorem requires a special limiting argument as it cannot be directly converted to homogeneous coordinates.

THEOREM 3.7 (Right-Half Plane). *Let p be a nonnegative quadratic polynomial on the right-half plane (Example 2.6). Then there exists a nonnegative constant α and a quadratic polynomial q which is nonnegative on all of $\mathbb{R}^2$ such that*

$$p(x) = \alpha x_1 + q(x).$$

PROOF. For every $t \geq 0$, p is nonnegative on the cone $x_1 \geq 0$, $x_1 + tx_2 \geq 0$. Therefore we can (after a linear change of variables) use Theorem 3.3 to express p in the form

$$p(x) = \alpha_t x_1 + \beta_t(x_1 + tx_2) + \gamma_t x_1(x_1 + tx_2) + q_t(x) \tag{3.5}$$

where $\alpha_t, \beta_t, \gamma_t \geq 0$ and q_t is a quadratic polynomial nonnegative on all of $\mathbb{R}^2$. We claim that

$$\delta_t = \alpha_t + \beta_t + \gamma_t + \max_{(x,x)=1} q_t(x)$$

remains bounded as $t \rightarrow 0^+$. If not we would divide both sides of (3.5) by δ_t and pass to the limit through a subsequence to obtain

$$0 = \alpha' x_1 + \beta' x_1 + \gamma' x_1^2 + q'(x)$$

where α', β', γ' and q' are not all zero. Since all these quantities are nonnegative we easily obtain a contradiction. Thus δ_t must remain bounded as $t \rightarrow 0^+$. We now pass to a limit in (3.5) through a subsequence, add the β term to the α term, and the γ term to the q term. ∎

ACKNOWLEDGEMENT. We thank Professor W. Dahmen for conversations pertaining to Theorem 2.6.

REFERENCES

Berg, C., J. P. R. Christensen, and C. U. Jensen (1979), A remark on the multidimensional moment problem, *Mathematische Annalen* **243**, 163–169.

Choi, M.-D. (1975), Positive semidefinite biquadratic forms, *Linear Algebra and its Applications*, **12**, 95–100.

Cottle, R. W., G. J. Habetler, and C. E. Lemke (1970), On classes of copositive matrices, *Linear Algebra and its Applications*, **3**, 295–310.

Dahmen, W., and C. A. Micchelli (1988), Convexity and Bernstein polynomials on k-simploids, IBM Research Report.

Diananda, P. H. (1962), On nonnegative forms in real variables some or all of which are nonnegative, *Proceedings of the Cambridge Philosophy Society*, **58**, 17–25.

Edelman, A., and C. A. Micchelli (1987), Admissible slopes for monotone and convex interpolation, *Numerische Mathematik*, **51**, 441–458.

Erdélyi, T. and J. Szabados (1988), On polynomials with positive coefficients, *Journal Approximation Theory*, **54**, 107–122.

Garsia, A. (1964), Remarks about copositive forms, working paper, Department of Mathematics, University of California, San Diego.

Hadeler, K. P. (1983), On copositive matrices, *Linear Algebra and its Applications*, **49**, 79–89.

Hall, M. Jr. (1986), *Combinatorial Theory*, Second edition, John Wiley and Sons, New York.

Hilbert, D. (1888), Über die Darstellung definiter Formen als Summe von Formenquadraten, *Mathematische Annalen*, **32**, 342–350.

Hilbert, D. (1893), Über ternäre definite Formen, *Acta Mathematica Scientia*, **17**, 169–197.

Karlin, S. (1963), Representation theorems for positive functions, *Journal of Mathematics and Mechanics*, **12**, 599–618.

Karlin, S., and L. S. Shapley (1953), Geometry of moment spaces, *Memoirs of the American Mathematical Society*, **12**.

Karlin, S., and W. J. Studden (1966), *Tchebycheff Systems: With Applications in Analysis and Statistics*, Interscience Publishers, New York.

Martin, D. H. (1981), Finite criteria for conditional definiteness of quadratic forms, *Linear Algebra and its Applications*, **39**, 9–21.

Martin, D. H., and D. H. Jacobson (1981), Copositive matrices and definiteness of quadratic forms subject to homogeneous linear inequality constraints, *Linear Algebra and its Applications*, **35**, 227-258.

Martin, D. H., M. J. D. Powell and D. H. Jacobson (1981), On a decomposition of conditionally positive-semidefinite matrices, *Linear Algebra and its Applications*, **39**, 51–59.

Motzkin, T. S. (1952), Copositive quadratic forms, *National Bureau of Standards Report* 1818, 11–22.

Motzkin, T. S. (1965), Quadratic forms positive for nonnegative variables not all zero, *Notices of the American Mathematical Society*, **12**, 224.

Motzkin, T. S. (1967), The arithmetic-geometric inequality, *Inequalities* (O. Shisha, ed.), Academic Press, New York, 205–224.

Nadler, E. (1988), On nonnegative bivariate quadratic polynomials on a triangle, talk at conference *Mathematical Aspects of CAGD*, Oslo, Norway.

Pólya, G., and G. Szegö (1976), *Problems and Theorems in Analysis, Volume II*, Springer-Verlag, New York.

Robinson, R. M. (1973), Some definite polynomials which are not sums of squares of real polynomials, *Selected Questions of Algebra and Logic*, Academic Sciences USSR, 264–282.

Schmüdgen, K. (1979), An example of a positive polynomial which is not a sum of squares of polynomials. A positive, but not strongly positive functional, *Mathematische Nachrichten*, **88**, 385–390.

Szegö, G. (1939), *Orthogonal Polynomials*, American Mathematical Society Colloqium Publishers, Volume XXII.

Väliaho, H. (1986), Criteria for copositive matrices, *Linear Algebra and its Applications*, **81**, 19–34.

Väliaho, H. (1988), Testing the definiteness of matrices on polyhedral cones, *Linear Algebra and its Applications*, **101**, 135–165.

The Fundamental Period of the Queue with Markov-Modulated Arrivals

M. F. Neuts*
University of Arizona

1. INTRODUCTION

Stimulated by applications in communications engineering, as discussed e.g. in Heffes and Lucantoni (1986), queues with Markov-modulated arrivals have recently received much attention. Such queueing models have earlier been discussed in Naor and Yechiali (1971), Neuts (1971, 1978ab, 1986ab), Purdue (1974ab, 1979), Ramaswami (1980), Sengupta (1987a, 1989a), Yechiali (1973), and a comprehensive treatment may be found in Neuts (1989). More recently, Lucantoni, Meier-Hellstern and Neuts (1988) have considered queues with a highly tractable class of arrival processes which include the Markov-modulated arrival process as a particular case. An $M/G/1$ queue in which the arrival rate is modulated by an alternating renewal process is treated in Sengupta (1987b). For the single server queue

AMS 1980 Subject Classifications: Primary 60K25, Secondary 60K20.
Key words and phrases: Queues, Markov-modulated arrivals, fundamental period.
*Research supported in part by the National Science Foundation Grant ECS-8803061 and the Air Force Office of Scientific Research Grant AFOSR-88-0076.

PROBABILITY, STATISTICS,
AND MATHEMATICS
Papers in Honor of Samuel Karlin

Copyright © 1989 by Academic Press, Inc.
All rights of reproduction in any form reserved.
ISBN-0-12-058470-0

with Markov-modulated arrivals and general service times, the $MMPP/G/1$ queue, the nucleus of the theoretical discussion is a matrix generalization of the classical Takács equation (1962), for the joint distribution of the number of services dispensed during and the duration of the busy period in the $M/G/1$ queue.

One such generalization is now well-known and plays, more generally, an essential role in the study of Markov renewal processes of $M/G/1$ type, treated in Neuts (1989). Properties of the resulting equation will be reviewed in this paper. A recent article by Sengupta (1987), on a different model, has suggested an alternative methodological approach, which leads to a second, quite different non-linear matrix equation. The derivation of that equation is the subject of Section 3. It is based on the consideration of an embedded Markov process. The presence of such a Markov process appears to require somewhat more restrictive assumptions on the model than the first approach. We shall illustrate this by considering a model where the first generalization, but not the second, holds. Finally, we verify the consistency of both approaches by showing that certain important derived analytical results agree.

A *Markov-modulated Poisson process* is the doubly stochastic Poisson process whose arrival rate is given by $\lambda^*[J(t)]$, where $J(t)$, $t \geq 0$, is an $m-$state irreducible Markov process. The arrival rate therefore takes on only m values $\lambda_1, \ldots, \lambda_m$, and is equal to λ_j whenever the Markov process is in the state j, $1 \leq j \leq m$. The $MMPP$ is parametrized by specifying the initial probability vector ϕ, the infinitesimal generator Q of the (modulating) Markov process and the vector $\lambda = (\lambda_1, \ldots, \lambda_m)$ of arrival rates. It is notationally useful to define the diagonal matrix Λ with $\Lambda_{jj} = \lambda_j$, $1 \leq j \leq m$. For brevity, we shall often refer to the $MMPP$ with generator Q and rate vector λ as a $(Q, \Lambda)-$*source.*

The random variables $N(t)$ and $J(t)$, $t \geq 0$, with $N(0) = 0$, are respectively defined as the number of arrivals in $(0, t]$ in the $MMPP$ and as the state of the Markov process Q at time t. It is straightforward to verify that the process $\{N(t), J(t), t \geq 0\}$, which is of basic importance to this discussion, is a Markov process with the state space $\{k \geq 0\} \times \{1, \ldots, m\}$. Moreover, the increments of the process $\{N(t), t \geq 0\}$ are conditionally independent, given the process $\{J(t), t \geq 0\}$. Specifically for every choice of $t_0 = 0 < t_1 < \ldots < t_n$, and $n \geq 2$, the random variables $N(t_1), N(t_2) - N(t_1), \ldots, N(t_n) - N(t_{n-1})$ are conditionally independent, given the random variables $J(t_r)$, $0 \leq r \leq n$.

We consider the probabilities

$$P_{jj'}(k;t) = P\{N(t) = k, J(t) = j' \mid N(0) = 0, J(0) = j\},$$

for $k \geq 0$, $t \geq 0$, $1 \leq j, j' \leq m$, and obtain a system of difference-differential equations for the matrices $P(k;t) = \{P_{jj'}(k;t)\}$. These generalize the elementary equations for the Poisson process.

By considering all possible occurrences in the interval $(t, t+dt)$, we obtain the (forward) Chapman-Kolmogorov equations, which we immediately write in matrix notation as

$$P'(0;t) = P(0;t)(Q - \Lambda), \tag{1}$$

and

$$P'(k;t) = P(k;t)(Q - \Lambda) + P(k-1;t)\Lambda, \quad \text{for } k \geq 1.$$

It is readily shown that the matrix generating function

$$P^*(z;t) = \sum_{k=0}^{\infty} P(k;t)z^k,$$

is given by

$$P^*(z;t) = \exp\{[Q - \Lambda + \Lambda z]t\}. \tag{2}$$

We denote by $\mathbf{e}$, a column vector of ones, by θ, the stationary probability (row) vector of the generator Q and by λ^*, the inner product $\theta\Lambda\mathbf{e}$. We note that the matrix $\mathbf{e}\theta - Q$ is nonsingular, as is readily proved by contradiction. See Hunter (1982) for this and many related results. The components of the vector $\mathbf{V}_1(t) = P_z^*(1-;t)\mathbf{e}$, give the conditional mean numbers of arrivals during $[0,t)$, given the initial state of the Markov process Q. The vector $\mathbf{V}_1(t)$ is given by

$$\mathbf{V}_1(t) = \lambda^* t\mathbf{e} + [I - \exp(Qt)](\mathbf{e}\theta - Q)^{-1}\Lambda\mathbf{e}, \quad \text{for } t \geq 0. \tag{3}$$

Clearly $\theta\mathbf{V}_1(t) = \lambda^* t$, so that λ^* is the fundamental rate of the stationary version of the Markov-modulated Poisson process.

2. THE $MMPP/G/1$ QUEUE

Consider a single server queue with a (Q, Λ) source as its arrival process. In the *general* model, the service times of successive customers are allowed to depend on the state of the Markov process Q at the beginnings of these services, but are mutually conditionally independent, given these states. The service time distributions of all customers, starting service in the "phase" j of the process Q are identical, are denoted by $H_j(\cdot)$ and have finite non-zero means α_j, $1 \leq j \leq m$. In the *restricted* model, the service times of all customers are independent of the arrival process and are mutually independent with the common distribution $H(\cdot)$ of finite mean α. Some results in the sequel hold only for the restricted model.

We view the queue at successive departure epochs and, for convenience, choose the time origin at such an epoch. We define the trivariate sequence of random variables $\{(I_n, J_n, \tau_n), n \geq 0\}$ where $\tau_0 = 0$, the I_n are the numbers of customers in the system following the successive departures, and the J_n are the states of the Markov chain Q at these same epochs. For $n \geq 1$, the τ_n are the successive interdeparture times. Under the assumptions of the model, the sequence $\{(I_n, J_n, \tau_n), n \geq 0\}$ is readily seen to be a *Markov renewal sequence,* and its transition probability matrix is given by

$$
(4) \qquad Q(x) = \begin{pmatrix} B_0(x) & B_1(x) & B_2(x) & B_3(x) & B_4(x) & \cdots \\ A_0(x) & A_1(x) & A_2(x) & A_3(x) & A_4(x) & \cdots \\ 0 & A_0(x) & A_1(x) & A-2(x) & A_3(x) & \cdots \\ 0 & 0 & A_0(x) & A_1(x) & A_2(x) & \cdots \\ \vdots & \vdots & \vdots & \vdots & \vdots & \ddots \end{pmatrix}
$$

where the matrices $A_k(\cdot)$, $k \geq 0$, of probability mass-functions are given by

$$
(5) \qquad A_k(x) = \int_0^x d\Delta[\mathbf{H}(u)]P(k; u), \text{ for } x \geq 0.
$$

The matrix $\Delta[\mathbf{H}(\cdot)]$ is an $m \times m$ diagonal matrix with the probability distributions $H_j(\cdot)$ as its diagonal elements.

The boundary matrices $B_k(\cdot)$, $k \geq 0$, which describe the transitions from states corresponding to an empty queue, are given by

$$
(6) \qquad B_k(x) = \int_0^x \exp[(Q - \Lambda)u]\Lambda A_k(u)du, \text{ for } x \geq 0.
$$

The analytic expressions for the matrices $A_k(\cdot)$, $k \geq 0$, and $B_k(\cdot)$, $k \geq 0$, given in formulas (5) and (6), follow by the standard counting argument for the number of arriving customers between two successive departures. The matrix formalism implicitly keeps track of all possible changes in the state of the environmental Markov chain.

The transition probability matrix $\mathbf{Q}(\cdot)$ has the canonical form which characterizes the Markov renewal processes *of* $M/G/1$ *type*, studied in full generality in Neuts (1989). The nucleus of the theory of such Markov renewal processes, which arise frequently in the study of queues and related models, is the *fundamental period,* which is the first passage time (measured in real time or in number of transitions) from an initial state $(i+1, j)$ with $i \geq 0$ and $1 \leq j \leq m$, to the set of states $(i, j\prime)$, $1 \leq j\prime \leq m$. The joint probabilities of the durations in number of transitions and in real time of the fundamental period are described by a sequence of $m \times m$ matrices of mass-functions, of which, for the sake of brevity, we shall only consider the transform matrix $\tilde{G}(z; s)$. The variable z corresponds to the *number of transitions,* while s is the variable of Laplace-Stieltjes transforms corresponding to real time.

The following are a statements of a number of classical results on Markov renewal processes *of* $M/G/1$ *type*. It is not feasible to repeat the lengthy proofs here and we therefore refer the reader to the journal literature or to the detailed discussion in Neuts (1989). The transform matrix $\tilde{G}(z, s)$ satisfies the equation

$$\tilde{G}(z, s) = z \sum_{\nu=0}^{\infty} \tilde{A}_\nu(s) \tilde{G}^\nu(z, s), \tag{7}$$

where

$$\tilde{A}_\nu(s) = \int_0^\infty e^{-sx} dA_\nu(x), \text{ for } \nu \geq 0,$$

and $\tilde{G}^\nu(z, s)$ is the ν-th power of the matrix $\tilde{G}(z, s)$. As shown in Neuts (1989), the solution of probabilistic interest of (7) is, in a sense made precise there, the minimal nonnegative solution to that equation. The proof of the uniqueness of that solution may also be found there.

A necessary condition for stability of the $MMPP/G/1$ queue is that the substochastic matrix $G = \tilde{G}(1-; 0+)$, be *stochastic.* This is the case, if and only if the inner product

$$\rho = \pi \beta, \tag{8}$$

does not exceed one. The vector π is the invariant probability vector of the (positive) stochastic matrix

$$A = \sum_{k=0}^{\infty} A_k(\infty) = \int_0^{\infty} d\Delta[\mathbf{H}(u)] \exp(Qu),$$

and the column vector β is given by

$$\beta = \sum_{k=1}^{\infty} kA_k(\infty)\mathbf{e} = \int_0^{\infty} d\Delta[\mathbf{H}(u)]\mathbf{V}_1(u).$$

By use of formula (3), the vector β may be explicitly written as

$$\beta = \lambda^* \alpha + (I - A)(\mathbf{e}\theta - Q)^{-1}\Lambda\mathbf{e}, \tag{9}$$

so that the quantity ρ for the general version of the $MMPP/G/1$ queue is given by

$$\rho = \lambda^*(\pi\alpha). \tag{10}$$

The quantity π_j may be interpreted as the fraction of services which start with the arrival process in its phase j, while the quantity β_j is the expected number of new arrivals to the queue during such a service. The traffic intensity ρ is therefore the average number of new arrivals during a "typical" service. The inequality $\rho < 1$, is the "negative drift condition" to be anticipated for recurrence of a process that is essentially a random walk on the lattice points in a semi-infinite strip. See Asmussen (1987) and Neuts (1989).

If $\rho < 1$, the vectors of means

$$\tilde{\mu}_1 = \left[\frac{\partial \tilde{G}(z,s)}{\partial z}\right]_{z=1,s=0} \mathbf{e}, \qquad \tilde{\mu}_1^* = \left[-\frac{\partial \tilde{G}(z,s)}{\partial s}\right]_{z=1,s=0} \mathbf{e}, \tag{11}$$

are finite, whereas both are infinite when $\rho = 1$. The components of $\tilde{\mu}_1$ and $\tilde{\mu}_1^*$ are respectively the conditional mean numbers of customers served and the mean durations, given the state j of the Markov process Q at the start of the fundamental period. Denoting by $\mathbf{g}$, the invariant probability vector of the (positive) stochastic matrix G, the vectors $\tilde{\mu}_1$ and $\tilde{\mu}_1^*$ are explicitly given by the general formulas

$$\tilde{\mu}_1 = (I - G + \mathbf{e}\mathbf{g})[I - A + (\mathbf{e} - \beta)\mathbf{g}]^{-1}\mathbf{e}, \tag{12}$$

and

$$\tilde{\mu}_1^* = (I - G + \mathbf{eg})[I - A + (\mathbf{e} - \beta)\mathbf{g}]^{-1}\beta^*, \tag{13}$$

and the equalities

$$\mathbf{g}\tilde{\mu}_1 = (1-\rho)^{-1}, \quad \mathbf{g}\tilde{\mu}_1^* = (1-\rho)^{-1}\pi\beta^*, \tag{14}$$

hold. The vector β^* is defined by

$$\beta^* = -\sum_{\nu=0}^{\infty} \tilde{A}'_\nu(0+)\mathbf{e} = \alpha. \tag{15}$$

For the $MMPP/G/1$ queue, the formulas (12) and (13) may be written in simpler explicit forms. We record these in the following theorem, in which Z is defined as the inverse of the matrix $I - A + \mathbf{e}\pi$. It is clear that $Z\mathbf{e} = \mathbf{e}$.

THEOREM 1. *For the general version of the stable $MMPP/G/1$ queue, the vectors $\tilde{\mu}_1$ and $\tilde{\mu}_1^*$ are given by*

$$(1-\rho)\tilde{\mu}_1 = \mathbf{e} + (I - G)(\mathbf{e}\theta - Q)^{-1}\Lambda\mathbf{e} + \lambda^*(I - G)Z\alpha, \tag{16}$$

$$\tilde{\mu}_1^* = (\pi\alpha)\tilde{\mu}_1 + (1-\rho)^{-1}(I - G)Z\alpha, \tag{17}$$

For the restricted version, $\pi = \theta$, $\pi\alpha = \alpha$, and the last term in each of the formulas (16) and (17) vanishes since $Z\mathbf{e} = \mathbf{e}$.

PROOF. We set $\mathbf{v} = [I - A + (\mathbf{e} - \beta)\mathbf{g}]^{-1}\mathbf{e}$. Premultiplication by the row vector π in the correponding system of linear equations leads to $\mathbf{gv} = (1-\rho)^{-1}$. Adding the vector $\mathbf{e} \cdot \pi\mathbf{v}$ to both sides of the equation

$$(I - A)\mathbf{v} = (1-\rho)^{-1}(\mathbf{e} - \beta),$$

further elementary manipulations yield that

$$\mathbf{v} = (1-\rho)^{-1}Z\beta + C\mathbf{e},$$

where C is a constant. Formula (9) implies that

$$Z\beta = \lambda^* Z\alpha + (\mathbf{e}\theta - Q)^{-1}\Lambda\mathbf{e} + [\pi(\mathbf{e}\theta - Q)^{-1}\Lambda\mathbf{e}]\mathbf{e}.$$

Substitution into (12) and use of (14) readily give an expression for the constant C. Simplifications leads to (16). Entirely similar calculations produce an analogous formula for $\tilde{\mu}_1^*$, which leads to the interesting relation (17) between the two vectors of means. The particular forms of these equations for the restricted case are now obvious. We note that in the restricted case, $\tilde{\mu}_1^* = \alpha\tilde{\mu}_1$, as was to be anticipated. ∎

3. AN EMBEDDED MARKOV PROCESS

Next, we consider the continuous parameter process $\{(J(t), R(t)), t \geq 0\}$, where $J(t)$ is the state of the Markov process Q and $R(t)$ the *residual busy period* at time t. When, at time t, the queue is empty, we set $R(t) = 0$. For the *restricted* model, the process $\{(J(t), R(t)), t \geq 0\}$, is Markovian, but not for the general version of the $MMPP/G/1$. This is easily seen by observing that at an arrival, the residual busy period is augmented by the service time of the new customer. In the general, but not in the restricted version, the distribution of that increment depends on the future of the process through the state of the Markov process Q at the time that customer will enter service. Since for the $MMPP/G/1$, the busy period is also a fundamental period, it suffices to derive the distribution of the busy period.

To that end, we introduce the conditional probability $\Psi_{jj'}(x; k, y)$ that, given $J(0) = j$, $1 \leq j \leq m$, and $R(0) = x$, $x > 0$, the (current) busy period ends before time y, $y \geq x$, with the Markov process Q in the state j' and involves the service of k, $k \geq 0$, new customers. The $m \times m$ matrix $\Psi(x; k, y)$ has elements $\Psi_{jj'}(x; k, y)$ and we define the transform $\Psi^*(x; z, s)$ by

$$\Psi^*(x; z, s) = \sum_{k=0}^{\infty} \int_x^{\infty} e^{-sy} d\Psi(x; k, y) z^k \text{ for } |z| \leq 1, Re s \geq 0. \tag{18}$$

THEOREM 2. *The matrix $\Psi^*(x; z, s)$ is given by*

$$\Psi^*(x; z, s) = e^{-sx} \exp[Q - \Lambda + \Lambda\tilde{G}(z, s)]x, \tag{19}$$

and the matrix $\tilde{G}(z,s)$ *satisfies the equation*

$$\tilde{G}(z,s) = z \int_0^{\infty} \Psi^*(x;z,s)dH(x). \tag{20}$$

PROOF. By viewing time $t = 0$, as the beginning of a busy period and conditioning on the duration of the first service, equation (20) follows by the law of total probability. To prove (19), we note that, by the spatial homogeneity of the Markov process away from the boundary and by considering the first time the residual busy period reduces to the value x_1, that for $k \geq 0$, and $y \geq x_1 + x_2$,

$$\Psi(x_1 + x_2; k, y) = \sum_{r=0}^{k} \Psi(x_1; r, \cdot) * \Psi(x_2; k-r, y), \tag{21}$$

where $*$ denotes matrix convolution. Upon evaluating the transforms of both sides in (21), we obtain routinely that

$$\Psi^*(x_1 + x_2; z, s) = \Psi^*(x_1; z, s)\Psi^*(x_2; z, s), \text{ for } x_1 > 0, x_2 > 0, \tag{22}$$

and by continuity, we may set $\Psi^*(0; z, s) = 0$.

We further note that

$$\Psi(x; 0, y) = \exp[(Q - \Lambda)x], \text{ for } y \geq x. \tag{23}$$

For $0 \leq y < x$, $\Psi(x; 0, y) = 0$, so that there is an atom at $y = x$, as is to be expected. By conditioning on the length of the service time of the first arriving customer, for $k \geq 1$,

$$\Psi(x; k, y) = \int_{0(u)}^{x} du \int_{0(v)}^{y-x} \exp[(Q-\Lambda)u]\Lambda\Psi(x-u+v; k-1, y-u)dH(v). \tag{24}$$

Upon evaluating transforms, we obtain the equation

$$\Psi^*(x;z,s) = \exp[-(sI - Q + \Lambda)x] \tag{25}$$

$$+z\exp[-(sI-Q+\Lambda)x] \int_{0(u)}^{x} \int_{0(v)}^{\infty} \exp[-(sI-Q+\Lambda)u]$$

$$\cdot \Lambda du \; d\,\Lambda[H(\nu)]\Psi^*(u+v;z,s),$$

By the relations (20) and (22), the preceding equation may be rewritten as

$$\Psi^*(x;z,s) = \exp[-(sI - Q + \Lambda)x] \tag{26}$$

$$+\exp[-(sI-Q+\Lambda)x] \int_0^x \exp[-(sI-Q+\Lambda)u]\Lambda\tilde{G}(z,s)\Psi^*(u;z,s)du.$$

By premultiplying both sides of (26) by the matrix $\exp[(sI - Q + \Lambda)x]$, which is always nonsingular, and differentiating with respect to x, routine simplifications yield that $\Psi^*(x;z,s)$ satisfies the linear differential equation

$$\Psi^*_z(x;z,s) = -[sI - Q + \Lambda - \Lambda\tilde{G}(z,s)]\Psi^*(x;z,s), \quad \text{for } x \geq 0, \tag{27}$$

with the initial condition $\Psi^*(0;z,s) = I$. This immediately yields the equation (19). ∎

Since all its row sums are zero and its off-diagonal elements nonnegative, the matrix $Q - \Lambda + \Lambda G$ is clearly an infinitesimal generator. Moreover, the positivity of the matrix G implies that all its off-diagonal elements positive. We let the row vector $\mathbf{u}$ be its stationary probability vector. ¿From the equation

$$G = \int_0^\infty \exp[(Q - \Lambda + \Lambda G)x]dH(x),$$

it is apparent that the matrices G and $Q - \Lambda + \Lambda G$ commute. This accounts for several remarkable factorization properties of stationary distributions for the $MMPP/G/1$ queue, discussed in Lucantoni, Meier-Hellstern and Neuts (1988).

We shall now derive an expression for the vector

$$\Psi_1(x) = \left[\frac{\partial \Psi^*(x;z,s)}{\partial z}\right]_{z=1,s=0} \mathbf{e}, \tag{28}$$

whose components are the conditional mean numbers of new customers served during a residual busy period, starting with an amount of work x is the system, given the various states of the process Q at time 0.

THEOREM 3. *The vector* $\mathbf{\Psi}_1(x)$ *is given by*

$$\mathbf{\Psi}_1(x) = (1-\rho)^{-1}\lambda^* x\mathbf{e} + (1-\rho)^{-1}[I - \Psi^*(x;1,0)](\mathbf{e}\theta - Q)^{-1}\Lambda\mathbf{e}. \tag{29}$$

The expressions for $\tilde{\mu}_1$*, derived from the equations (7) and (20) agree.*

PROOF. Setting $s = 0$, and differentiation with respect to z in (19) leads to

$$\mathbf{\Psi}_1(x) = \sum_{\nu=1}^{\infty} \frac{x^\nu}{\nu!}(Q - \Lambda + \Lambda G)^{\nu-1}\Lambda\tilde{\mu}_1. \tag{30}$$

Again appealing to a classical property of irreducible infinitesimal generators for Markov processes, we see that the matrix $\mathbf{eu} - Q + \Lambda - \Lambda G$ is nonsingular and this leads, by routine matrix calculations to

$$\sum_{\nu=1}^{\infty} \frac{x^\nu}{\nu!}(Q - \Lambda + \Lambda G)^{\nu-1} = x\mathbf{eu} + [I - \Psi^*(x;1,0)](\mathbf{eu} - Q + \Lambda - \Lambda G)^{-1}.$$

Substitution into (30) yields

$$\mathbf{\Psi}_1(x) = x(\mathbf{u}\Lambda\tilde{\mu}_1)\mathbf{e} + [I - \Psi^*(x;1,0)](\mathbf{eu} - Q + \Lambda - \Lambda G)^{-1}\Lambda\tilde{\mu}_1. \tag{31}$$

Evaluating the vector $\tilde{\mu}_1$ by differentiation in the equation (20), we obtain that

$$\tilde{\mu}_1 = \mathbf{e} + (\mathbf{u}\Lambda\tilde{\mu}_1)\alpha\mathbf{e} + (I - G)(\mathbf{eu} - Q + \Lambda - \Lambda G)^{-1}\Lambda\tilde{\mu}_1. \tag{32}$$

Premultiplication in that equation by Λ yields the equality

$$(\mathbf{eu} - Q + \Lambda - \Lambda G)^{-1}\Lambda\tilde{\mu}_1 = [1 + (\mathbf{u}\Lambda\tilde{\mu}_1)\alpha](\mathbf{eu} - Q)^{-1}\Lambda\mathbf{e}. \tag{33}$$

Premultiplying in (33) by $\mathbf{u}$ readily shows that

$$\mathbf{u}\Lambda\tilde{\mu}_1 = (1-\rho)^{-1}\lambda^*.$$

Combining that equality with the equations (32) and (33) leads to

$$\tilde{\mu}_1 = (1-\rho)^{-1}\mathbf{e} + (1-\rho)^{-1}(I-G)(\mathbf{eu}-Q)^{-1}\Lambda\mathbf{e},$$

and it is elementary to verify that this expression agrees with the one given in (16) for the restricted model. A final substitution into formula (31) yields the expression for $\Psi_1(x)$, stated in (29). ∎

REMARKS. *a*. There does not appear to be a direct analytic way to show that for the restricted model, the two nonlinear equations (7) and (20) are equivalent in the sense that one can be derived from the other purely by matrix manipulations.

b. It is interesting that the arguments based on the embedded Markov process can be extended to the case where the service time distribution of a customer depend on the environmental state *at the time of the customer's arrival.* For that case, there is no longer an embedded Markov renewal process of $M/G/1$ type, so that the derivations in Section 2 can no longer be carried out. We shall leave the derivation of the analogues of equations (19) and (20) for that model to the initiative of the reader.

ACKNOWLEDGEMENTS

The author's renewed interest in the properties of the $MMPP/G/1$ queue was stimulated by joint work with David M. Lucantoni and Kathleen S. Meier-Hellstern of Bell Laboratories on a related model with server vacations. The idea to consider the embedded Markov process to derive a new set of equations for the fundamental period was suggested by the work of Bhaskar Sengupta of Bell Laboratories, who kindly shared his research ideas with us at the earliest stage. With deep appreciation, the contributions of these colleagues are acknowledged.

REFERENCES

Asmussen, Søren (1987), *Applied Probability and Queues*, New York, John Wiley and Sons.

Heffes, Harry and Lucantoni, David M. (1986), A Markov modulated characterization of packetized voice and data traffic and related statistical multiplexer performance, *IEEE Journal on Selected Areas in Communication, Special Issue on Network Performance Evaluation*, **SAC-4**, 6, 856-868.

Hunter, Jeffrey J. (1982), Generalized inverses and their application to applied probability problems, *Linear Algebra and Applications*, 45, 157-98.

Lucantoni, David M.; Meier-Hellstern, Kathleen S. and Neuts, Marcel F. (1988), A single server queue with server vacations and a class of non-renewal arrival processes, *Advances in Applied Probability*, forthcoming.

Naor, Paul, and Yechiali, Uri (1971), Queueing problems with heterogeneous arrivals and service, *Operations Research*, **19**, 722-34.

Neuts, Marcel F. (1971), A queue subject to extraneous phase changes, *Advances in Applied Probability*, **3**, 78-119.

Neuts, Marcel F. (1978a), The $M/M/1$ queue with randomly varying arrival and service rates, *Opsearch*, **15**, 139-57.

Neuts, Marcel F. (1978b), Further results on the $M/M/1$ queue with randomly varying rates, *Opsearch*, **15**, 158-68.

Neuts, Marcel F. (1981), *Matrix-Geometric Solutions in Stochastic Models: An Algorithmic Approach*, Baltimore: The Johns Hopkins University Press.

Neuts, Marcel F. (1986a), The caudal characteristic curve of queues, *Advances in Applied Probability*, **18**, 221-54.

Neuts, Marcel F. (1986b), Generalizations of the Pollaczek-Khinchin integral equation in the theory of queues, *Advances Applied Probability*, **18**, 952-90.

Neuts, Marcel F. (1989), *Structured Stochastic Matrices of $M/G/1$ Type and their Applications*, New York: Marcel Dekker, Publishers, in press.

Purdue, Peter, (1974a), The single server queue in a Markovian environment, *Proceedings of the Conference on Mathematical Methodology in the Theory of Queues*, Kalamazoo, Michigan, Springer–Verlag, 359-65.

Purdue, Peter, (1974b), The $M/M/1$ queue in a Markovian environment, *Operations Research*, **22**, 562-69.

Purdue, Peter, (1979), The single server queue in a random environment, *Operations Research Verfahren*, **33**, 363-72.

Ramaswami, Vadyanathan, (1980), The $N/G/1$ queue and its detailed analysis, *Advances Applied Probability,* **12**, 222-61.

Sengupta, Bhaskar, (1987a), Sojourn time distributions for the $M/M/1$ queue in a Markovian environment, *European Journal of Operational Research,* **32**, 140-9.

Sengupta, Bhaskar, (1987b), A queue with service interruptions in an alternating random environment, Manuscript, *AT&T Bell Laboratories,* submitted for publication.

Sengupta, Bhaskar, (1989a), A perturbation method for solving some queues with processor sharing discipline, *Journal of Applied Probability,* to appear.

Sengupta, Bhaskar, (1989b), Markov processes whose steady state distribution is matrix- exponential with an application to the $GI/PH/1$ queue, *Advances Applied Probability,* **21**, forthcoming.

Takács, Lajos, (1962), *Introduction to the Theory of Queues,* Oxford: Oxford University Press.

Yechiali, Uri, (1973), A queueing-type birth-and-death process defined on a continuous-time Markov chain, *Operations Research,* **21**, 604-9.

Some Remarks on a Limiting Diffusion for Decomposable Branching Processes

Peter Ney*
University of Wisconsin, Madison

1. INTRODUCTION

In branching process theory, as in many areas of probability, diffusion approximations frequently shed light on the behavior of a process. Professor Karlin has himself made extensive contributions to this subject; so it is appropriate that we touch on an aspect of it here.

The grandfather of the processes we will consider is the "Feller diffusion." Consider a sequence $\{Z_k^{(n)};\ k = 0, 1, \ldots\}$, $n = 0, 1, \cdots$, of critical Galton-Watson processes with the same particle production, mean offspring per parent = m = 1, variance = $\sigma^2 < \infty$ and with $Z_0^{(n)} = u_0 n$; i.e. start the $n'th$ process with $u_0 n$ particles. Let

$$U_n(t) = \frac{Z_{[nt]}^{(n)}}{n}, \qquad 0 \le t \le 1.$$

AMS 1980 Subject Classifications: Primary 60J80, Secondary 60J70.
Key words and phrases: Branching processes, limiting diffusions.
*Research supported in part by the National Science Foundation.

PROBABILITY, STATISTICS, AND MATHEMATICS
Papers in Honor of Samuel Karlin

Copyright © 1989 by Academic Press, Inc.
All rights of reproduction in any form reserved.
ISBN-0-12-058470-0

Then [Feller (1951)]

$$\{U_n(t)\} \underset{n\to\infty}{\overset{weakly}{\longrightarrow}} \{U(t)\},$$

where $U(\cdot)$ is a diffusion with state space $[0,\infty)$ and generator

$$Af(u) = \frac{\sigma^2}{2} u f_{uu}(u), \tag{1.1}$$

and

$$Ee^{-sU(t)} = exp\left\{-u_0 \frac{s}{1+\frac{\sigma^2}{2}ts}\right\}. \tag{1.2}$$

Letting W be the well-known limit r.v.

$$W = lim\left\{\frac{Z_n}{n}|Z_n > 0\right\}$$

with

$$\varphi(s) = Ee^{-sW} = \frac{1}{1+s\frac{\sigma^2}{2}},$$

we note (with $u_0 = 1$)

$$Ee^{-sU(1)} = e^{\frac{2}{\sigma^2}[\varphi(s)-1]},$$

i.e. $U(1)$ is a Poisson compounding of W.

In this note we will observe that similar relations hold for somewhat more complicated processes, namely decomposable critical processes. In the interest of simplicity we consider a 3-type process with mean matrix

$$M = \begin{bmatrix} 1 & m_{12} & m_{13} \\ 0 & 1 & m_{23} \\ 0 & 0 & 1 \end{bmatrix} \tag{1.1a}$$

where $0 < m_{12}, m_{13}, m_{23} < \infty$. Thus type $1's$ can produce type $1's, 2's$, and $3's$, type $2's$ can produce $2's$ and $3's$, type $3's$ only $3's$. This already includes all the interesting aspects of the general case.

Let $\tilde{Z}_n = (Z_{n_1}, Z_{n_2}, Z_{n_3})$, where Z_{n_i} = number of type i particles in the nth generation. We assume throughout that

$$(1.1b) \qquad E[Z_{1_i} Z_{i_j} | \tilde{Z}_0] < \infty,$$

and let $\sigma_i^2 = var(Z_{1_i} | \tilde{Z}_0 = 1$ type i particle).

As for a single type, conditioned limit laws are available for $\{\tilde{Z}_n\}$, but now several different types of conditioning are possible, and they lead to different limit laws. We focus attention on the two extremes, other cases being similarly treatable. Assuming second moments exist we have

(i) Savin and Chistyakov (1962)

$$(1.2) \qquad \left\{ \frac{\tilde{Z}_n}{n} | \tilde{Z}_n \neq \tilde{O} \right\} \xrightarrow{D} \tilde{W} = (O, O, W_3)$$

(convergence in distribution), and

(ii) Foster and Ney (1978), Ogura (1975)

$$(1.3) \qquad \left\{ \frac{Z_{n_1}}{n}, \frac{Z_{n_2}}{n^2}, \frac{Z_{n_3}}{n^3} | Z_{n_1} > 0 \right\} \xrightarrow{D} \tilde{U} = (U_1, U_2, U_3)$$

where U_i are non-degenerate. Intuitively the conditioning in (i) helps the type $3's$ to survive and $\frac{Z_{n_3}}{n}$ to converge, while the type 1^s and 2^s become extinct. The conditioning in (ii) leads to different normalizations, and a non-degenerate limit law in all three components.

We will use diffusion approximations to tell us what the relation between these limits is.

2. LIMITING DIFFUSION

The basic limit theorem we will need has recently been proved by S. Goldstein (1988). Let $\{\tilde{Z}(n|z_0)\}$ denote a critical decomposable process with mean matrix M as above. Let $\tilde{e}_i$, $i = 1, 2, 3$ be unit coordinate vectors and

$$\tilde{U}_n(t) = \left(\frac{Z_1(nt|\tilde{e}_1 n)}{n}, \frac{Z_2(nt|\tilde{e}_1 n)}{n^2}, \frac{Z_3(nt|\tilde{e}_1 n)}{n^3} \right).$$

THEOREM. (S. Goldstein). *assume that* (1.1a) *and* (1.1b) *hold. Then*

$$tildeU_n(t) \xrightarrow{W} \tilde{U}(t) = (U_1(t), U_2(t), U_3(t)), \qquad 0 \le t \le 1,$$

where $\tilde{U}(\cdot)$ *is a diffusion with state space* $([0,\infty))^3$ *and generator*

$$Af(u_1, u_2, u_3) = \frac{1}{2}\sigma_1^2 u_1 f_{u_1 u_1} + m_{12} u_1 f_{u_2} + m_{23} u_2 f_{u_3}. \tag{2.1}$$

REMARK. For a discussion of the domain of A, see Goldstein (1988). It follows that $(U_1(\cdot), U_2(\cdot), U_3(\cdot))$ satisfies the stochastic integral equations

$$\begin{aligned} U_1(t) &= U_1(0) + \sigma_1 \int_0^t \sqrt{U_1(s)} dB, \\ U_2(t) &= U_2(0) + m_{12} \int_0^t U_1(s) ds, \\ U_3(t) &= U_3(0) + m_{23} \int_0^t U_2(s) ds. \end{aligned} \tag{2.2}$$

where $B(t)$ is an appropriately chosen standard Brownian motion. Thus, once the diffusion $U_1(\cdot)$ is determined, U_2 and U_3 can be obtained in terms of U_1.

The properties of $U_1(t)$ and $U_2(t)$ are easy to determine quite explicitly, and will be useful later. Write $m_{12} = m$. Then

$$Ef(U_1(t), U_2(t)) = f(U_1(0), U_2(0)) + \int E_{\tilde{U}(0)} Af(U_1(s), U_2(s)) ds,$$

where

$$Af(u_1, u_2) = \frac{1}{2}\sigma_1^2 u_1 f_{u_1 u_1} + m u_1 f_{u_2}.$$

Letting

$$\theta(t; \lambda_1, \lambda_2) = E e^{-\lambda_1 U_1(t) - \lambda_2 U_2(t)}$$

we see that this function must satisfy

$$\theta_t(t; \lambda_1, \lambda_2) + (\frac{1}{2}\sigma_1^2 \lambda_1^2 - m\lambda_2)\theta_\lambda(t; \lambda_1, \lambda_2) = 0. \tag{2.3.}$$

This can be solved explicitly (see (3.5) below), and thereby we can determine the distribution of $(U_1(1), U_2(1))$ and relate it to that of (U_1, U_2) in (1.2). A similar analysis is presumably possible for (U_1, U_2, U_3). We will not go into these rather tedious details, but turn to the more interesting question of how W_3 and (U_1, U_2, U_3) are related.

3. RELATION BETWEEN DIFFERENT CONDITIONS

In the conditioned process $\{Z_{n_1}, Z_{n_2}, Z_{n_3} | Z_{n_3} > 0\}$, types 1 and 2 become extinct long before type 3. (Types $3's$ of course survive up to n, but die out if the process is extended past n.) Hence to analyze W_3 we need to extend our constructions beyond the extinction times of types 1 and 2. Arguing heuristically, at the extinction time of type $1's$, we are faced with a two type process, initiated by order of n^2 particles of the first type. We must therefore change our normalization and scaling from n to n^2. Similarly after the death of the next type the scaling goes to n^4, etc. We will carry out this construction later.

We also have to extend the diffusion $\tilde{U}(t)$ in a suitable way past the time the first components hit 0. Let $\tau_1 = inf\{t : U_1(t) = 0\}$. Consider a process $\tilde{V}(t) = (V_1(t), V_2(t), V_3(t))$ with generator

$$A_1 f(u, v, w) = \frac{\sigma_1^2}{2} u f_{uu} + m_{12} u f_v + m_{23} v f_w \qquad \text{for} \quad 0 \le t \le \tau_1,$$

$$A_2 f = \frac{\sigma_2^2}{2} v f_{vv} + m_{23} v f_w \qquad \text{for} \quad \tau_1 < t \quad .$$

Now let $\tau_2 = inf\{t : V_2(t) = 0\}$ and consider the process $\tilde{X}(t)$ with generator

$$(3.1) \qquad \begin{array}{lll} A_1 f & \text{for} & 0 \le t \le \tau_1 \\ A_2 f & \text{for} & \tau_1 < t \le \tau_2 \\ A_3 f = \frac{\sigma_3^2}{2} w f_{ww} & \text{for} & \tau_2 < t. \end{array}$$

Recall $\sigma_i^2 = var(Z_{1_i} | \tilde{Z}_0 = \tilde{e}_i)$. Let $U(t; \sigma^2, u_0)$ be a Feller diffusion with variance σ^2 and $U(0; \sigma^2, u_0) = u_0$. Here

$$(3.2) \qquad E e^{-\beta U(t; \sigma^2, u_0)} = exp\left\{ -u_0 \frac{\beta}{1 + \beta \frac{\sigma^2}{2} t} \right\}.$$

Let $U^{(i)}(t, \sigma_i^2, u_0)$ be independent Feller processes, and define

$$\tau_i(u_0) = inf\{\tau : U^{(i)}(t, \sigma_i^2, u_0) = 0\}.$$

Also let

$$X = m_{12} \int_0^{\tau_1(u_0)} U^{(1)}(s; \sigma_1^2, u_0) ds,$$

and

$$Y = m_{23} \int_{\tau_1(u_0)}^{\tau_2(X)} U^{(2)}(s - \tau_1; \sigma_2^2, X) ds.$$

Now construct the process $\tilde{X}(t) = (X_1(t), X_2(t), X_3(t))$ as indicated in the following table:
(3.3)

For	$X_1(t)$	$X_2(t)$	$X_3(t)$
$0 \leq t \leq \tau_1(u_0)$	$U^{(1)}(t; \sigma_1^2, u_0)$	$m_{12} \int_0^t U^{(1)}(s) ds$	$m_{12} m_{23} \int_0^t (t-s) U^{(1)}(s) ds$
$\tau_1 < t \leq \tau_2(X)$	0	$U^{(2)}(t - \tau_1; \sigma_2^2, X)$	$m_{23} \int_{\tau_1}^t U^{(2)}(s; \sigma_2^2, X) ds$
$\tau_2 < t$	0	0	$U^{(3)}(t - \tau_2; \sigma_3^2, Y)$

The generator of $X(t)$ will be of the form (3.1). We can now relate $X(1)$ to W_3. Savin and Chistyakov (1962) have shown that

$$E\left[E^{-\lambda_1 \frac{Z_{n_1}}{n} - \lambda_2 \frac{Z_{n_2}}{n} - \lambda_3 \frac{Z_{n_3}}{n}} | Z_{n_3} > 0\right]$$

$$\longrightarrow \quad 1 - \left(\frac{\lambda_3}{1 + \lambda_3}\right)^{1/4} = H(\lambda_3) \quad (say).$$

PROPOSITION. *Assume* (1.1a and b). *Then*

$$Ee^{-\gamma U(1;\sigma_3^2,Y)} = e^{u_0 K[H(\gamma)-1]} \tag{3.4}$$

where

$$K = \left(\frac{m_{12}}{\frac{\sigma_1^2}{2}}\right)^{1/2} \left(\frac{m_{23}}{\frac{\sigma_2^2}{2}}\right)^{1/4} \left(\frac{1}{\frac{\sigma_3^2}{2}}\right)^{1/4}.$$

Thus, as in the Feller case, the appropriate diffusion at time 1 is a Poisson compounding of the conditioned limit law.

PROOF.

$$\begin{aligned} Ee^{-\alpha X} &= Ee^{-\alpha m_{12}\int_0^\infty U^{(1)}(s;\sigma_1^2,u_0)ds} \\ &= \lim_{t\to\infty} \theta(t;0,\alpha), \end{aligned}$$

where θ *is given by (2.3). One can show that the solution is*

$$\theta(t;\lambda_1,\lambda_2) = e^{-\psi(t;\lambda_1,\lambda_2)} \tag{3.5}$$

where $\psi(t;\lambda_1,\lambda_2) =$

$$\left(\frac{m\lambda_2}{\lambda}\right)^{1/2} \frac{(\lambda_2^{1/2}+\lambda_1(\lambda/m)^{1/2})e^{(m\lambda_2\lambda)^{1/2}t} - (\lambda_2^{1/2}-\lambda_1(\lambda/m)^{1/2})e^{-(m\lambda_2\lambda)^{1/2}t}}{(\lambda_2^{1/2}+\lambda_1(\lambda/m)^{1/2})e^{(m\lambda_2\lambda)^{1/2}t} + (\lambda_2^{1/2}-\lambda_1(\lambda/m)^{1/2})e^{-(m\lambda_2\lambda)^{1/2}t}}$$

and $\lambda = \frac{\sigma_1^2}{2}$. *Thus*

$$\begin{aligned} Ee^{-\alpha X} &= exp\left\{-u_0\left(\frac{m_{12}\alpha}{\frac{\sigma_1^2}{2}}\right)^{1/2}\right\} \\ &= \int_0^\infty f(x)e^{-\alpha x}dx, \end{aligned} \tag{3.6}$$

where

$$f(x) = \frac{b}{2\sqrt{\pi}}x^{-3/2}e^{-b^2/4x}, \quad b = u_0\sqrt{\frac{2m_{12}}{\sigma_1^2}}.$$

By (3.2)

$$\begin{aligned} Ee^{-\beta U(t;\sigma_2^2,X)} &= \int_0^\infty exp\left\{-x\frac{\beta}{1+\beta\frac{\sigma_2^2}{2}t}\right\} f(x)dx \\ &= exp\left\{-u_0\left(\frac{m_{12}}{\frac{\sigma_1^2}{2}}\right)^{1/2}\left(\frac{\beta}{1+\beta\frac{\sigma_2^2}{2}t}\right)^{1/2}\right\}. \end{aligned}$$

Similarly

$$Ee^{-\alpha Y} = Ee^{-\alpha_{23}\int_0^\infty U^{(2)}(s;\sigma_2^2,X)ds},$$

(note $\int_{\tau_1}^{\tau_2} U^{(2)}(s-\tau_1)ds = \int_0^\infty U(s)ds$*) and hence by (3.6)*

$$Ee^{-\alpha Y} = Ee^{-X\left(\frac{m_{23}}{\sigma_2^2/2}\right)^{1/2}\alpha^{1/2}} \tag{3.7}$$

$$= exp\left\{-u_0\left(\frac{m_{12}}{\sigma_1^2/2}\right)^{1/2}\left(\frac{m_{23}}{\sigma_2^2/2}\right)^{1/4}\alpha^{1/4}\right\}.$$

But now by (3.2)

$$Ee^{-\gamma U(t;\sigma_3^2,Y)} = Ee^{-\theta Y}, \tag{3.8}$$

with

$$\theta = \frac{\gamma}{1+\gamma\frac{\sigma_3^2}{2}t}$$

Thus by (3.7)

$$Ee^{-\theta\gamma} = \exp\left\{-u_0\left(\frac{m_{12}}{\sigma_1^2/2}\right)^{1/2}\left(\frac{m_{23}}{\sigma_2^2/2}\right)^{1/4}\left(\frac{\gamma}{1+\gamma\frac{\sigma_3^2}{2}t}\right)^{1/4}\right\}$$

$$= e^{u_0K[H_t(\gamma)-1]}$$

where $H_t(\gamma) = [\gamma/(1+\gamma\frac{\sigma_3^2}{2}t)]^{1/4}$. *Setting* $t = 1$ *we have (3.4).* ∎

We conclude by describing the sequence of discrete processes of which $X(t)$ is the limit.

Recall $\tilde{Z}(n|v) = (Z_1(n), Z_2(n), Z_3(n))$ denotes the population vector with $Z(0|v) = v$. Let $N_1^{(n)} = inf\{k : Z_1(k|n) = 0\}$. Then $N_1^{(n)}/n \longrightarrow a$ r.v. τ_1. Let $Z_2(N_1^{(n)}|\tilde{e}_1 n) = \xi_n^{(2)}$; also

$$N_2^{(n)} = inf\{k : Z_2(k|\tilde{e}_2\xi_n^{(2)}) = 0\} \quad \text{and} \quad \xi_n^{(3)} = Z_3(N_2^{(n)}|\tilde{e}_2\xi_n^{(2)}).$$

Note that $\frac{N_2^{(n)}}{n^2} \longrightarrow \tau_2$ (a r.v.).

Now define the sequence of processes

$$\tilde{X}_n(t) = (X_n^{(1)}(t), (X_n^{(2)}(t), (X_n^{(3)}(t))$$

by the following table, in which $Z^{(1)}, Z^{(2)}, Z^{(3)}$ are independent processes distributed as $Z(\cdot|\cdot)$.

For	$X_n^{(1)}(t)$	$X_2^{(n)}(t)$	$X_3^{(n)}(t)$
$0 \leq t \leq \tau_1$	$\frac{1}{n} Z_1^{(1)}(nt\|\tilde{e}_1 n)$	$\frac{1}{n^2} Z_2^{(1)}(nt\|\tilde{e}_1 n)$	$\frac{1}{n^3} Z_3^{(1)}(nt\|\tilde{e}_1 n)$
$\tau_1 < t \leq \tau_2$	0	$\frac{1}{n^2} Z_2^{(2)}(n^2(t-\tau_1)\|\xi_n^{(2)})$	$\frac{1}{n^4} Z_3^{(2)}(n^2(t-\tau_1)\|\xi_n^{(3)}$
$\tau_2 < t$	0	0	$\frac{1}{n^4} Z_3^{(3)}(n^4(t-\tau_2)\|\xi_n^{(3)}$

Successive application of Goldstein's theorem shows that $\tilde{X}_n(t) \xrightarrow{w} \tilde{X}(t)$.

REFERENCES

Feller, W. (1951), Diffusion processes in genetics, *Proceedings of the Second Berkeley Symposium*, University of California Press, Berkeley, 227-246.

Foster, J. and Ney, P. (1978), Limit theorems for decomposable branching processes *Zeitschrift für Wahrscheinlichkeitstheorie und Verwandte Gebiete*, **46**, 13-43.

Goldstein, S. (1988), *Multi-type branching processes: diffusion approximation for critical decomposable processes.* Ph.D. Thesis, University of Wisconsin, Madison, WI.

Ogura, Y. (1975), Asymptotic behavior of multi-type GW processes. *Journal of Mathematics Kyoto University*, **15**, 251-302.

Savin, A. A. and Chistyakov, V. P. (1962), Some theorems for branching processes with several types of particles. *Theory of Probability and Applications*, **7**, 93-100.

Some Results on Repeated Risktaking

John W. Pratt*
Harvard University

1. INTRODUCTION

Repeated risktaking raises many interesting questions about rational behavior. This paper answers a few of them. A summary of these answers follows. As will be indicated, the paper was prompted by and complements Samuelson (1989), but neither depends on nor subsumes it.

We will consider mainly independent, identically distributed gambles. As Samuelson has pointed out (1989, end of Section 1, restating 1963, p. 156), an expected utility maximizer "who is risk averse enough to refuse a specified favorable bet *at every* [wealth] level cannot ever rationally accept a set of 2,3, ..., or $N, \ldots$ such favorable bets." Samuelson's condition is very strong, however, as his emphasis and wording indicate – he could have said simply "who would refuse a specified bet at every wealth level," and had he been considering sequential decisions, his conclusion would still

AMS 1980 Subject Classification: Primary 90A10, Secondary 62C99.
Key words and phrases: Risk aversion, repeated gambles, utility theory, proper risk aversion, exponential utility, constant risk aversion.

*I am grateful to the Harvard Business School for research support, to Paul Samuelson and Richard Zeckhauser for highly stimulating and useful comments, and to Samuel Karlin for teaching me much of what I know in the way of mathematical technique and repeatable tricks.

Copyright © 1989 by Academic Press, Inc.
All rights of reproduction in any form reserved.
ISBN-0-12-058470-0

have applied (Pratt and Zeckhauser, 1987, hereafter *P&Z*). Matters are more subtle, however, for bets that are undesirable at present wealth but desirable at greater wealth levels. Even if a bet is undesirable, and would be still more undesirable after a loss, it could rationally be desirable enough after a gain so that two independent bets are better than none, and indeed also 3 better than 2, 4 than 3, and so on up to any finite number. Thus, as Samuelson says, his colleague Brown may be quite rational in rejecting one favorable bet while embracing 100, even though these preferences can also be explained less charitably as resulting from a common misinterpretation of the Law of Large Numbers and the Central Limit Theorem, whose true economic import Samuelson elucidates so eloquently in his papers.

Can a preference ordering of the sets of repetitions of a favorable bet ever be demonstrably irrational? For bounded gambles and bounded repetition the answer is No. Let X be any bounded, favorable gamble and let S_n be the sum of n independent repetitions of X. Given any finite N and any ranking of $S_1, \ldots, S_N$, there exists a decreasingly risk-averse utility function that yields this ranking. Thus rationality perhaps calls for further conditions, and avoidance of extreme irregularity certainly requires them.

One intuitive condition that precludes such complicated preferences is that utility be proper in the sense of *P&Z*, that is, besides decreasing risk aversion, if X, Y, and W are independent and neither $W + X$ nor $W + Y$ is better than W, then neither is $W + X + Y$; a sum of independent, unattractive risks X and Y is not attractive. This condition implies immediately for $W = S_n$ that if S_{n+1} is no better than S_n then neither is S_{n+2} or, therefore, S_{n+3}, etc., and less immediately that S_{m+1} is no better than S_m for every $m \geq n$. Indeed it implies that the certainty equivalent of a further bet X in the presence of S_m is nondecreasing in m for $m \geq n$. This in turn suggests but does not prove that the certainty equivalent of S_m is concave in m for $m \geq n$. (All risk-averse utilities of exponential, power, and logarithmic form and all mixtures thereof are proper.)

Properness says that if you would reject each of two independent gambles you would reject their sum, and indeed implies that the sum is worse than either. It does not say the same for acceptance. This asymmetry is natural if you feel that two gambles might each be attractive alone but so risky that one is enough. The increased risk of combined gambles strengthens the case for rejection but weakens the case for acceptance.

It is familiar and easily verified that if utility is exponential and the available gambles are independent, then a combination of gambles is optimal if and only if it includes all individually desirable gambles and excludes all individually undesirable gambles. (It is indifferent which, if any, of the individually indifferent gambles are included.) Samuelson (1989) asks the converse question for identically distributed gambles, and answers it using 6th order Taylor expansions for small 50-50 gambles with possible gain just enough larger than the possible loss to yield indifference. He shows that utility must be exponential if, for all w and all 50-50 gambles X such that $w + X$ is indifferent to w, $w + X + Y$ is also indifferent to w, where Y is an independent repetition of X. This paper gives a quite short proof, without expansions, of the same conclusion when the gambles in the hypothesis have two possible values $\pm x$ but have probabilities adjusted for indifference rather than equal.

The local condition for properness at a nonrandom wealth $W = w$ is $r''(w) \geq r'(w)r(w)$ where $r(w) = -u''(w)/u'(w)$ is the local risk aversion function ($P\&Z$ equation (28)). The same local condition is obtained even if X and Y are restricted to have identical 2-point distributions. (One way to show this is to allow inequality in Samuelson's derivation of $r'' = r'r$ as one step on his path to exponential utility.) Unfortunately this local condition is not sufficient globally, apparently not even for identically distributed 2-point gambles restricted to a small interval. It will be shown that if $r'' > r'r$ in some neighborhood of w then there exists an $\epsilon > 0$ such that accepting a gamble X in $(-\epsilon, \epsilon)$ increases risk aversion in some neighborhood of w unless either X is desirable and $r'(w) < 0$ or X is undesirable and $r'(w) > 0$. The same statement holds with the inequalities reversed and "increases" replaced by "decreases." Unfortunately the second neighborhood of w depends on X, so that we cannot infer proper preference or the opposite even in a small neighborhood of w. This may be why the statement can be symmetric in acceptance and rejection, and it indicates that the boundary of properness is complicated. Other indications include the global insufficiency of the local condition $r'' \geq r'r$, and Samuelson's need for a step beyond the equality $r'' = r'r$, which is satisfied by non-exponential utilities, one such being the basis of Lindsey's example as reported by $P\&Z$. (The exponential utilities are the extreme points of a convex set of proper utilities that $P\&Z$ call "completely proper" because of their relation to completely monotone functions, but it is not known whether the set of all proper utilities is convex.)

Life presents many needs or opportunities for taking repeated risks that are not purely monetary. In conversation, Samuelson and Zeckhauser have suggested a variety of interesting, repeatable risks, ranging from speeding tickets to accidental death. Unfortunately time has permitted only a quick look at just one of these – whether a husband and wife should travel together or separately. The next section gives a simple argument leading to a possibly surprising answer to this question, along with indications and illustrations of the need for some easily overlooked assumptions, but attempting to reduce the assumptions to a nontautological minimum does not appear fruitful.

Subsequent sections provide further details and proofs. Section 3 deals with sets of repetitions of a favorable bet under decreasing risk aversion, Section 4 with related certainty equivalents under proper risk aversion, Section 5 with the derivation of exponential utility from bets on $\pm x$, and Section 6 with the implications of $r'' >$ or $< r'r$ at w.

2. SHOULD WE BOTH GO TOGETHER WHEN WE GO?

A husband and wife who often travel to a common destination must decide before each trip whether to travel together or separately. Their concern is only with life and death; pleasure or tedium in transit is negligible by comparison, and any injury can be treated as either negligible or equivalent to death. It may be important that at least one spouse survive, perhaps for the sake of children. Yet a life alone may be little better than death. Could it be that the couple should travel together if the number of trips they will make is small but separately if it is large, or vice versa? That they should "hedge their bets" by traveling together on some trips and separately on others? That the risk of each trip should influence their decision?

It turns out that, under suitable assumptions, the answer to all such questions is "No," and the decision should depend only on the desirability of 1 survivor relative to 0 or 2. The argument is very simple: on any trip, the decision matters only if both spouses would survive all other trips, and only if exactly one of the two transport choices contemplated on this trip turns out to be fatal. In this case, if they travel separately, exactly one is sure to survive, while if they travel together, it is (by assumption) 50-50 whether both survive or both die. They should therefore travel separately always if they prefer one survivor for sure to a 50-50 gamble between 0 and 2 survivors, together always if they prefer the gamble, and it doesn't matter how they travel if they are indifferent. This is true regardless of the number of trips to be made or how risky each is, and regardless of the relative desirability of wife or husband surviving alone, since a sole survivor is equally likely to be either one. (This argument may have appeared somewhere; it would be hard to find.)

One implicit assumption of this argument is that, when there is a serious choice of transport, the couple agrees on what the two safest choices are and that they are equally safe. Another is that the couple has a joint utility function, or at least agrees on preference between one survivor for sure and a 50-50 gamble between 0 and 2 survivors. (Otherwise they might make different decisions on different trips as an interpersonal compromise, not because some expected utility is maximized thereby. If their criterion is a nonlinear preference functional, or some other exotic beast, their problem is difficult or even undefined; as others have noted, they face a serious conceptual question of consistency over time, between the static or normal form and the dynamic, decision-tree, or extensive form, and between stages of the latter.)

A third assumption is that no relevant learning is affected by transport choice. This has at least two possible justifications. One is that the risk probabilities *per se* are not relevant, only the first assumption above. Another is that taking a trip does not yield significant information about its safety that cannot be obtained otherwise.

The argument implicitly uses an assumption that the same fate befalls all who travel together, but this assumption is not needed. If there is a probability p that exactly one of two given passengers will survive, then compared to a situation with the same chance of survival for each passenger but a common fate, the probability that a couple traveling together will have 1 survivor is increased by p while the probabilities of 0 and 2 survivors must each be reduced by $p/2$. This change is exactly the trade-off between a certainty and a gamble that determined the choice before, and hence does not change the choice, unless p is so large that the probability of 1 survivor is larger than it would be under independence, an implausible case.

One set of assumptions that clearly suffices is that, on each trip, the available choices are exchangeable and the chance that both survive is greater for travel together than separately, and that outcomes on different trips are independent. In this case the probability B that both survive all trips is clearly maximized by travel together and minimized by travel apart, while the marginal probabilities H and W that the husband and wife individually survive are unaffected by their choice. The probability of exactly 1 survivor is $H + W - 2B$ and the probability of no survivor is $1 - H - W + B$. If their utility for k survivors is U_k, then their expected utility is

$$BU_2 + (H + W - 2B)U_1 + (1 - H - W + B)U_0.$$

Thus they should maximize B (travel together) if $U_2 + U_0 > 2U_1$ and minimize B (travel separately) if $U_2 + U_0 < 2U_1$.

The following cxamples show that the implicit assumptions mentioned above are not sufficient and that capturing their spirit correctly is not a matter for casual verbalization. Weaker sufficient conditions will not be investigated here. We note, however, that if sufficient conditions hold given each possible value of a random variable (the state of nature), this suffices, since a strategy is obviously optimal if it is optimal in every state of nature. Thus the examples must violate the sufficient conditions above (and all others) in some state of nature. The first example shows that the probability of surviving two trips is not always maximized by choosing the safest alternative on each trip. The second example shows that traveling together may not maximize the probability of joint survival even if the fates of those who travel together are positively correlated and no choice affects either learning or the individual probability of survival.

EXAMPLE 1. Two alternatives are available for Trip 1, only one for Trip 2. Nature has two equally likely states. The probabilities of survival are given in Table 1. Alternative B is safer on Trip 1 alone, but the probability of surviving both trips is larger if Alternative A is chosen on Trip 1. The same would also be true if Alternative B afforded any probability of survival between .9 and .82/.9 = .911 in each state. The story might be that Carrier

A is available for both trips but its safety is very uncertain, while Carrier B is available only for Trip 1 but there is little uncertainty about its safety. The best hope may then be that Carrier A is in fact very safe, even though this hope may not be good enough to make B a better bet for Trip 1 alone.

Table 1. Probability of Survival

	State 1	State 2	Average
Trip 1, Alt. A	0.8	1.0	0.9
Trip 1, Alt. B	1.0	0.82	0.91
Trip 2	0.8	1.0	0.9
Both trips, Alt. A	0.64	1.0	0.82
Both trips, Alt. B	0.8	0.82	0.81

EXAMPLE 2. Nature has two equally likely states. For Trip 1 there are two carriers. In State 1, each carrier affords survival probability a, with a common fate for all. In State 2, each carrier affords each passenger survival probability b but never kills more than one passenger per trip. For Trip 2, only one carrier is available, with a common fate for all and survival probability c in State 1 and 1 in State 2. In each state, the carriers and trips are independent. Straightforward calculation shows that travel together gives positively correlated fates on Trip 1 iff

$$a + 2b - 1 > (a + b)^2/2; \tag{1}$$

the probability of jointly surviving Trip 1 is greater for travel together than separately iff

$$a + 2b - 1 > a^2 + b^2; \tag{2}$$

and the probability of jointly surviving both trips is greater for taking Trip 1 together than separately iff

$$ac + 2b - 1 > a^2c + b^2. \tag{3}$$

It is easy to see that (2) implies (1) and that (2) holds but (3) does not whenever

$$a(1-a)c < (1-b)^2 < a(1-a). \tag{4}$$

Thus, under condition (4), separate travel on Trip 1 maximizes the probability of jointly surviving both trips, although on Trip 1 alone, travel together maximizes the probability of joint survival and gives positively correlated fates. Condition (4) is satisfied for some values of b provided that $c < 1$ and $0 < a < 1$.

3. HOW BROWNIAN CAN RATIONAL RANKING OF REPEATED GAMBLES BE?

Let S_n be the sum of n independent repetitions of any gamble X with positive mean and finite extrema $a, b, a < 0 < b$. Let any preference ranking of $S_0, S_1, S_2, \ldots, S_N$ be given; ties are allowed. Then a decreasingly risk averse utility function that yields this ranking can be constructed recursively as follows. Let u_0 have constant risk aversion c, say $u_0(x) = -e^{-cx}$, with c chosen so that $Eu_0(X) = u_0(0)$. Then $Eu_0(S_n) = u_0(0)$ for all n. Suppose u_n yields the required ranking of $S_0, S_1, \ldots, S_n$. If u_n ranks S_{n+1} correctly among these, let $u_{n+1} = u_n$. If u_n ranks S_{n+1} too low, let $v_n(x) = u_0(x)$ for $x \leq nb$ but $v_n(x) > u_0(x)$ for $x > nb$, with v_n decreasingly risk averse. Then $Ev_n(S_{n+1}) > u_0(0) = Ev_n(S_i)$ for $i \leq n$. Let $u_{n+1} = u_n + d_n v_n$ with $d_n > 0$. Then u_{n+1} is decreasingly risk averse (Pratt, 1964, Theorem 5), ranks $S_0, \ldots, S_n$ just as u_n does, and ranks S_{n+1} correctly among them for some d_n. If u_n ranks S_{n+1} too high, the same construction works with $v_n(x) = u_0(x)$ for $x \geq na$ but $v_n(x) < u_0(x)$ for $x < na$.

4. PROPER CERTAINTY EQUIVALENTS FOR REPEATED GAMBLES

The results of the previous section imply that a stronger assumption than decreasing risk aversion is needed to infer any regularity in the ranking of sets of repeated gambles. One such assumption is properness. Define

$C(X,W)$ as the certainty equivalent of X in the presence of W, the certain amount such that $W+X$ and $W+C(X,W)$ are indifferent, and let $C(X)=C(X,0)$. One implication (or definition) of properness given by $P\&Z$ (Theorem 1, equation (7)) is that if $X,Y,$ and W are independent, $C(X,W)\leq 0$, and $C(Y,W)\leq 0$, then $C(X,Y+W)\leq C(X,W)$. For $W=S_m$ and Y distributed like X, it follows immediately that if $C(X,S_m)\leq 0$, then $C(X,S_{m+1})\leq C(X,S_m)$. Hence $c_m=C(X,S_m)$ is decreasing in m beyond the first m for which it is nonpositive. In particular, it never becomes positive again, that is, S_{m+1} is no better than S_m. Therefore S_m improves monotonically as m increases up to its optimum value n (if any), and thereafter is monotonically nonimproving.

The monotonicity of c_m would imply the concavity of $s_m=C(S_m)$ as a function of m for $m\geq n$ if s_{m+1} were identical to s_m+c_m, but the correct relation is

$$s_{m+1}=C(S_m+c_m)=C(S_m,c_m)+c_m. \tag{5}$$

(The more general relation

$$\begin{aligned} C(X+Y,W) &= C(Y+C(X,Y+W),W) \\ &= C(X,Y+W)+C(Y,C(X,Y+W)+W) \end{aligned} \tag{6}$$

is easily verified.) By (5) the second difference is

$$s_{m+2}-2s_{m+1}+s_m=C(S_m+d_m)-2C(S_m+c_m)+C(S_m) \tag{7}$$

where $d_m=c_{m+1}+C(X,S_m+c_{m+1})$. For $c_m<0$, we have seen that properness implies $c_{m+1}\leq c_m$, whence $d_m\leq 2c_m$. Negativity of the second difference would then follow for $m\geq n$ if $C(S_n+t)$ were concave in t for $t<0$. (Decreasing risk aversion implies $C(S_n,t)$ is nondecreasing in t and hence, by (5), $s_{m+1}\geq s_m+c_m$ for $c_m\geq 0$, and similarly for $\leq$.)

A natural strengthening of properness is that the reduction in certainty equivalent resulting from adding an undesirable gamble is increased by the addition of another undesirable gamble, that is, if X,Y and W are independent and $C(X,W)\leq 0$ and $C(Y,W)\leq 0$, then

$$C(W)-C(W+X)\leq C(W+Y)-C(W+Y+X). \tag{8}$$

This condition implies directly both concavity of $C(S_m)$ for $m\geq n$ and concavity of $C(W+t)$ for $t\leq 0$ and all W. The latter in turn implies $r''\geq 0$, even if W is restricted to gambles of the form S_n. Since properness does not imply $r'' \geq 0$, it follows that (8) is strictly stronger than properness and that properness does not imply concavity of $C(S_n + t)$ for $t < 0$, but this does not resolve the question whether or not properness implies concavity of $C(S_m)$ for $m \geq n$.

5. WHEN INDIFFERENCE ONCE IMPLIES INDIFFERENCE TWICE

We prove here that

THEOREM. *If a utility function on wealth has the property that indifference to a gamble that gains or loses some fixed amount always implies indifference to two independent repetitions of the gamble, then it has constant absolute risk aversion, that is, it is linear or exponential.*

Specifically, let u be a utility function and suppose that a gamble gains b with probability p and loses b with probability $q = 1 - p$. At wealth w, this gamble is a matter of indifference if

$$pu(w + b) + qu(w - b) = u(w) \tag{9}$$

while two independent repetitions are a matter of indifference if

$$p^2u(w + 2b) + 2pqu(w) + q^2u(w - 2b) = u(w). \tag{10}$$

The hypothesized property of u is that every w, b, p, and q that satisfy (9) with $0 < p = 1 - q < 1$ also satisfy (10). PROOF. Given any a and b, let

$u_i = u(a + ib)$ for all i and define p_i so that at an initial wealth of $a + ib$, a gamble gaining b with probability p_i and losing b with probability $q_i = 1-$ p_i is a matter of indifference, namely

$$\frac{p_i}{q_i} = \frac{u_i - u_{i-1}}{u_{i+1} - u_i}, \quad p_i = \frac{u_i - u_{i-1}}{u_{i+1} - u_{i-1}}, \quad q_i = \frac{u_{i+1} - u_i}{u_{i+1} - u_{i-1}}. \tag{11}$$

Two such independent gambles are likewise a matter of indifference iff

$$\frac{p_i^2}{q_i^2} = \frac{u_i - u_{i-2}}{u_{i+2} - u_i}. \tag{12}$$

Dividing (12) by (11) gives

$$\frac{p_i}{q_i} = \frac{p_{i+1}}{q_{i-1}}. \tag{13}$$

By (12), a gamble gaining $2b$ with probability P_i and losing $2b$ with probability Q_i is a matter of indifference iff $P_i/Q_i = p_i^2/q_i^2$ and hence $P_i = p_i^2/(p_i^2 + q_i^2)$. For such gambles (13) becomes

$$\frac{p_i^2}{q_i^2} = \frac{p_{i+2}^2}{p_{i+2}^2 + q_{i+2}^2} \frac{p_{i-2}^2 + q_{i-2}^2}{q_{i-2}^2}. \tag{14}$$

From (13) with $i = 1, 2, 3$, we obtain successively

$$\frac{1}{q_0} + \frac{1}{p_2} = \frac{1}{p_2 q_1} = \frac{1}{p_3 q_2} = \frac{1}{p_4} + \frac{1}{q_2}. \tag{15}$$

Subtracting 1 gives

$$\frac{p_0}{q_0} + \frac{1}{p_2} = \frac{q_4}{p_4} + \frac{1}{q_2}. \tag{16}$$

From (14) with $i = 2$ we obtain

$$\frac{p_2^2}{q_2^2}(1 + \frac{q_4^2}{p_4^2}) = \frac{p_0^2}{q_0^2} + 1. \tag{17}$$

Letting $z = q_4/p_4$, we have by (17) and (16)

$$\frac{p_2^2}{q_2^2}(1 + z^2) - 1 = \frac{p_0^2}{q_0^2} = \left(z + \frac{1}{q_2} - \frac{1}{p_2}\right)^2. \tag{18}$$

Multiplying by q_2^2 gives

$$(p_2^2 - q_2^2)(z^2 + 1) = 2zq_2(1 - \frac{q_2}{p_2}) + (1 - \frac{q_2}{p_2})^2. \tag{19}$$

Either $p_2 = q_2$ or we can divide by $p_2 - q_2$ to obtain

$$z^2 - 2z\frac{q_2}{p_2} + \frac{q_2^2}{p_2^2} = 0, \tag{20}$$

in which case the definition of z, (20), and (17) give successively

$$\frac{q_4}{p_4} = z = \frac{q_2}{p_2} = \frac{q_0}{p_0}. \tag{21}$$

Replacing a by $a + kb$ for integer k, we see that the differences $u_{i+1} - u_i$ must form either a geometric or, if $p_2 = q_2$ for every k, an arithmetic sequence, and hence must agree with some exponential or linear function. Since the spacing was arbitrary and since u cannot agree with different exponential or linear functions on different spacings, u must be a single exponential or linear function. ∎

6. IMPLICATIONS OF STRICT INEQUALITY IN THE LOCAL CONDITION FOR PROPERNESS

The local condition $r'' \geq r'r$ is not sufficient for properness, as remarked earlier. Furthermore, equality on an interval does not imply borderline properness in any simple sense, as the earlier discussion and the beginning of Lindsey's argument (*P&Z*, p. 152) indicate. We therefore consider the case of strict inequality at a point w. This implies strict inequality in a neighborhood of w, provided r'' is continuous, which we assume. We now show what can be inferred in this case.

If a gamble X is accepted, the derived utility $U(w) = Eu(w + X)$ has local absolute risk aversion

$$R(w) = -Eu''(w + X)/Eu'(w + X). \tag{22}$$

The difference $R(w) - r(w)$ has the same sign at $w = 0$ as

$$-Eu''(X) - r(0)Eu'(X) = E[r(X) - r(0)]u'(X). \tag{23}$$

We can compare this to $r'(0)E[u(X) - u(0)]$ as follows. Expanding r around 0 and u around x gives

$$
\begin{aligned}
(24) \qquad [r(x) &- r(0)]u'(x) + r'(0)[u(0) - u(x)] \\
&= [xr'(0) + \frac{1}{2}x^2 r''(0)]u'(x) + \\
&\qquad r'(0)[-xu'(x) + \frac{1}{2}x^2 u''(x)] + O(x^3) \\
&= \frac{1}{2}x^2 u'(x)[r''(0) - r'(0)r(x)] + O(x^3).
\end{aligned}
$$

If $r'' > r'r$ at 0 then $\exists \epsilon > 0$ such that (24) is positive for $| x | \leq \epsilon$ and hence, if $| X | \leq \epsilon$ with probability 1,

$$
(25) \qquad E[r(X) - r(0)]u'(X) > r'(0)E[u(X) - u(0)].
$$

If $r'(0)$ and $E[u(X) - u(0)]$ do not have opposite signs, then the right-hand side of (25) is nonnegative and hence $R(0) > r(0)$. If $r'' < r'r$ at 0, the inequality in (25) is reversed and $R(0) < r(0)$ if $r'(0)$ and $E[u(X) - u(0)]$ do not have the same sign. Since $w = 0$ was arbitrary, the statements made in Section 1 follow. The particular form of this derivation was suggested by *P&Z* (equation (27)), which is, not coincidentally, the step preceding the local condition for properness $r'' \geq r'r$. Specifically, the left-hand side of (24) is the difference between the two sides of *P&Z* equation (27) for $w = 0$.

REFERENCES

Pratt, J.W. (1964), Risk aversion in the small and in the large, *Econometrica*, **32**, 122-136.

Pratt, J.W. and R.J. Zeckhauser (1987), Proper risk aversion, *Econometrica*, **55**, 143-154.

Samuelson, P.A. (1963), Risk and uncertainty: a fallacy of large numbers, *Scientia*, **98**, 108-113.

Samuelson, P.A. (1989), The $\sqrt{N}$ law and repeated risktaking. This volume.

The Rate of Escape Problem for a Class of Random Walks

William E. Pruitt*
University of Minnesota

1. INTRODUCTION

Let X, $X_1, X_2, \ldots$ be a sequence of i.i.d. random variables and define $S_n = X_1 + \ldots + X_n$. The problem that will interest us is the minimal rate of growth of $|S_n|$. Suppose that Z is a nonnegative stable random variable of index $\alpha \in (0,1)$ and we choose the scale factor so that

$$Ee^{-\lambda Z} = e^{-\lambda^\alpha}, \qquad \lambda \geq 0.$$

For $x > 0$, let $P\{X > x\} = P\{Z > x\}/2$. If we place the remaining mass of $\frac{1}{2}$ at zero and define $\beta_n = n^{1/\alpha}(\ell\ell\, n)^{-(1-\alpha)/\alpha}$ where $\ell\ell\, n$ is the iterated logarithm, then

$$\liminf_{n\to\infty} \frac{|S_n|}{\beta_n} = \alpha(1-\alpha)^{(1-\alpha)/\alpha}2^{-1/\alpha}. \tag{1.1}$$

AMS 1980 Subject Classification: Primary 60J15, Secondary 60F15.
Key words and phrases: Probability estimates, domain of attraction, lim inf, integral test.
* This work was partially supported by NSF Grant DMS 86-03437.

PROBABILITY, STATISTICS,
AND MATHEMATICS
Papers in Honor of Samuel Karlin

Copyright © 1989 by Academic Press, Inc.
All rights of reproduction in any form reserved.
ISBN-0-12-058470-0

If, on the other hand, we keep the positive tail of the distribution of X as above but make the distribution symmetric, then there is no nice sequence $\{\beta_n\}$ that will make the lim inf in (1.1) a positive finite constant. But if

$$\beta_n(\varepsilon) = n^{1/\alpha}(\log n)^{-1/(1-\alpha)}(\ell\ell\, n)^{-(1+\varepsilon)/(1-\alpha)},$$

then

$$\liminf_{n\to\infty} \frac{|S_n|}{\beta_n(\varepsilon)} = 0 \text{ or } \infty \quad \text{according as} \quad \varepsilon = 0 \text{ or } \varepsilon > 0. \tag{1.2}$$

Thus there are two differences between the two situations. In the symmetric case, the sums are smaller due to cancellation but they are also more irregular making it impossible to normalize. The goal here is to see how the transition between these two cases occurs as the mass on the negative axis is changed. To do this we will consider the above question for the family of distributions for X with density f which is defined as above on the positive axis, i.e. $f(x) = f_Z(x)/2$ for $x \geq 0$, f is to be zero on $(-e^e, 0)$, and

$$f(x) = \frac{d}{|x|^{\alpha+1}(\log|x|)^\gamma(\ell\ell|x|)^\delta}, \quad x \leq -e^e.$$

$d > 0$ is chosen so that f integrates to one and $\gamma > 0$, $\delta \in \mathbb{R}^1$. For comparison, we note that the asymptotic behavior of f as $x \to \infty$ is given by

$$f(x) \sim \frac{\alpha}{2\Gamma(1-\alpha)x^{\alpha+1}},$$

so that the negative tail is small compared to the positive tail. We will see that there is indeed a smooth transition from one case to the other as we pass through this family of distributions. Here is the result:

THEOREM 1. (a) *If* $\gamma > 1$ *or if* $\gamma = 1$ *and* $\delta > 1 - \alpha^{-1}(1-\alpha)^2$, *then* (1.1) *holds.*

(b) *If* $\gamma < 1$ *or if* $\gamma = 1$ *and* $\delta \leq 1 - \alpha^{-1}(1-\alpha)^2$, *then there is no normalizer* β_n *which makes* $\liminf \beta_n^{-1}|S_n|$ *positive and finite and has the property that* $n^{-1/\alpha}\beta_n$ *decreases; however, in this case, if*

$$\beta_n(\varepsilon) = n^{1/\alpha}(\log n)^{-(1-\gamma)/(1-\alpha)}(\ell\ell\, n)^{-(1-\delta+\varepsilon)/(1-\alpha)},$$

then (1.2) *holds.*

The main ingredient in the proof is a probability estimate for $P\{|S_n| \leq x\}$ when this probability is small. We will use the notation $A(n,x) \approx B(n,x)$ to mean that the ratio of A to B is bounded above and below by positive finite constants for large n, uniformly in x.

THEOREM 2. *For all* $1 < x < n^{1/\alpha}$, *we have*

$$P\{|S_n| \leq x\} \approx \max\left(\frac{x}{n^{1/\alpha}(\log n)^{\gamma}(\ell\ell\, n)^{\delta}}\ ,\ \left(\frac{x}{n^{1/\alpha}}\right)^{\frac{\alpha}{2(1-\alpha)}} \exp\left(-\theta\left(\frac{n^{1/\alpha}}{x}\right)^{\frac{\alpha}{1-\alpha}}\right)\right)$$

where $\theta = (1-\alpha)\alpha^{\alpha/(1-\alpha)}(1/2)^{1/(1-\alpha)}$.

Note that $n^{-1/\alpha}S_n$ converges weakly to a stable law supported on $[0,\infty)$ so that this probability is comparable to one when $x \geq n^{1/\alpha}$.

The results of Theorem 1 are true whenever

$$P\{X > x\} \sim \frac{1}{2\Gamma(1-\alpha)x^{\alpha}} \qquad \text{as } x \to \infty.$$

We have chosen the distribution of X, conditioned to be positive, to be exactly that of Z in order to simplify the proofs. The results of Theorem 2 are actually more delicate and require

$$P\{X > x\} = \frac{1}{2\Gamma(1-\alpha)x^{\alpha}}\left(1 + O\left(\frac{1}{\ell\ell\, x}\right)\right) \qquad \text{as } x \to \infty.$$

Thus, both results are valid, for example, if one takes

$$f(x) = \frac{c}{b + x^{\alpha+1}}, \quad x > 0,$$

where $c = \alpha/(2\Gamma(1-\alpha))$ and b is chosen to make f integrate to one.

A few words on the history of the problem are in order. The rate of escape problem was first considered for simple random walk in dimension $d \geq 3$ by Dvoretzky and Erdös (1951). The lim inf result for nonnegative stable random variables is due to Lipschutz (1956). An alternative approach is in Fristedt (1964) and Breiman (1968). Results for more general distributions are in Fristedt and Pruitt (1971), Zhang (1986) and Pruitt (1990). The result for symmetric stable random variables is due to Takeuchi (1964) with generalizations by Taylor (1967), Erickson (1976) and Griffin (1983). Since the symmetric distribution we have used above is not actually stable, Erickson's result is needed for (1.2). The only known result that applies to all distributions for the summands is due to Kesten (1978) who proved Erickson's conjecture that in dimension $d \geq 3$, any random walk escapes at least as fast as simple random walk.

2. PROOF OF THEOREM 2

It will be convenient to denote the two terms in the probability estimate in Theorem 2 by $P_1(n,x)$ and $P_2(n,x)$ respectively. In order for $|S_n|$ to be unusually small, there must be cancellation. But since we have made the negative tail small compared to the positive tail, this means that either the sum of the negative terms is unusually large in absolute value and the sum of the positive terms about the right size (which leads to $P_1(n,x)$), the sum of the positive terms is unusually small and the sum of the negative terms about the right size (which leads to $P_2(n,x)$), or they could meet in some middle ground (which has probability at most $P_1(n,x)$). In order to make use of the fact that X, conditioned to be positive, is stable, we introduce three independent i.i.d. sequences $\{V_i\}$, $\{P_i\}$, and $\{N_i\}$ where V_i is Bernoulli with parameter $\frac{1}{2}$, and

$$P\{P_i > x\} = 2P\{X > x\}, \quad x > 0, \quad P\{N_i \leq x\} = 2P\{X \leq x\}, \quad x \leq 0.$$

Then X has the same distribution as $V_i P_i + (1 - V_i)N_i$. Next let

$$T_n = \sum_{i=1}^{n} P_i, \quad U_n = \sum_{i=1}^{n} N_i, \quad W_n = \sum_{i=1}^{n} V_i.$$

Now fix a sequence of ones and zeros corresponding to the V's and let $m = W_n$. Then, conditionally, S_n has the same distribution as $T_m + U_{n-m}$. We will estimate $P\{|T_m + U_{n-m}| \leq x\}$ assuming only that $|m - n/2| \leq n^{\xi}$ where $\frac{1}{2} < \xi < 1$. Since

$$P\left\{\left|W_n - \frac{n}{2}\right| > n^{\xi}\right\} \leq \exp(-n^{2\xi-1})$$

(see e.g. page 193 of Feller (1968)) and this bound is of smaller order than $P_1(n,x)$ for any $x \geq 1$ and any ξ, this assumption on m is harmless. We will need the following bounds on the lower tail of a nonnegative stable law:

$$(2.1)\qquad \begin{aligned} P\{T_n \leq x\} &= P\{T_1 \leq n^{-1/\alpha}x\} \\ &\approx (xn^{-1/\alpha})^{\alpha/2(1-\alpha)}\exp(-\theta_1(n^{1/\alpha}x^{-1})^{\alpha/(1-\alpha)}) \end{aligned}$$

for $x \leq n^{1/\alpha}$ where $\theta_1 = (1-\alpha)\alpha^{\alpha/(1-\alpha)}$. This is given in this form in Example 4.1 in Jain and Pruitt (1987). Let h_n be the density of T_n, h the density of T_1. Then

$$(2.2)\qquad h_n(x) = n^{-1/\alpha}h(n^{-1/\alpha}x) \quad \text{and} \quad h(u) = O(\exp(-cu^{-\alpha/(1-\alpha)}))$$

as $u \to 0$ where c is some positive constant. The exact asymptotic behavior is available in Skorokhod (1954) but (2.2) is sufficient for our needs here. Define

$$\begin{gathered} A_n = \{|T_m + U_{n-m}| \leq x\},\ B_n = \{U_{n-m} \leq -n^{1/\alpha}\}, \\ C_n = \{-n^{1/\alpha} < U_{n-m} \leq -(x/n)^{1/(1-\alpha)}\}, \\ D_n = \{U_{n-m} > -(x/n)^{1/(1-\alpha)}\}. \end{gathered}$$

We will now prove the following:

$$\begin{aligned} &(2.3) && P(A_nB_n) \approx P_1(n,x), \\ &(2.4) && P(A_nC_n) = O(P_1(n,x)), \\ &(2.5) && P(A_nD_n) = O(P_2(n,x)) && \text{if } x \geq n^{(\xi+\alpha-\xi\alpha)/\alpha}, \\ &(2.6) && P_2(n,x) = O(P(A_nD_n)) && \text{if } x \geq cn^{1/\alpha}(\ell\ell\, n)^{-(1-\alpha)/\alpha}, \end{aligned}$$

where c is a positive constant. This will be enough to complete the proof of the theorem. To see this, observe that if $1 \leq x \leq n^{(\xi+\alpha-\xi\alpha)/\alpha}$, then using the monotonicity of A_n and D_n in x and (2.5) for $x = n^{(\xi+\alpha-\xi\alpha)/\alpha}$, we obtain

$$P(A_nD_n) = O(P_2(n, n^{(\xi+\alpha-\xi\alpha)/\alpha})) = O(\exp(-\theta n^{1-\xi})) = o(P_1(n,x))$$

which is sufficient for the upper bound. For the lower bound, simply note that if $x \leq cn^{1/\alpha}(\ell\ell\, n)^{-(1-\alpha)/\alpha}$ for an appropriate c and n is large, then $P_2(n,x) \leq P_1(n,x)$.

PROOF OF (2.3): If $y \in [b_n, n^{1/\alpha}]$, where

$$b_n = n^{1/\alpha}(\log n)^{-\gamma/\alpha}(\ell\ell\, n)^{-\delta/\alpha},$$

then

$$\begin{aligned} P\{U_{n-m} \le -y\} &\approx P\{-3y \le U_{n-m} \le -y\} \\ &\approx ny^{-\alpha}(\log n)^{-\gamma}(\ell\ell\, n)^{-\delta}. \end{aligned} \tag{2.7}$$

To see the upper bound, truncate the summands at $-y$ and use Chebyshev; see e.g. the Lemma in Pruitt (1981b). For the lower bound, consider the disjoint union of the events that one of the summands is in the interval $(-2y, -y]$ and the sum of the remaining terms is in the interval $(-b_n, 0]$. Since $b_n^{-1}U_{n-m-1}$ converges weakly to a stable law supported on $(-\infty, 0]$, we have

$$P\{-b_n < U_{n-m-1} \le 0\} \approx 1. \tag{2.8}$$

Now to prove (2.3), we apply (2.7) with $y = n^{1/\alpha}$. Then we observe that $P\{|T_m - z| \le x\} = O(xn^{-1/\alpha})$ since $m^{-1/\alpha}T_m$ is stable and has a bounded density. This gives the upper bound. For the lower bound, we use the fact that $P\{|T_m - z| \le x\} \approx xn^{-1/\alpha}$ when $z \in [n^{1/\alpha}, 3n^{1/\alpha}]$ since this condition restricts $m^{-1/\alpha}T_m$ to its support.

PROOF OF (2.4): Here we split C_n further into the disjoint union of

$$C_{nk} = \left\{-\frac{n^{1/\alpha}}{k} < U_{n-m} \le -\frac{n^{1/\alpha}}{k+1}\right\}, \; k = 1, 2, \ldots, \left[\left(\frac{n^{1/\alpha}}{x}\right)^{1/(1-\alpha)}\right].$$

Then by (2.7),

$$P(C_{nk}) = O(k^\alpha(\log n)^{-\gamma}(\ell\ell\, n)^{-\delta}). \tag{2.9}$$

This is true for all k for if $n^{1/\alpha}k^{-1} < b_n$, then $k^\alpha(\log n)^{-\gamma}(\ell\ell\, n)^{-\delta} > 1$. Now suppose first that $n^{1/\alpha}k^{-1} \le x$. Then

$$P(A_nC_{nk}) \le P\{T_m \le x + n^{1/\alpha}k^{-1}\}P(C_{nk}) \le P\{T_m \le 2x\}P(C_{nk}).$$

For the first factor, we use (2.1):

$$P\{T_m \le 2x\} = O(\exp(-c(n^{1/\alpha}x^{-1})^{\alpha/(1-\alpha)})).$$

Then

$$n^{1/\alpha}x^{-1}P\{T_m \le 2x\} = O(n^{1/\alpha}x^{-1}\exp(-c(n^{1/\alpha}x^{-1})^{\alpha/(1-\alpha)}))$$
$$\le Ck^{1-\alpha}\exp(-ck^{\alpha})$$

since the bound on k implies that $n^{1/\alpha}x^{-1} \ge k^{1-\alpha}$. Thus, by (2.9)

$$P(A_n C_{nk}) \le Ck\exp(-ck^{\alpha})xn^{-1/\alpha}(\log n)^{-\gamma}(\ell\ell\, n)^{-\delta}$$
$$= Ck\exp(-ck^{\alpha})P_1(n,x).$$

Now we want to obtain a similar bound when $x \le n^{1/\alpha}k^{-1}$. Here we need to use (2.2). For $n^{1/\alpha}(k+1)^{-1} \le z \le n^{1/\alpha}k^{-1}$,

$$P\{z - x \le T_m \le z + x\} \le 2xm^{-1/\alpha}\sup_{0\le u\le\zeta}h(u)$$

where $\zeta = (z+x)m^{-1/\alpha}$. Now

$$\zeta \le 2n^{1/\alpha}k^{-1}m^{-1/\alpha} = O(k^{-1})$$

so that by (2.2) and (2.9)

$$P(A_n C_{nk}) \le Cxn^{-1/\alpha}\exp(-ck^{\alpha/(1-\alpha)})P(C_{nk})$$
$$\le C_1 k^{\alpha}\exp(-ck^{\alpha/(1-\alpha)})P_1(n,x).$$

Now we have two different bounds for $P(A_n C_{nk})$ depending on the relative sizes of x, n, and k but both are of the form $P_1(n,x)$ times a factor that is summable on k. This is enough for (2.4).

PROOF OF (2.5): Here we note that by using the bounds on m, we have

$$m^{-1/\alpha}\left(x + \left(\frac{x}{n}\right)^{1/(1-\alpha)}\right) = \frac{2^{1/\alpha}x}{n^{1/\alpha}}\left(1 + \left(\frac{x}{n^{1/\alpha}}\right)^{\alpha/(1-\alpha)} + O(n^{\xi-1})\right).$$

Then

$$P(A_n D_n) \le P\left\{T_m \le x + \left(\frac{x}{n}\right)^{1/(1-\alpha)}\right\}$$
$$= P\left\{T_1 \le m^{-1/\alpha}\left(x + \left(\frac{x}{n}\right)^{1/(1-\alpha)}\right)\right\}$$
$$\le C\left(\frac{x}{n^{1/\alpha}}\right)^{\alpha/2(1-\alpha)}\exp\left(-\theta_1\left(\frac{n^{1/\alpha}}{2^{1/\alpha}x}\right)^{\alpha/(1-\alpha)}\right) = CP_2(n,x).$$

For the second error term in the exponent, we have used the fact that

$$x \geq n^{(\xi+\alpha-\xi\alpha)/\alpha} \quad \text{implies} \quad (n^{1/\alpha}x^{-1})^{\alpha/(1-\alpha)}n^{\xi-1} = O(1).$$

PROOF OF (2.6): First note that $x \leq n^{1/\alpha}$ implies that $(xn^{-1})^{1/(1-\alpha)} \leq x$. Thus we have $P(A_n D_n) \geq P\{T_m \leq x\}P(D_n)$. The fact that $P\{T_m \leq x\} \geq cP_2(n, x)$ follows as above. (Although we no longer have to deal with the error term from $(xn^{-1})^{1/(1-\alpha)}$ we still have the one from $n^{\xi-1}$.) The fact that $P(D_n) \approx 1$ follows from (2.8) since even

$$x \geq \frac{n^{1/\alpha}}{(\log n)^{\gamma(1-\alpha)/\alpha}(\ell\ell\, n)^{\delta(1-\alpha)/\alpha}} \quad \text{implies} \quad \left(\frac{x}{n}\right)^{1/(1-\alpha)} \geq b_n,$$

the normalizer for U_{n-m}. ∎

3. PROOF OF THEOREM 1

If $\gamma > 1$ or if $\gamma = 1$ and $\delta > 1$, then it follows from Lemma 8.1 of Pruitt (1981a) that

$$\frac{\sum_{i=1}^n X_i^-}{\sum_{i=1}^n X_i^+} \to 0 \quad \text{a.s.} \tag{3.1}$$

and so the result is a consequence of the known result for $\sum X_i^+$. But (3.1) is not valid if $\gamma = 1$ and $\delta \leq 1$ so a different proof is needed. The proof we give will also work in the above cases. We use Theorem 3 of Kesten (1970). It asserts that b is a limit point of the sequence $\{\beta_n^{-1}S_n\}$ iff

$$\sum_{n=1}^{\infty} \frac{P\{|\beta_n^{-1}S_n - b| < \varepsilon\}}{\sum_{i=0}^{n-1} P\{|S_i| < \beta_n\}} = \infty \qquad \text{for every } \varepsilon > 0.$$

One may estimate the denominator by using Theorem 2 but it is easier to deal with it directly. We will prove for $x > 1$,

$$\sum_{i=0}^{n} P\{|S_i| < x\} \approx \min(x^\alpha, n). \tag{3.2}$$

For the lower bound, observe that since $n^{-1/\alpha}S_n$ is attracted to a stable law, there is an M such that if $x \geq Mi^{1/\alpha}$, then $P\{|S_i| \geq x\} < \frac{1}{2}$. Thus the summands in (3.2) corresponding to $i \leq (x/M)^\alpha$ are at least $\frac{1}{2}$. This proves the lower bound. For the upper bound, the ordinary local limit theorem for S_n implies that the summands in (3.2) are at most $Cxi^{-1/\alpha}$. Using this bound for $i > x^\alpha$ and 1 as an upper bound for $i \leq x^\alpha$ completes the proof of (3.2). Since we will only consider β_n such that $1 \leq \beta_n \leq n^{1/\alpha}$, Kesten's test becomes

$$\sum_{n=1}^{\infty} \beta_n^{-\alpha} P\{|\beta_n^{-1}S_n - b| < \varepsilon\} = \infty. \tag{3.3}$$

Now suppose we consider $\beta_n = n^{1/\alpha}(\ell\ell\, n)^{-(1-\alpha)/\alpha}$ and $b = \theta^{(1-\alpha)/\alpha}$. Then $P_2(n, b\beta_n) \approx (\ell\ell\, n)^{-1/2}(\log n)^{-1}$ and so

$$\sum_{n=1}^{\infty} \beta_n^{-\alpha} P\{|S_n| \leq b\beta_n\} = \infty. \tag{3.4}$$

On the other hand,

$$P_2(n, b(1-\varepsilon)\beta_n) \approx (\ell\ell\, n)^{-1/2}(\log n)^{-\eta} \tag{3.5}$$

where $\eta = (1-\varepsilon)^{-\alpha/(1-\alpha)} > 1$ and

$$P_1(n, b(1-\varepsilon)\beta_n) \approx (\ell\ell\, n)^{-\delta-(1-\alpha)/\alpha}(\log n)^{-\gamma}. \tag{3.6}$$

Thus

$$\sum_{n=1}^{\infty} \beta_n^{-\alpha} P\{|S_n| \leq b(1-\varepsilon)\beta_n\} < \infty \tag{3.7}$$

by (3.5) and (3.6) provided $\gamma > 1$ or $\gamma = 1$ and $\delta > 1 - \alpha^{-1}(1-\alpha)^2$. Then by (3.4) and (3.7)

$$\sum_{n=1}^{\infty} \beta_n^{-\alpha} P\{b(1-\varepsilon)\beta_n < |S_n| \leq b\beta_n\} = \infty.$$

Then by (3.3), either b or $-b$ must be a limit point of $\beta_n^{-1}S_n$ and so b is a limit point of $\beta_n^{-1}|S_n|$. On the other hand, by (3.7) there can be no limit points that are smaller in absolute value than b and this proves (1.1). Next we consider

$$\beta_n(\varepsilon) = n^{1/\alpha}(\log n)^{-(1-\gamma)/(1-\alpha)}(\ell\ell\, n)^{-(1-\delta+\varepsilon)/(1-\alpha)}.$$

Then

$$(\beta_n(0))^{-\alpha} P_1(n, \varepsilon\beta_n(0)) \approx n^{-1}(\log n)^{-1}(\ell\ell\, n)^{-1},$$

which means by (3.3) that 0 is a limit point of $(\beta_n(0))^{-1}S_n$ and this proves (1.2) in case $\varepsilon = 0$. Next, observe that if $\varepsilon > 0$, then

$$(\beta_n(\varepsilon))^{-\alpha} P_1(n, M\beta_n(\varepsilon)) \approx n^{-1}(\log n)^{-1}(\ell\ell\, n)^{-1-\varepsilon}. \tag{3.8}$$

Furthermore, if $\gamma < 1$,

$$(\beta_n(\varepsilon))^{-\alpha} P_2(n, M\beta_n(\varepsilon)) = O(n^{-1}\exp(-(\log n)^{\eta})) \tag{3.9}$$

for some $\eta > 0$. If $\gamma = 1$, we obtain

$$(\beta_n(\varepsilon))^{-\alpha} P_2(n, M\beta_n(\varepsilon)) = O(n^{-1}\exp(-(\ell\ell\, n)^{1+\eta})) \tag{3.10}$$

for some $\eta > 0$, provided $\delta \le 1 - \alpha^{-1}(1-\alpha)^2$. Thus, whenever $\gamma < 1$ or $\gamma = 1$ and $\delta \le 1 - \alpha^{-1}(1-\alpha)^2$ we have by (3.8)-(3.10)

$$\sum_{n=1}^{\infty} (\beta_n(\varepsilon))^{-\alpha} P\{|S_n| \le M\beta_n(\varepsilon)\} < \infty.$$

Using (3.3) again, this proves (1.2) when $\varepsilon > 0$. It remains to prove that there is no "correct nice normalizer" in these last cases. Suppose that $\beta_n n^{-1/\alpha}$ is nonincreasing and that $\beta_n \ge cn^{1/\alpha}(\ell\ell\, n)^{-(1-\alpha)/\alpha}$ infinitely often. Choose an n where this last inequality holds and let $k \le n$. Then

$$\begin{aligned}\frac{1}{\beta_k^{\alpha}} P\{|S_k| \le \varepsilon\beta_k\} \ge \frac{1}{\beta_k^{\alpha}} P_1(k, \varepsilon\beta_k) &= \varepsilon \left(\frac{\beta_k}{k^{1/\alpha}}\right)^{1-\alpha} \frac{1}{k(\log k)^{\gamma}(\ell\ell\, k)^{\delta}} \\ &\ge \varepsilon \left(\frac{\beta_n}{n^{1/\alpha}}\right)^{1-\alpha} \frac{1}{k(\log k)^{\gamma}(\ell\ell\, k)^{\delta}},\end{aligned}$$

and so when we sum on k we obtain for any fixed M,

$$\sum_{k=M}^{n} \frac{1}{\beta_k^{\alpha}} P\{|S_k| \le \varepsilon\beta_k\} \ge c_1 \left(\frac{\beta_n}{n^{1/\alpha}}\right)^{1-\alpha} (\ell\ell\, n)^{1-\delta} \ge c_1 c^{1-\alpha}(\ell\ell\, n)^{\eta}$$

where $\eta = 1 - \delta - (1-\alpha)^2/\alpha$. (This is what happens if $\gamma = 1$; if $\gamma < 1$, we obtain a factor of $(\log n)^{1-\gamma}$ which makes the bound tend to ∞ regardless of the value of δ.) Since $\eta \geq 0$, the series must diverge and hence by (3.3) $\liminf \beta_n^{-1}|S_n| = 0$. If, on the other hand, $\beta_n = o(n^{1/\alpha}(\ell\ell\, n)^{-(1-\alpha)/\alpha})$, then $P\{|S_n| \leq \varepsilon\beta_n\} = P_1(n, \varepsilon\beta_n)$ and so the series in (3.3) with $b = 0$ is either convergent for all $\varepsilon > 0$ or is divergent for all $\varepsilon > 0$. This means that $\liminf \beta_n^{-1}|S_n| = 0$ or ∞ for any sequence $\{\beta_n\}$. ∎

4. FINAL REMARKS

The symmetric case corresponds to $\gamma = 0$ and $\delta = 0$ in Theorem 2. In this case, it is easy to see that $P_1(n, x)$ always dominates and so the exponential term $P_2(n, x)$ disappears. This makes it impossible to find a sequence $\{\beta_n\}$ with $\liminf \beta_n^{-1}|S_n|$ positive and finite even under very mild conditions: $\beta_n \uparrow \infty$ and $\beta_{2n}/\beta_n \leq C$. (In fact, this is probably true even with no restrictions on β_n.) But once $\gamma > 0$, it is possible to make $P_2(n, x)$ play a role for short intervals of n by making β_n comparable to $n^{1/\alpha}(\ell\ell\, n)^{-(1-\alpha)/\alpha}$ there and then making β_n relatively small in between. Thus, even though there is no sequence β_n with $n^{-1/\alpha}\beta_n \downarrow$ that gives a positive, finite lim inf, it is possible to find such a sequence which satisfies $n^{-\xi}\beta_n \uparrow$ and $n^{-\eta}\beta_n \downarrow$ when ξ and η are given with $\xi < \alpha^{-1} < \eta$.

It is probably now possible to prove results similar to Theorems 1 and 2 when the distribution for the summands is much more general than the stable case considered here. This would involve using the bounds on the lower tail of the distribution of the sum given in Jain and Pruitt (1987) in place of the exact behavior of the stable distribution used here and also using the Esscher transform and a more general local limit theorem to replace the bounds that used (2.2). It may also be possible to obtain an integral test (even in this more general context) to determine whether $\{|S_n| \leq \beta_n \text{ i.o.}\}$ has probability 0 or 1. This would be the analogue of the result obtained by Lipschutz (1956) in the nonnegative stable case. It would require the use of some method other than Kesten's Theorem such as that in Pruitt (1990). We have not attempted any of these generalizations here in order to keep the proofs as simple as possible.

REFERENCES

Breiman, Leo (1968), A delicate law of the iterated logarithm for non-decreasing stable processes, *Annals of Mathematical Statistics*, **39**, 1818-1824. (Correction, **41** (1970), 1126-1127.)

Dvoretzky, A., and P. Erdös (1951), Some problems on random walk in space, *Proceedings of the Second Berkeley Symposium on Mathematical Statistics and Probability* (Jerzy Neyman, ed.), University of California, Berkeley, 353-367.

Erickson, K. Bruce (1976), Recurrence sets of normed random walk in $\mathbb{R}^d$, *Annals of Probability*, **4**, 802-828.

Feller, William (1968), *An Introduction to Probability Theory and Its Applications*, **1** (3rd ed.), Wiley, New York.

Fristedt, Bert E. (1964), The behavior of increasing stable processes for both small and large times, *Journal of Mathematics and Mechanics*, **13**, 849-856.

Fristedt, Bert E., and William E. Pruitt (1971), Lower functions for increasing random walks and subordinators, *Zeitschrift für Wahrscheinlichkeitstheorie und Verwandte Gebiete*, **18**, 167-182.

Griffin, Philip S. (1983), An integral test for the rate of escape of d-dimensional random walk, *Annals of Probability*, **11**, 953-961.

Jain, Naresh C., and William E. Pruitt (1987), Lower tail probability estimates for subordinators and nondecreasing random walks, *Annals of Probability*, **15**, 75-101.

Kesten, Harry (1970), The limit points of a normalized random walk, *Annals of Mathematical Statistics*, **41**, 1173-1205.

Kesten, Harry (1978), Erickson's conjecture on the rate of escape of d-dimensional random walk, *Transactions of the American Mathematical Society*, **240**, 65-113.

Lipschutz, Mariam (1956), On strong bounds for sums of independent random variables which tend to a stable distribution, *Transactions of the American Mathematical Society*, **81**, 135-154.

Pruitt, William E. (1981a), General one-sided laws of the iterated logarithm, *Annals of Probability*, **9**, 1-48.

Pruitt, William E. (1981b), The growth of random walks and Lévy processes, *Annals of Probability*, **9**, 948-956.

Pruitt, William E. (1990), The rate of escape of random walk, *Annals of Probability*, to appear.

Skorohod, A. V. (1954), Asymptotic formulas for stable distribution laws, *Selected Translations in Mathematical Statistics and Probability*, **1**, 157-161 (1961). (Original in Russian.)

Takeuchi, J. (1964), On the sample paths of symmetric stable processes in space, *Journal of the Mathematical Society of Japan*, **16**, 109-127.

Taylor, S. J. (1967), Sample path properties of a transient stable process, *Journal of Mathematics and Mechanics*, **16**, 1229-1246.

Zhang, C. H. (1986), The lower limit of a normalized random walk, *Annals of Probability*, **14**, 560-581.

Recent Advances on the Integrated Cauchy Functional Equation and Related Results in Applied Probability

C. Radhakrishna Rao
Pennsylvania State University

and

D. N. Shanbhag
University of Sheffield

1. INTRODUCTION

The integrated Cauchy functional equation (ICFE)

$$H(x) = \int_S H(x+y)\, \sigma(dy), \qquad x \in S \tag{1.1}$$

with S as a certain semigroup with identity, H as a nonnegative real valued continuous function on S and σ as a measure on (the Borel σ-field of) S has generated considerable interest among the research workers in characterization theory of probability distributions during recent years. The

AMS 1980 Subject Classifications: Primary 62H10, Secondary 60E05.
Keywords and phrases: Deny's theorem, point processes, integrated Cauchy functional equation, exponential and geometric distributions.

Copyright © 1989 by Academic Press, Inc.
All rights of reproduction in any form reserved.
ISBN-0-12-058470-0

origin of this problem goes back to the work of Choquet and Deny (1960) and Deny (1961). Choquet and Deny (1960) proved that in the special case when S is a locally compact second countable Hausdorff Abelian group with the additional restriction that the smallest closed subgroup of S containing the support of σ is S itself, H is bounded and σ is a probability measure, the equation (1.1) is satisfied if and only if H is a constant. (Observe that the specialized result remains valid even when the restriction that H is nonnegative is relaxed.) Deny (1961) extended this result by showing that if the restriction that H is bounded is replaced by that H is a nonnegative function not identically zero and the restriction that σ is a probability measure is relaxed, then H has an integral representation as a (constant multiple of) weighted average of the σ-harmonic exponential functions on S. (A function $e: S \to \mathbf{R}_+$ is called an exponential function if it is continuous and satisfies the condition $e(x+y) = e(x)e(y)$ for all $x, y \in S$; an exponential function e is called σ-harmonic if the integral of e over S with respect to the measure σ in (1.1) is unity.) A special case of the Choquet-Deny theorem for $S = \mathbf{R}$ with an important application in renewal theory has appeared in Feller (1966, p.351).

In many problems of characterization of probability distributions one has a functional equation of the type (1.1) where S is a certain semigroup (not a group as assumed in the theorems of Choquet and Deny and Deny). The first successful attempts to solve the equation with S as a semigroup were made by Shanbhag (1977) in the case $S = \{0, 1, \cdots\}$ and by Lau and Rao (1982) in the case $S = \mathbf{R}_+ (= [0, \infty))$. Davies and Shanbhag (1987) obtained the solution in the more general case when S is a Polish Abelian semigroup satisfying some mild conditions. Some variants and extensions of these results have been obtained among others by Alzaid, Lau, Rao and Shanbhag (1988) and Rao and Shanbhag (1989).

Applications of the results mentioned here in characterization theory of probability distributions have been discussed by Rao (1983), Rao and Shanbhag (1986) and several others. There also exist instances in the stochastic processes where in the integrated Cauchy functional equation (ICFE) plays a key role. It is known, for example, that the equation (1.1) in a specialized form appears in problems dealing with Markov chains (see, for example, Alzaid, Rao and Shanbhag (1987)) and its variant considered by Alzaid, Lau, Rao and Shanbhag (1988) has applications in queueing and storage theories. Futhermore, special cases of the equation (1.1) or its variants of the type in Rao and Shanbhag (1986) lead us to characterization results of Poisson processes (see, for example, Rao and Shanbhag (1986) and Gupta and Gupta (1986)).

The purpose of the present paper is to give a brief account of the various general results available on (1.1) and its variants with improvements and unification wherever possible. In the course of doing this, we shall discuss some further applications of the results in epidemiological research (Daniels (1975)), characterization of normal distribution (Letac (1981)), and stochastic processes (Shanbhag (1973)).

2. THEOREMS ON THE ICFE AND SOME OF ITS VARIANTS

We shall state here only the general theorems concerning the solution of the ICFE (integrated Cauchy functional equation) and some of its variants and make some remarks on their inter connections and their implications. Special cases of the results existing in the literature such as those due to Shanbhag (1977) and Lau and Rao (1982) have been discussed earlier indicating various methods for arriving at them by Rao and Shanbhag (1986), Alzaid, Lau,Rao and Shanbhag (1988) and Rao and Shanbhag (1989).

Consider now the case where S is a Polish Abelian semigroup with zero element, with H and σ as defined in (1.1). If $H(0) > 0$, as pointed out by Davies and Shanbhag (1987), we have a unique probability measure P_H defined on $\mathcal{B}(S^{\mathbf{N}})$, the Borel σ-field of $S^{\mathbf{N}}$, where $\mathbf{N} = \{1, 2, \cdots\}$, such that for each $n \geq 1$ and $B_1, B_2, \cdots, B_n \in \mathcal{B}(S)$

$$P_H(B_1 \times \cdots \times B_n \times \prod_{n+1}^{\infty} S) = \frac{1}{H(0)} \int_{B_n} \cdots \int_{B_1} H(x_1 + \cdots + x_n)\sigma(dx_1) \cdots \sigma(dx_n),$$

where $\mathcal{B}(S)$ is the Borel σ-field of S. The probability measure P_H obtained here is symmetric in the sense of Hewitt and Savage (1955). (Incidentally, even when σ is not σ-finite, one can verify P_H to be a well-defined symmetric measure.) On the corresponding probability space $(S^{\mathbf{N}}, B^{\mathbf{N}}, P_H\ )$ Davies and Shanbhag (1987) define $\{\xi_n(x, \cdot)\}$ such that

$$\xi_n(x, \cdot) \;=\; \begin{cases} H(x + Y_n)/H(Y_n) & \text{if } H(Y_n) \text{ and } H(x) > 0 \\ 0 & \text{otherwise,} \end{cases}$$

where $Y_n = \sum_{m=1}^{n} X_m$ with X_m, $m \geq 1$ as projection maps $S^{\mathbf{N}} \rightarrow S$, and point out that this is a supermatingale. Defining then $\xi(x,w)$ to be the limit of $\xi_n(x,w)$ as $n \rightarrow \infty$ whenever it exists and is finite, and zero otherwise, these authors establish the following theorem:

THEOREM 2.1. *Let $S^*(\sigma)$ be the smallest closed subsemigroup of S containing the zero element and* supp*(σ) and $H(0) > 0$. Let $\xi(x,\cdot)$ be as defined above for each $x \in S$. Then*

(2.1) $$H(x) \geq H(0)E(\xi(x,\cdot)), \; x \in S, \; H(x) = H(0)E(\xi(x,\cdot)), \; x \in S^*(\sigma),$$

(2.2) $$\xi(x+y,\cdot) = \xi(x,\cdot)\xi(y,\cdot) \;\; a.s. \; [P_H] \text{ for every } x, y \in S^*(\sigma),$$

(2.3) $$\int_S \xi(x,\cdot)\sigma(dx) = 1 \;\; a.s. \; [P_H],$$

(2.4) $$\xi(x,\cdot) \equiv 0 \;\text{ if } H(x) = 0.$$

The proof of the theorem given by the authors involves among other things Fatou's lemma and the observation that for each $x \in S^*(\sigma)$, $\{\xi_n(x,\cdot)\}$ is a martingale converging in L_2 to $\xi(x,\cdot)$. (Indeed, these authors establish that $\{\xi_n(x,\cdot)\}$ is a martingale converging in L_p for each $p > 0$ to $\xi(x,\cdot)$ for each $x \in S^*(\sigma)$.)

Using the present theorem, Rao and Shanbhag (1989) have extended the Choquet-Deny theorem to the case where the integral equation corresponds to a certain semigroup and has a certain signed measure in place of a probability measure. See also Davies and Shanbhag (1987) for a relevant result.

REMARK 2.1. If S is countable and $S^*(\sigma) = S$, then Theorem 2.1 implies that under discrete topology H has an integral representation of the type in Deny's theorem. This specialized result is due to Ressel (1985) and it yields, among other things, Shanbhag's(1977) result referred to earlier as a corollary.

REMARK 2.2. At this stage one might raise a relevant question as to whether Theorem 2.1 remains valid with $S^*(\sigma)$ in (2.1) and (2.2) replaced by S without any further assumptions except that $\sigma(\{0\}) < 1$. The following example shows that the answer to this question is in the negative.

EXAMPLE. Let $S = \mathbf{N}_0^2$, where $\mathbf{N}_0 = \{0, 1, \cdots\}$ and define σ by $\sigma(\{0\} \times \{n\}) = 2^{-1}$, $n = 0, 1, \cdots, \sigma(\{1, \cdots, m\} \times \mathbf{N}_0) = 0$ and $\sigma(\{n_1\} \times \{n_2\}) = 1$, $n_1 = m+1, m+2, \cdots, n_2 = 0, 1, \cdots$, where m is a fixed positive integer ≥ 2. Define $q: \mathbf{N}_0 \to \mathbf{R}_+$ such that $q(n) = 0$ for all $n > m$ and $q(n) \neq 0$ otherwise. The H on S such that $H(n_1, n_2) = q(n_1)(\frac{1}{2})^{n_2}$, $n_1, n_2 = 0, 1, \cdots$ satisfies (1.1). In this case, $\{\xi_n(x, \cdot)\}$ turns out to be a bounded martingale for each $x \in S$ and we have (2.1) with $S^*(\sigma)$ replaced by S satisfied. However, we cannot have here (2.2) with $S^*(\sigma)$ replaced by S satisfied since the H has $H(1,1) > 0$ and for some x, $H(x) = 0$.

It is now clear in view of what is revealed in Remark 2.2 that we need some further conditions (in addition to $\sigma(\{0\}) < 1$) to have the validity of Theorem 2.1 with $S^*(\sigma)$ in (2.1) and (2.2) replaced by S. Davies and Shanbhag have essentially proved using an extended version of the martingale argument considered in their proof of Theorem 2.1 that if the following structural condition is satisfied the modified theorem is valid:

CONDITION A. Let there be subsets B and D of S such that i) $B \subset \text{supp}(\sigma)$, $B \in \mathcal{B}(S)$, $\sigma(S \setminus B) = 0$, ii) the smallest closed sub-semigroup with identity containing D equals S, iii) for every $x \in D + (B \setminus \{0\})$, we have $k \geq 1$ and $y_1, \cdots, y_k$ in $S^*(\sigma)$ such that $x + \sum_1^{r-1} y_i \in S + y_r$, $r = 1, 2, \cdots, k$, and $x + \sum_1^k y_i \in S^*(\sigma)$.

Observe that if S is a group and the smallest closed subgroup of S containing $\text{supp}(\sigma)$ equals S then the structural Condition A is trivially satisfied and hence in this case we have the validity of Theorem 2.1 with $S^*(\sigma)$ in (2.1) and (2.2) replaced by S; it is also worth noting here that in view of essentially the same reason as in Remark 7 of Davies and Shanbhag (1987) the argument given by Davies and Shanbhag (1987) for establishing the modified version of Theorem 2.1 simplifies in the present case. One could obviously find several other simplified situations such as the ones satisfying

i) and ii) of Condition A together with $D+(B\setminus\{0\})\subset S^*(\sigma)$ in which the requirements of Condition A are met.

REMARK 2.3. If in the above modified version of Theorem 2.1 we have S to be countable, then it follows that under discrete topology the function H appearing in the theorem has an integral representation as a weighted average of σ-harmonic exponential functions. It also follows easily that the function H in the theorem has an integral representation of the type just mentioned if we specialize to $S=\prod_{i=1}^{n} S_i$ with $n\geq 1$ and $S_i = \mathbf{Z}(=\{0,\pm 1,\pm 2,\cdots\})$ or $\mathbf{N}_0(=\{0,1,2,\cdots\})$ or $-\mathbf{N}_0$ or $\mathbf{R}$ or $\mathbf{R}_+$ or $-\mathbf{R}_+$. This latter fact is evident from the proof of Theorem 2 of Davies and Shanbhag (1987) and could also be seen via a some what different approach observing in particular that if $x\in S$ and $\{x_n : \text{n=1,2,}\cdots\}$ is a sequence of members of S with rational components converging to x, then, in view of the modified versions of (2.1) and (2.2), $\{\xi(x_n,.)\}$ converges in L_2 to $\xi(x,.)$ and hence

$$\xi(x,\cdot)=\prod_{i=1}^{n}(a_i(\cdot))^{x(i)} \quad a.s.\ [P_H]$$

for some nonnegative a_i's with these to be positive if S_i is uncountable or a group, where $x(i)$ is the ith component of x (and $0^0=1$).

Extending the argument of Davies and Shanbhag (1987) further, Rao and Shanbhag (1989) have established Theorem 2.2 given below; to state the Theorem, we require the following condition A^* in which S stands for the semigroup of Theorem 2.1, μ and ν denote some measures on S and $\hat{S}$ denotes the smallest closed subsemigroup of S with zero element such that it includes $\text{supp}(\mu)\cup\big(\text{supp}(\nu)+\text{supp}(\nu)\big)$.

CONDITION A^*. $\text{Supp}(\nu)\subset\hat{S}$ and there exist subsets B and D of S satisfying (i), (ii) and (iii) of Condition A with σ replaced by $\mu+\nu$.

We can now state the theorem as follows:

THEOREM 2.2. *Suppose S, $\hat{S}$, μ and ν are as in Condition A^* and Condition A^* is met. Assume that $\mu(\{0\})+\nu(\{0\})<1$ and the following* *equations are satisfied with H_1 and H_2 as continuous nonnegative functions on S.*

$$H_i(x) = \int_S H_i(x+y)\mu(dy) + \int_S H_j(x+y)\nu(dy), \ x \in S, \ i,j = 1,2, \ i \neq j.$$

Then either $H_1(x) = H_2(x) = 0$ for all $x \in S$ or we have a probability space (Ω, F, P) and product measurable functions $\xi_i : S \times \Omega \to \mathbf{R}_+$, $i = 1,2$ such that

$$H_i(x) = H_i(0)E(\xi_i(x,\cdot)), \ x \in S, \ i = 1,2$$

and the following conditions are met:

(i) $\xi_i(x+y,\cdot) = \xi_i(x,\cdot)\xi_i(y,\cdot)$ *a.s.* $[P]$ *for every* $x, y \in S$, $i = 1,2$.

(ii) $\int_S \xi_i(x,\cdot)(\mu+\nu)(dx) = 1$ *a.s.* $[P]$, $i = 1,2$.

(iii) $\xi_i(x,\cdot) \equiv 0$ *if* $H_i(x) = 0$ *and* $H_i(0) \neq 0$, $i = 1,2$

REMARK 2.4. Suppose S is a closed subsemigroup of a locally compact second countable Hausdorff Abelian group and we have a compact subset K of S, which is the closure of a nonempty open subset of the group, and a subset D^* dense in supp$(\mu+\nu)$ with the property that for every $x \in D^* \setminus \{0\}$ there exists an $x' \in K$ such that $x \in S + x'$. (Observe in particular that if $0 \in K$, then the existence of D^* meeting the requirement is assured.) Then Theorem 2.2 remains valid with a.s. $[P]$ in (i) and (ii) deleted and also $\xi_i(\cdot,\omega)$ as $(\mu+\nu)$-harmonic exponential functions on S for each $\omega \in \Omega$. This follows from arguments on pages 27 and 28 in the proof of Theorem 2 of Davies and Shanbhag (1987).

Using essentially an argument implied in Remark 4 of Davies and Shanbhag (1987), Rao and Shanbhag (1989) have established the following corollary of Theorem 2.2 (subject to the modification of Remark 2.4):

COROLLARY 2.1. *Let $n \geq 1$, $S = \prod_{i=1}^{n} S_i$ with $S_i = \mathbf{Z}(= \{0, \pm 1, \pm 2, \cdots\})$ or $\mathbf{N}_0(= \{0,1,2,\cdots\})$ or $-\mathbf{N}_0$ or $\mathbf{R}$ or $\mathbf{R}_+$ or $-\mathbf{R}_+$ and let λ be the restriction to S of a Haar measure on the smallest subgroup of $\mathbf{R}^n$ containing S. Let $h_1 : S \to \mathbf{R}_+$ and $h_2 : S \to \mathbf{R}_+$ be Borel measurable functions that are locally integrable with respect to λ, and μ and ν be σ-finite measures on S with $\mu(\{0\}) + \nu(\{0\}) < 1$. Assume that Condition A^* is met. Then*

$$h_i(\cdot) = \int_S h_i(\cdot + y)\mu(dy) + \int_S h_j(\cdot + y)\nu(dy) \quad a.e. \ [\lambda], \ i,j = 1,2, \ i \neq j$$

implies that

$$(2.5) \qquad h_i(\cdot) = \int_{[-\infty,\infty]^n} e^{<\cdot,x>} \alpha_i(dx) \quad a.e.\ [\lambda],\ i = 1,2\ ,$$

where α_i, *i=1,2 are measures on* $[-\infty,\infty]^n$ *such that*

$$\alpha_i = (\{x \in [-\infty,\infty]^n : \textstyle\int_S e^{<x,y>}(\mu+\nu)(dy) \neq 1 \quad or$$
$$< x, y > \ is\ undefined\ for\ some\ y \in S\}) = 0,\ i = 1,2$$

REMARK 2.5. In corollary 2.1 $e^{-\infty}$, e^{∞} and $0 \cdot (\pm\infty)$ are defined as usual to be 0, ∞ and 0 respectively: also, it is implicit from the statement of the corollary that the measures α_i, $i = 1, 2$ are such that the integrals in (2.5) are λ-locally integrable on S.

REMARK 2.6. Theorem 2.2 and Corollary 2.1 yield in particular on taking $\nu \equiv 0$ and $H_1 \equiv 0$ (or $= H_2$) respectively the Davies-Shanbhag (1987) modified version of Theorem 2.1 mentioned above and a stronger version of Corollary 2 of Theorem 2 of Davies and Shanbhag (1987); the latter of the two results yields the Lau-Rao theorem as an obvious corollary. It is also worth pointing out that using essentially the argument mentioned in Remark 2.3, one could produce a method slightly different to the one appearing in Rao and Shanbhag (1989) for arriving at Corollary 2.1.

3. SOME APPLICATIONS OF THE ICFE

A number of examples have been given in previous papers (Rao (1983) and Rao and Shanbhag (1986), (1989) among others) where the solution of the ICFE is used to characterize probability distributions. We propose to discuss a few other applications of the ICFE to demonstrate its applicability in solving a variety of problems.

3.1 A Problem in Epidemiology.

In a problem of deterministic spread of a simple epidemic, Daniels (1975) derived the following equation, with f nonnegative, not identically zero and everywhere differentiable,

$$-cf'(z) = \int_{-\infty}^{\infty} f(z-u)dF(u),\ z \in \mathbf{R}, \tag{3.1.1}$$

where F is a probability distribution function. There is no loss of generality in assuming that c > 0 in the equation (3.1.1) and hence we consider this to be so. Then in view of the nonnegativity of f, the equation implies that f is decreasing with $\lim f(z) = 0$ as $z \to \infty$. In that case, integrating both sides of (3.1.1) from x to ∞ and dividing by c, we have

$$f(x) = \int_{-\infty}^{\infty} f(z+x)c^{-1}(1-F(-z))dz \quad \forall x \in \mathbf{R}, \tag{3.1.2}$$

which is an ICFE. In view of Deny's theorem for $S = \mathbf{R}$, it follows that we have either two real roots θ_1, θ_2 of

$$\int_{-\infty}^{\infty} c^{-1}e^{\theta u}(1-F(u))du = 1 \tag{3.1.3}$$

(or equivalently of $\int_{-\infty}^{\infty} e^{\theta u}dF(u) = c\theta$) giving the solution for f as

$$f(x) = p_1 \exp(-\theta_1 x) + p_2 \exp(-\theta_2 x),\ x \in \mathbf{R} \tag{3.1.4}$$

with $p_1,\ p_2 \geq 0$ and at least one $p_i > 0$, or just one real root θ_0 of (3.1.3) giving

$$f(x) = p\ \exp(-\theta_0 x),\ x \in \mathbf{R} \tag{3.1.5}$$

with $p\ > 0$. Daniels arrived at this result in an intuitive way assuming apriori the existence of the moment generating function F on some interval which does not appear to be necessary.

3.2 A Characterization of the Normal Distribution.

Letac (1981) established the result that if X_1, X_2 and X_3 are three independent real random variables such that $P\{X_j = 0\} = 0,\ j = 1, 2, 3$ then $(R^{-1}X_1,\ R^{-1}X_2,\ R^{-1}X_3)$, where $R = (X_1^2 + X_2^2 + X_3^2)^{1/2}$ is uniformly distributed on the sphere $\{x \in R^3 \colon ||x|| = 1\}$ if and only if X_j's are all distributed as $N(0, \sigma^2)$ for some σ^2. (Indeed the statement of Letac's result is more general than this; however, the general result in question can easily be arrived at using the result we have stated.) We now show in what follows that this can be arrived at by a direct application of Deny's theorem:

Since the 'if' part is trivial, it is sufficient if we prove just the 'only if' part of the result. Assume then that the vector has the spherical distribution. We have, in view of the structure of the spherical distribution satisfying $|X_j| > 0$ a.s., for every $-1 < \alpha < 1$

$$E\left\{\frac{X_1|X_1|^\alpha}{(|X_2||X_3|)^{\frac{1+\alpha}{2}}}\right\} = 0$$

with the function under expectation well defined a.s., which implies that $E\{X_1|X_1|^\alpha\} = 0$. Consequently, we have X_1 and hence by symmetry all X_i to be symmetric. The assumption of sphericity (on appealing to $|X_j| > 0$ a.s.) yields

$$E\left\{\frac{|X_1|^\alpha}{|X_3|^\alpha}\right\} = E\left\{\frac{|X_2|^\alpha}{|X_3|^\alpha}\right\} < \infty, \quad -1 < \alpha < 1, \tag{3.2.1}$$

(with the functions under the expectations well defined a.s.,) and hence that $E\{|X_1|^\alpha\} = E\{|X_2|^\alpha\} < \infty$, $-1 < \alpha < 1$, implying X_1 and X_2 to be identically distributed. By symmetry then we have X_1 and X_3 and hence X_1, X_2, X_3 to be identically distributed. Observe now that (3.2.1) remains valid with X_2 replaced by $(2X_1 + X_2)\,/\sqrt{5}$; we can hence claim that

$$X_1 \stackrel{\text{d}}{=} \frac{2X_1 + X_2}{\sqrt{5}}, \tag{3.2.2}$$

which implies that the characteristic function f of X_1 satisfies the equation

$$f(t) = f(\frac{2t}{\sqrt{5}})f(\frac{t}{\sqrt{5}}), \quad -\infty < t < \infty, \tag{3.2.3}$$

which in turn implies in view of the symmetry of X_1 that f is a nonvanishing and positive (real) even function. Defining $g(x) = -\log f(e^x)$, $x \in \mathbf{R}$, we can then see from Deny's theorem that $f(t) = e^{-at^2}$, $t \in \mathbf{R}$ for some $a > 0$. Consequently, we have X_i to be independent identically distributed normal random variables with zero mean.

As revealed in Letac (1981), the present result yields a result due to Zinger as a corollary. (For the details and the reference, see Letac (1981).)

3.3 The Poisson Process.

Gupta and Gupta (1986) proved the following theorem.

THEOREM 3.1. *Let $\{S_n, n \geq 1, F(x)\}$ be a renewal process where $F(x)$ is continuous. Let $G(x)$ be a monotone non-decreasing function having a non-lattice support on $x \geq 0$ with $G(0) = 0$ and $\int_0^\infty \exp(-\xi x)dG(x) < \infty$ for all $\xi > 0$. Suppose further that $g(t) = E(G(v(t))) < \infty$ (where $v(t)$ is the forward recurrence time at t) for all t. If $g(t)$ is a constant, then $\{S_n\}$ is a Poisson process.*

It may be noted that if we replace the assumption that F is continuous by the weaker assumption $F(0) = 0$ subject to an alteration that G is now assumed right continuous and if we drop the condition that $\int_0^\infty \exp(-\xi x)dG(x) < \infty$ for all $\xi > 0$, the theorem still remains valid. This follows because the equation $g(t) = c$ now reduces to a special case of the equation in the Lau-Rao theorem (1982); indeed, the equation in the present case is equivalent to

$$c(1 - F(x)) = \int_{\mathbf{R}_+} (1 - F(x + y))dG(y), \quad x \in \mathbf{R}_+ \tag{3.3.1}$$

showing that F is exponential. Further, the differentiability of G, implicitly assumed in the proof given by Gupta and Gupta (1986), does not seem to be necessary.

3.4 The Yule Process.

In this section, we provide a characterization of the Yule process through the solution of an ICFE.

Suppose that $\{X_n : n = 1, 2, \cdots\}$ is a sequence of independent positive random variables. Define $N(t) = \sup\{n : X_1 + \cdots + X_n \leq t\}$, $t > 0$. Let $n_0 (\geq 2)$ be a fixed positive integer and let $\{X_n : n = 1, 2, \cdots\}$ be such that it satisfies additionally $P\{N(t) = n\} > 0$, $n = 1, 2, \cdots, n_0$. We have then the following theorem:

THEOREM 3.2. *The conditional distribution of $N(y)$ given $N(t) = n$ for each $0 < y < t$, $t > 0$, and $n = 1, 2, \cdots, n_0$ is nondegenerate binomial with* *index n and success probability parameter independent of n if and only if for some $\lambda_0 > 0$ and $\lambda \geq 0$,*

$$P\{X_i > x\} = e^{-\{\lambda_0+(i-1)\lambda\}x}, \quad x \geq 0, \quad i = 1, 2, \cdots, n_o + 1.$$

PROOF. The 'if' part of the result follows essentially from Neuts and Resnick (1971). We shall now prove the 'only if' part. Denote by $p(y,t)$ the p-parameter of the binomial distribution. It follows that unless the distribution of X_1 has full support on $\mathbf{R}_+$, we have a contradiction to the implicit assumption that $0 < p(y,t) < 1$. We have for $0 < y < t < \infty$.

$$(3.4.1) \qquad \frac{1}{\Psi_n(t)} \int_0^y [B_{n-1}(t-x) - B_n(t-x)]dA(x) = 1 - \{q(y,t)\}^n, \quad n = 1, 2, \cdots, n_0,$$

where $\Psi_n(t) = P\{N(t) = n\}$, $B_r(t) = P\{X_2 + \cdots + X_{r+1} \leq t\}$, $B_0(t) = 1$, and A is the distribution function of X_1. If $y \in (0,t)$, then it follows from (3.4.1) that

$$(3.4.2) \qquad \frac{B_{n-1}\big((t-y)-\big) - B_n\big((t-y)-\big)}{\Psi_n(t)} = n(q(y,t))^{n-1}q^*(y,t), \quad n = 1, 2, \cdots, n_0,$$

where $q^*(y,t) = \lim_{m\to\infty} \left(\frac{q(y,t)-q(y+\frac{1}{m},t)}{A(y+\frac{1}{m})-A(y)}\right)$ (which is easily seen to exist). Also, it follows from (3.4.1) with $n = 1$ that

$$(3.4.3) \qquad A(y)(1 - B_1(t-y)) \geq p(y,t)\Psi_1(t) > 0,$$

implying that $(1 - B_1(t-y)) > 0$ and hence $1 - B_1((t-y)-) > 0$ for all $0 < y < t < \infty$. Dividing (3.4.2) by its special case for $n = 1$, we can then conclude that $q(y,t)$ is of the form $\Psi^*(t-y)/\Psi(t)$. Since, given $t > 0$, each continuity point y of A in $(0,t)$ is, in view of (3.4.1), a continuity point of $q(y,t)$, it follows that Ψ^* is continuous on $(0,\infty)$. Furthermore, in view of the fact that $\lim_{y\to 0_+} q(y,t) = 1$, we can then immediately conclude that $q(y,t)$ is of the form $\Psi(t-y)/\Psi(t)$ with Ψ continuous and strictly increasing on $(0,\infty)$ with $\Psi(0_+) = 0$. We have now for $0 < y < t < \infty$

$$(3.4.4) \qquad \frac{\int_0^y \big(1 - C_{n+1}(t-x)\big)\, dA_n(x)}{\Psi_n(t)(\Psi(t))^{-n}} = (\Psi(t) - \Psi(t-y))^n, \quad n = 1, 2, \cdots, n_0,$$

where A_n and C_{n+1} are d.f.'s. of $X_1+\cdots+X_n$ and X_{n+1} respectively, which implies in view of the properties of Ψ that each A_n is continuous with support $\mathbf{R}_+$. From the observation immediately below (3.4.3) we have, in view of (3.4.2), $q^*(y,t) \neq 0$. Consequently from (3.4.4), in view of (3.4.2), we can easily see that for some function ξ_n

$$\xi_n(t-y)\, a_n^*(y) \;=\; \Psi(t) \;-\; \Psi(t-y), \tag{3.4.5}$$

where $a_n^*(y) = \lim_{m\to\infty} \left\{ \left(A_n(y+\frac{1}{m}) - A_n(y)\right) / \left(A(y+\frac{1}{m}) - A(y)\right) \right\}^{1/(n-1)}$ (which is clearly seen to exist). In (3.4.5) we should have clearly ξ_n and a_n^* to be positive and continuous on $(0,\infty)$ and indeed $a_n^*(y) \;\propto \Psi(y)$. Writing in (3.4.5) for y and t respectively t and $y+t$, and dividing both sides by $\xi_n(y)$, we get

$$K\Psi(t) \;=\; \frac{\Psi(t+y)-\Psi(y)}{\xi_n(y)}, \qquad 0<y,t<\infty, \tag{3.4.6}$$

where K is a positive constant. Using Theorem 2 of Rao and Shanbhag (1986) (which in turn follows from the Lau-Rao theorem) or directly via (3.4.5) we can conclude that $\Psi(t)$ is either proportional to t or to 1-$e^{-\lambda t}$ for some λ. In either of the two situations, we have in (3.4.4), A_n and Ψ to be differentiable on $(0,\infty)$. (To see the differentiability of A_n, first choose t such that $t-y$ is a continuity point of $C_{n+1}(t-y)$ given y.) Differentiating both sides of (3.4.4) with respect to y, we get

$$\frac{\left(1-C_{n+1}(t-y)\right)a_n(y)}{\Psi_n(t)\left(\Psi(t)\right)^{-n}} = n\left(\Psi(t)-\Psi(t-y)\right)^{n-1}\Psi'(t-y),\ 0<y<t<\infty,$$

where $a_n(y) \;=\; \frac{dA_n(y)}{dy}$. Substituting the form of Ψ, we see by induction on n here (or somewhat more easily using in this equation an observation based on (3.4.2) that $\Psi_n(t)\left(\Psi(t)\right)^{-n} \;\propto\; a_1(t)$ or exclusively from (3.4.2)) that $X_1,\cdots,X_{n_0+1}$ are exponential random variables of the form in the theorem. (It may be mentioned here that the situation $\lambda=0$ arises when Ψ is linear, otherwise, we get $\lambda>0$.) ∎

COROLLARY 3.1. *If we assume,* $P\{N(t)=n\}>0$ *for every* $n\geq 1$ *and every* $t>0$*, then the conditional distribution of* $N(y)$ *given* $N(t)=n$ *is nondegenerate binomial with parameters as stated in Theorem 3.2 for* *every* $0 < y < t < \infty$ *and every* $n \geq 1$ *if and only if the process* $\{N(t)\}$ *is Yule. (The process constructed with intervals such that* $P\{X_n > x\} = e^{-\{\lambda_0+(n-1)\lambda\}x}$, $x \geq 0$ *is referred to as Yule.)*

REMARK 3.1. A version of Liberman's (1985) characterization of a Poisson process in the class of renewal processes is a special case of the above corollary.

REFERENCES

Alzaid, A. A., Rao, C. R. and Shanbhag, D. N. (1987), Solution of the integrated Cauchy functional equation using exchangeability, *Sankhyā*, A, **49**, 189–194.

Alzaid, A. A., Lau Ka-Sing, Rao, C. R. and Shanbhag, D. N. (1988), Solution of Deny convolution equation restricted to a half line, *Journal Multivariate Analysis*, **24**, 309–329.

Choquet, G and Deny, J. (1960), Sur l'equation de convolution $\mu = \mu * \sigma$, *Comptes Rendus Hebdomadaires des Séances de l'Académie des Sciences, Paris*, **250**, 799–801.

Daniels, H. E. (1975), The deterministic spread of a simple epidemic, *In honor of M. S. Bartlett* (J. Gani, ed.), *Perspectives in Probability and Statistics*, 373–386.

Davies, P. L. and Shanbhag, D. N. (1987), A generalization of a theorem of Deny with applications in characterization theory, *Quarterly Journal of Mathematics, Second Series, Oxford*, **38**, 13–34.

Deny, J. (1961), Sur l'equation de convolution $\mu * \sigma$, *Semin. Theor. Potent. M. Brelot, Fac. Sci. Paris*, 1959–60, **4e** ann.

Feller, W. (1966), *An introduction to Probability Theory and its Applications*, **Vol. 2**, John Wiley and Sons Inc., New York.

Gupta, P. L. and Gupta, R. C. (1986), A characterization of the Poisson process, *Journal of Applied Probability*, **23**, 233–235.

Hewitt, E. and Savage, L. J. (1955), Symmetric measures on cartesian products, *American Mathematical Society Transactions*, **80**, 470–501.

Lau, Ka-Sing and Rao, C. Radhakrishna (1982), Integrated Cauchy functional equation and characterizations of the exponential law, *Sankhyā*, A, **44**, 72–90.

Letac, G. (1981), Isotropy and sphericity: Some characterizations of the normal distribution, *Annals of Statistics*, **9**, 408–417.

Liberman, U. (1985), An order statistic characterization of the Poisson renewal process, *Journal of Applied Probability* **22**, 717–722.

Neuts, M. F. and Resnick, S. I. (1971), On the times of births in a linear birth process, *Journal of the Australian Mathematical Society*, **12**, 473–475

Rao, C. Radhakrishna (1983), An extension of Deny's theorem and its application to characterization of probability distributions. *In a Festschrift for Eric Lehmann* (P. J. Bickel et al., eds.), Wadsworth, Monterey, 348–365.

Rao, C. Radhakrishna and Shanbhag, D. N. (1986), Recent results on characterization of probability distributions: A unified approach through extensions of Deny theorem, *Advances in Applied Probability*, **18**, 660–678.

Rao, C. Radhakrishna and Shanbhag, D .N. (1989), Further extensions of the Choquet-Deny and Deny theorems with applications in characterization theory, *Quarterly Journal of Mathematics, Oxford*, **40**, to appear.

Ressel, P. (1985), De Finetti-type theorems: An analytic approach, *Annals of Probability*, **13**, 898–922.

Shanbhag, D. N. (1973), Characterization of Yule process, unpublished manuscript.

Shanbhag, D. N. (1977), An extension of Rao-Rubin characterization of the Poisson distribution, *Journal of Applied Probability*, **14**, 540–546.

A Note on Maximum Entropy

Murray Rosenblatt*
University of California, San Diego

The maximum entropy method is claimed to be effective in detecting sharp spectral features. In this paper we want to see to what extent it is useful in isolating discrete harmonics in the case of a simple model. I shall follow in part the notation given in a discussion in the excellent book on computational procedures of Press, Flannery, Teukolsky, and Vetterling (1986) from page 430 on.

Suppose we have a power spectrum of a stationary sequence of random variables

$$P(z) = \frac{a_0}{\left|1 + \sum_{j=1}^{M-1} a_j z^j\right|^2}, \quad z = \exp(2\pi i \lambda), \tag{1}$$

with

$$P(z) \approx \sum_{j=-M+1}^{M-1} \phi_j z^j \tag{2}$$

AMS 1980 Subject Classifications: Primary 60G35, Secondary 62M10, 62M15.
Key words and phrases: Maximum entropy, discrete harmonic, power spectrum.
*Research supported in part by ONR Contract No. N00014-81-K-0003 and NSF Grant DMS 83-12106.

PROBABILITY, STATISTICS,
AND MATHEMATICS
Papers in Honor of Samuel Karlin

Copyright © 1989 by Academic Press, Inc.
All rights of reproduction in any form reserved.
ISBN-0-12-058470-0

where the symbol $\approx$ in (2) means that the series expansion of $P(z)$ in z agrees with that on the right side from z^{-M+1} to z^{M-1}. If the ϕ_j are the covariances indexed from $-M+1$ to $M-1$, the expansion of $P(z)$ in z provides an extrapolation of the covariance sequence for lags greater than $M-1$. Among all Gaussian stationary processes with covariances ϕ_j, $j=-M+1,\ldots,M-1$, the process with this extrapolation for the covariance sequence has maximal entropy. The $(M-1)^{\text{st}}$ order autoregressive estimator $P(z)$ is the maximal entropy approximation since it maximizes

$$\int_{-1/2}^{1/2} \log g(\lambda)\, d\lambda$$

among all power spectra $g(\lambda)$ satisfying

$$\int_{-1}^{1} e^{2\pi i\lambda h}\, g(\lambda)\, d\lambda = \phi_h, \quad h = 0, \pm 1, \ldots, \pm(M-1).$$

The idea of using autoregressive approximations of the power spectrum with this interpretation is due to J. P. Burg (1967). A discussion of the maximum entropoy method can be found in Priestley.

Relation (2) implies that

$$\begin{bmatrix} \phi_0 & \phi_1 & \phi_2 & \cdots & \phi_{M-1} \\ \phi_1 & \phi_0 & \phi_1 & \cdots & \phi_{M-2} \\ \phi_2 & \phi_1 & \phi_0 & \cdots & \phi_{M-3} \\ \vdots & \vdots & \vdots & \ddots & \vdots \\ \phi_{M-1} & \phi_{M-2} & \phi_{M-3} & \cdots & \phi_0 \end{bmatrix} \begin{bmatrix} 1 \\ a_1 \\ a_2 \\ \vdots \\ a_{M-1} \end{bmatrix} = \begin{bmatrix} a_0 \\ 0 \\ 0 \\ \vdots \\ 0 \end{bmatrix}. \tag{3}$$

The harmonics are specified so that the Toeplitz forms are circulants and this allows the formal computations to be carried out. We shall consider the special case of a sequence of random variables with covariance sequence

$$\phi_\tau = \cos \omega_M \tau + \epsilon \delta_{0,\tau}, \quad \tau = 0, 1, \ldots, M-1, \tag{4}$$

with the harmonic

$$\omega_M = \frac{2\pi k_M}{M}$$

and $\epsilon > 0$. Here k_M is an integer $0 \le k_M \le M-1$ and we may assume (though it is not necessary) that

$$k_M/M \to \mu > 0 \quad \text{as} \quad M \to \infty.$$

Here $\delta_{0,\tau}$ is the Kronecker δ

$$\delta_{0,\tau} = \begin{cases} 1 & \text{if } \tau = 0 \\ 0 & \text{if } \tau \neq 0. \end{cases}$$

Notice that (4) can be seen to be the covariance sequence of the random variables

$$V_\tau = U \cos \omega_M \tau + V \sin \omega_M \tau + \sqrt{\epsilon}\, W_\tau,$$

$\tau = 0, 1, \ldots, M-1$, where U, V, W_τ are independent identically distributed normal random variables with mean zero and variance one. The sequence has a discrete harmonic (with random amplitudes) at $\omega_M = 2\pi k_M/M$ (and $\omega = 2\pi(M - k_M)/M$) added to white noise of variance $\epsilon > 0$. With the covariances (4) the matrix in the covariances ϕ_j on the left of (3) is a circulant matrix and we can solve the corresponding system of equations in closed terms.

The right eigenvectors of the basic circulant matrix

$$J = \begin{bmatrix} 0 & 1 & & & \\ & 0 & 1 & \bigcirc & \\ & & \ddots & \ddots & \\ & \bigcirc & & & 1 \\ 1 & & & & 0 \end{bmatrix}$$

are

$$x^{(j)} = \frac{1}{\sqrt{M}} \begin{bmatrix} 1 \\ \exp(2\pi i j/M) \\ \vdots \\ \exp(2\pi i(M-1)j/M) \end{bmatrix}$$

with eigenvalue $\lambda_j = \exp(2\pi i j/M)$, $j = 0, 1, \ldots, M-1$. Let Q be the $M \times M$ circulant on the left of (3). Let U be the unitary matrix with the jth column given by $x^{(j)}$. Then

$$Q = U \Lambda U^{-1}, \qquad U^{-1} = \overline{U}^T$$

with the eigenvalues λ_s of Q (the diagonal entries of the diagonal matrix Λ) such that

$$\lambda_s = \begin{cases} \epsilon & \text{if } s \neq k,\ M-k \\ \epsilon + \sqrt{M}/2 & \text{if } s = k,\ M-k. \end{cases}$$

Then

$$\begin{aligned} Q^{-1} &= U\Lambda^{-1}U^{-1} \\ &= \epsilon^{-1}I + UDU^{-1} \end{aligned}$$

where D is the diagonal matrix with diagonal entries

$$c = (\epsilon + \sqrt{M}/2)^{-1} - \epsilon^{-1}$$

at the kth and $(M-k)^{\text{th}}$ diagonal locations and zeros everywhere else. The $(u,v)^{\text{th}}$ element of UDU^{-1}, $u, v = 0, 1, \ldots, M-1$, is given by

$$\frac{2c}{M} \cos\left(2\pi(u-v)\frac{k}{M}\right).$$

We can then solve (3) and find that

$$\begin{aligned} 1 &= \left[\frac{2c}{M} + \epsilon^{-1}\right] a_0 \\ a_j &= \frac{2c}{M} a_0 \cos\left(2\pi\frac{jk}{M}\right), \qquad j = 1, \ldots, M-1. \end{aligned} \tag{5}$$

Now

$$\frac{2c}{M} + \epsilon^{-1} = \epsilon^{-1}\left(1 - \frac{2}{M}\right) + \frac{2}{M}\left(\epsilon + \frac{\sqrt{M}}{2}\right)^{-1}$$

and so

$$a_0 = \epsilon\left[\left(1 - \frac{2}{M}\right) + \frac{2}{M}\frac{\epsilon}{\epsilon + \sqrt{M}/2}\right]^{-1}.$$

Since we are interested in the case in which M is large, one has

$$a_0 \simeq \epsilon.$$

Also

$$\frac{2c}{M}a_0 = -\frac{2}{M} + 4\epsilon/M^{3/2} - \frac{4}{M^2}(1 + 2\epsilon^2) + O(M^{-5/2}).$$

Our object is now to compute (1) at the values $z_s = \exp(-2\pi i s/M)$, $s = 0, 1, \ldots, M-1$. Notice that

$$\sum_{j=0}^{M-1} z_s^j = 1 + \sum_{j=1}^{M-1} z_s^j = \begin{cases} M & \text{if } s = 0 \\ 0 & \text{if } s = 1, \ldots, M-1. \end{cases}$$

This implies that as $M \to \infty$,

$$1 + \sum_{j=1}^{M-1} a_j z_s^j \cong \begin{cases} \frac{2\epsilon}{M^{1/2}} - \frac{1+4\epsilon^2}{M}, & \text{if } s = k \text{ or } M-k \\ 1 + \frac{2}{M}, & \text{if } s = 0, 1, \ldots, M-1 \\ & \text{but } s \neq k,\ M-k \end{cases}$$

Using (1), this implies that our power spectral estimate at z_s is

$$f(z_s) \cong \begin{cases} \frac{M}{4\epsilon} & \text{if } s = k, M-k \\ \epsilon & \text{if } s = 0, 1, \ldots, M-1 \\ & \quad s \neq k, M-k. \end{cases} \tag{6}$$

We have the following proposition.

PROPOSITION. *Let X_τ, $\tau = 0, 1, \ldots, M-1$, be a sequence of random variables with covariances given by (4). If one employs the solution $a_0, \ldots, a_{M-1}$ of the system of equations (3) to obtain an estimate of the power spectrum by using (1) at $z = z_s = \exp(-2\pi i s/M)$, $s = 0, 1, \ldots, M-1$ then asymptotically as $M \to \infty$, the power spectral estimate is given to first order by (6).*

By comparing the magnitudes of the estimates at the points z_s, $s = 0, 1, \ldots, M-1$ one can isolate the spectral peaks at $s = k, M-k$. It is clear that this estimation procedure can be straightforwardly applied if one has several (a few) harmonics. This indicates that under some circumstances the maximum entropy method is effective in determining the location of harmonics.

Related questions are taken up in the papers of Pisarenko (1973), Sakai (1984), and Newton and Pagano (1983).

REFERENCES

Burg, J. P., (1967), "Maximum entropy numerical analysis." Paper presented at the 37th Annual International S. E. G. Meeting, Oklahoma City, Oklahoma.

Burg, J. P. (1972), The relationship between maximum entropy spectra and maximum likelihood spectra, *Geophysics*, **37**, 375–376.

Newton, H. J. and Pagano, M. (1983), A method for determining periods in time series, *Journal of the American Statistical Association*, **78**, 152–157.

Pisarenko, V. F. (1973), The retrieval of harmonies from a covariance function, *Geophysics Journal of the Royal Astronomical Society*, **33**, 347–366.

Press, W., B. Flannery, S. Teukolsky, and W. Vetterling (1986), *Numerical Recipes*, Cambridge University Press.

Priestley M. B. (1981), *Spectral Analysis and Time Series*, Academic Press.

Sakai, H. (1984), Statistical analysis of Pisarenko's method for sinusoidal frequency estimation, *IEEE Transactions on Acoustics, Speech, and Signal Processing*, **32**, 95–101.

The Various Linear Fractional Lévy Motions

G. Samorodnitsky and M.S. Taqqu*
Boston University

1. INTRODUCTION

We treat in this paper a problem raised by Cambanis and Maejima (1988). Linear Fractional Lévy motions are α-stable self-similar processes with stationary increments and a "moving average" representation. The representation involves two real parameters a and b. When $\alpha = 2$, the processes are identical to the Gaussian Fractional Brownian motion for all values of a and b. We will show that different values of the parameters a and b yield different processes when $0 < \alpha < 2$ and when the skewness intensity is not necessarily zero.

AMS 1980 Subject Classifications: Primary 60G99, Secondary 60H05.

Key words and phrases: Self-similar processes, Fractional Brownian motion, stable processes, long-range dependence.

*Supported in part by the National Science Foundation grant DMS-88-05627 and the Air Force Office of Scientific Research grant 89-0115 at Boston University. The first author was also supported by a grant of the Dr. Haïm Weizmann Foundation for Scientific Research and the second author by a grant of the Guggenheim Foundation. Part of the research for this paper was carried out while the second author was visiting Harvard University.

Copyright © 1989 by Academic Press, Inc.
All rights of reproduction in any form reserved.
ISBN-0-12-058470-0

We define in this introduction self-similar processes, fractional Brownian motion and the α-stable integral. In the following section, we introduce the linear fractional Lévy motions and establish our result.

A process $\{X(t), -\infty < t < \infty\}$ is *self-similar* with index $H > 0$ if $X(ct) \stackrel{d}{=} c^H X(t)$, that is, if the finite-dimensional distributions of $X(ct)$ are identical to those of $c^H X(t)$ for all $c > 0$. We say that $\{X(t), -\infty < t < \infty\}$ is $H-$sssi if it is self-similar with index H and has stationary increments. Standard Brownian motion for example, is $1/2-$sssi. $H-$sssi processes have been extensively studied in the literature. See Taqqu (1986) for a bibliography. They satisfy $X(0) = 0$ and $EX(t) = 0$ (if the mean exists and $H \neq 1$). If $X(t)$ has also finite second moments, then

$$E(X(t_1) - X(t_2))^2 = |t_1 - t_2|^{2H} EX^2(1)$$

and consequently

$$EX(t_1)X(t_2) = \{\frac{1}{2}(|t_1|^{2H} + |t_2|^{2H} - |t_1 - t_2|^{2H})\}EX^2(1). \tag{1.1}$$

The expression in brackets in the R.H.S. of (1.1) is positive definite if $0 < H \leq 1$. The case $H = 1$ will be ignored because it corresponds to the trivial process $tX(1)$.

DEFINITION. *Fractional Brownian motion is the Gaussian $H-$sssi process, $0 < H < 1$. It is denoted $\{B_H(t), -\infty < t < \infty\}$.*

Fractional Brownian motion, which was discovered by Kolmogorov (1940) and studied by Hunt (1951) and Mandelbrot and Van Ness (1968), plays a fundamental role in the theory of self-similar processes. It is standard Brownian motion when $H = 1/2$. When $H \neq 1/2$, it can be represented as

$$B_H(t) = \int_{-\infty}^{+\infty} f_{2,H}(a,b;t,s)B(ds), \; -\infty < t < \infty, \tag{1.2}$$

where

(1.3)

$$f_{2,H}(a,b;t,s) = a[(t-s)_+^{H-1/2} - (-s)_+^{H-1/2}] + b[(t-s)_-^{H-1/2} - (-s)_-^{H-1/2}].$$

Here a, b are real constants, x_+ and x_- denote respectively the positive and negative part of x, and $B(ds)$ is Gaussian white noise. More precisely, $B(\;)$ is an independently scattered random measure with Lebesgue control measure, that is, for disjoint Borel sets $A_1, \ldots, A_n$, the random variables $B(A_1), \ldots, B(A_n)$ are independent and, moreover, the random variable $B(A)$ is $N(0, |A|)$ for any Borel set A. Now $B(cA) \stackrel{d}{=} c^{1/2}B(A)$ where cA denotes the set A expanded by c. Therefore the R.H.S. of (1.2) is self-similar with index H. Since it is also Gaussian and has stationary increments, we conclude that the R.H.S. of (1.2) does indeed represent $B_H(t)$.

What role do the coefficients a and b play in (1.3)? They play no significant role because they only affect the multiplicative constant $EB_H^2(1)$ in the covariance (1.1), and consequently different constants give rise to the *same* process, once we normalize that process by the standard deviation $\sqrt{EB_H^2(1)}$. Typically one chooses $a = 1$ and $b = 0$ or $a = b = 1$. In the latter case, the representation (1.2) becomes

$$B_H(t) = \int_{-\infty}^{+\infty} (|t-s|^{H-1/2} - |s|^{H-1/2}) B(ds).$$

What happens if we replace the Gaussian white noise $B(ds)$ by an α-stable white noise $M(ds)$, $0 < \alpha < 2$, or, more precisely by an independently scattered α-stable random measure M? An independently scattered α-stable random measure M with $0 < \alpha < 2$, Lebesgue control measure and skewness intensity $\beta : \mathbf{R}^1 \to [-1, 1]$ is such that:

(1) For any disjoint Borel sets $A_1, \ldots, A_n$, the random variables $M(A_1), \ldots, M(A_n)$ are independent.

(2) For any Borel set A with finite Lebesgue measure, the random variable $M(A)$ has characteristic function

$$E\exp\{i\theta M(A)\} = \begin{cases} \exp\{-\lambda(A)|\theta|^\alpha(1 - i\beta(A)\text{sign}(\theta)\tan\frac{\pi\alpha}{2})\} & \text{if } \alpha \neq 1 \\ \exp\{-\lambda(A)|\theta|(1 + i\beta(A)\frac{2}{\pi}\text{sign}(\theta)\ln|\theta|)\} & \text{if } \alpha = 1, \end{cases}$$

where λ denotes the Lebesgue measure and where $\beta(A) = \dfrac{1}{\lambda(A)} \displaystyle\int_A \beta(s)ds$.

The α-stable integral

$$X(t) = \int_{-\infty}^{+\infty} f(t,s) M(ds)$$

is well-defined if, for all t,

$$\int_{-\infty}^{+\infty} |f(s,t)|^\alpha ds < \infty$$

and, if $\alpha = 1$, also

$$\int_{-\infty}^{+\infty} |f(s,t)\ (\ln_+ |f(t,s)|)\beta(s)|ds < \infty,$$

where $\ln_+ a$ equals $\ln a$ if $a > 1$ and 0 if $0 < a \leq 1$. The integral defines a stochastic process whose finite-dimensional distribution $(X(t_1), \ldots, X(t_d))$ are multivariate stable, with joint characteristic function

$$E \exp\{i \sum_{j=1}^{d} \theta_j X(t_j)\}$$

equal to

$$\exp\{-\int_{-\infty}^{+\infty} |\sum_{j=1}^{d} \theta_j f(t_j, s)|^\alpha (1 - i\beta(s)\text{sign}(\sum_{j=1}^{d} \theta_j f(t_j, s)) \tan \frac{\pi\alpha}{2})ds\}$$

if $\alpha \neq 1$, and

$$\exp\{-\int_{-\infty}^{+\infty} |\sum_{j=1}^{d} \theta_j f(t_j, s)|(1 + i\beta(s)\frac{2}{\pi}\text{sign}(\sum_{j=1}^{d} \theta_j f(t_j, s))$$

$$\cdot \ln |\sum_{j=1}^{d} \theta_j f(t_j, s)|)ds\}$$

if $\alpha = 1$. (See for example Hardin (1984).)

For example, if $\beta(s) \equiv 0$, $t \geq 0$, and $f(t,s) = 1_{[0,t]}(s)$, then $X(t)$ is the α-stable Lévy motion, a process with independent stationary increments which is the α-stable counterpart to standard Brownian motion. It satisfies $X(ct) \stackrel{d}{=} c^{1/\alpha} X(t)$, and hence is a $1/\alpha$-sssi process.

The characteristic function can be complex when $\beta(s) \not\equiv 0$. Observe however that the relation

$$|E \exp\{i \sum_{j=1}^{d} \theta_j X(t_j)\}| = \exp\{-\int_{-\infty}^{+\infty} |\sum_{j=1}^{d} \theta_j f(t_j, s)|^\alpha ds\} \tag{1.4}$$

holds for all $0 < \alpha < 2$.

We now want to focus on (1.2), replace B by M and the exponent $H-1/2$ in (1.3) by $H - 1/\alpha$.

If the skewness intensity $\beta(s)$ is constant, then $M(cA) \stackrel{d}{=} c^{1/\alpha}M(A)$, and hence, if we replace, in (1.2), B by M and, in (1.3), the exponent $H - 1/2$ by $H - 1/\alpha$ with $0 < H < 1$, $H \neq 1/\alpha$, we will get an α-stable process which is H–sssi. Such a process is of interest in many applications because its increments can exhibit long-range dependence and high variability (see Taqqu (1987)). Do different values of a and b yield the same process? Cambanis and Maejima (1988) show that different constants a and b give rise to different processes when $1 < \alpha < 2$ and when $\beta \equiv 0$. We prove in the following section that this is also the case when $\alpha \in (0,2)$ and $\beta \not\equiv 0$.

2. THE VARIOUS LINEAR FRACTIONAL LÉVY MOTIONS

The α-stable counterpart to (1.2) will be called *Fractional Lévy motion* and we add the qualifier "*linear*" in order to distinguish it from other types of α-stable H–sssi processes. More precisely,

DEFINITION. *A Linear Fractional Lévy Motion (LFLM) is a stochastic process* $\{L_{\alpha,H}(a,b;t),\ -\infty < t < \infty\}$ *given by*

$$L_{\alpha,H}(a,b,t) = \int_{-\infty}^{+\infty} f_{\alpha,H}(a,b;t,s)M(ds) \tag{2.1}$$

where

$$\begin{aligned} &f_{\alpha,H}(a,b;t,s) \\ &= a[(t-s)_+^{H-1/\alpha} - (-s)_+^{H-1/\alpha}] + b[(t-s)_-^{H-1/\alpha} - (-s)_-^{H-1/\alpha}] \end{aligned}$$

with a,b real constants, $0 < \alpha \leq 2$, $0 < H < 1$, $H \neq 1/\alpha$ and where M is an α-stable random measure on $\mathbf{R}^1$ with Lebesgue control measure and with skewness intensity $\beta(x)$, $-\infty < x < \infty$, satisfying

(i) $\beta(x) \equiv 0$ *if* $\alpha = 1$

(ii) for all integers $d \geq 1$ and real $\theta_j, t_{1j}, t_{2j},\ j = 1, \ldots, d$,

$$\int_{-\infty}^{+\infty} \left[\sum_{j=1}^{d} \theta_j (f_{\alpha,H}(a,b;t_{2j},s) - f_{\alpha,H}(a,b;t_{1j},s)) \right]^{<\alpha>} \beta(cs+h)ds$$

is independent of $c > 0$ and $-\infty < h < \infty$.

The notation $x^{<\alpha>}$ stands for $|x|^\alpha \text{sign} x$. The LFLM (2.1) is well-defined because

$$\int_{-\infty}^{+\infty} |f_{\alpha,H}(a,b;t,s)|^\alpha ds < \infty.$$

Note that (ii) always holds if $\beta(\ )$ is constant.

NORMALIZED LFLM. This is the process

$$X(t) = \frac{1}{K_{\alpha,H}(a,b)} L_{\alpha,H}(a,b;t)\ ,\ -\infty < t < \infty$$

where $K_{\alpha,H}(a,b)$ is such that the scaling parameter of $X(1)$ equals 1.

The following theorem shows that the result of Cambanis and Maejima (1988) extends to $\alpha \leq 1$ and $\beta \not\equiv 0$. The proof is similar to theirs but the starting point is different.

THEOREM. *Let $0 < \alpha < 2$, $0 < H < 1$, $H \neq 1/\alpha$, and let a, a', b, b' be real numbers satisfying $|a| + |b| > 0$ and $|a'| + |b'| > 0$. Then the normalized LFSM's*

$$\frac{1}{K_{\alpha,H}(a,b)} L_{\alpha,H}(a,b;t)\ ,\ -\infty < t < \infty \tag{2.2}$$

and

$$\frac{1}{K_{\alpha,H}(a',b')} L_{\alpha,H}(a',b';t)\ ,\ -\infty < t < \infty \tag{2.3}$$

have identical finite-dimensional distributions, if and only if one of the following conditions holds:

(i) $a = a' = 0$,
(ii) $b = b' = 0$,
(iii) a, a', b', b' *are non zero and* $\frac{a}{b} = \frac{a'}{b'}$.

The proof makes use of the following lemma due to Kanter (1973).

LEMMA (Kanter). *Let* $f, g \in L^{\alpha}(\mathbf{R})$. *If for some* $0 < \alpha < 2$, *any integer* n, *and any reals* $\theta_1, \ldots, \theta_n$, $t_1, \ldots, t_n$,

$$\int_{-\infty}^{+\infty} |\sum_{j=1}^{n} \theta_j f(t_j + s)|^{\alpha} ds = \int_{-\infty}^{+\infty} |\sum_{j=1}^{n} \theta_j g(t_j + s)|^{\alpha} ds,$$

then there exists a number ϵ *equal to* -1 *or* $+1$ *and a real number* τ *such that*

$$f(s) = \epsilon g(s + \tau) \ a.e. \ s.$$

PROOF OF THE THEOREM. Suppose that one of the three conditions hold, for example (iii). Then

$$\frac{1}{K_{\alpha,H}(a,b)} L_{\alpha,H}(a,b;t) \stackrel{d}{=} \frac{b}{K_{\alpha,H}(a,b)} L_{\alpha,H}(\frac{a}{b},1;t)$$

has same finite dimensional distributions (f.d.d.) as

$$\frac{1}{K_{\alpha,H}(a',b')} L_{\alpha,H}(a',b';t) \stackrel{d}{=} \frac{b'}{K_{\alpha,H}(a',b')} L_{\alpha,H}(\frac{a}{b},1;t)$$

because they differ by a multiplicative constant which must be equal to 1 since both processes have been normalized.

To prove the converse, suppose $L_{\alpha,H}(a,b;t)$ and $L_{\alpha,H}(a',b';t)$ have the same f.d.d.. Then the stationary processes

$$Y(a,b,t) = K_{\alpha,H}^{-1}(a,b)[L_{\alpha,H}(a,b;t+h) - L_{\alpha,H}(a,b;t)], \ -\infty < t < \infty$$

and

$$Y(a',b',t) = K_{\alpha,H}^{-1}(a',b')[L_{\alpha,H}(a',b';t+h) - L_{\alpha,H}(a',b';t)], \ -\infty < t < \infty$$

will also have identical f.d.d., for any given $h > 0$. We want to show that this implies that one of the conditions (i)-(iii) holds.

The process $Y(a,b;t)$ can be represented as

$$Y(a,b;t) = K_{\alpha,H}^{-1}(a,b)\int_{-\infty}^{+\infty} g(a,b;t-s)M(ds)$$

where

$$g(a,b;t) = a[(t+h)_+^{H-1/\alpha} - t_+^{H-1/\alpha}] + b[(t+h)_-^{H-1/\alpha} - t_-^{H-1/\alpha}].$$

Since $Y(a,b;t)$ and $Y(a',b';t)$ have same f.d.d.,

$$|E\exp\{i\sum_{j=1}^{n}\theta_j Y(a,b;t_j)\}| = |E\exp\{i\sum_{j=1}^{n}\theta_j Y(a',b';t_j)\}|,$$

i.e., using (1.4),

$$\exp\{-\int_{-\infty}^{+\infty}|\sum_{j=1}^{n}\theta_j K_{\alpha,H}^{-1}(a,b)g(a,b;t_j-s)|^\alpha ds\}$$

$$= \exp\{-\int_{-\infty}^{+\infty}|\sum_{j=1}^{n}\theta_j K_{\alpha,H}^{-1}(a',b',)g(a',b';t_j-s)|^\alpha ds\},$$

for all integers n and reals $\theta_1,\ldots,\theta_n,t_1,\ldots,t_n$. Hence, by Kanter's Lemma, there exists a real number τ (depending on h,a,b,a',b') and an $\epsilon\in\{+1,-1\}$ such that

$$K_{\alpha,H}^{-1}(a,b)g(a,b;t) = \epsilon K_{\alpha,H}^{-1}(a',b')g(a',b';t-\tau) \text{ a.e. } t,$$

i.e.,

$$\text{(2.4)}\quad K_{\alpha,H}^{-1}(a,b)\Big[a[(t+h)_+^{H-1/\alpha} - t_+^{H-1/\alpha}] + b[(t+h)_-^{H-1/\alpha} - t_-^{H-1/\alpha}]\Big]$$

$$= \epsilon K_{\alpha,H}^{-1}(a',b')\Big[a'[(t+h-\tau)_+^{H-1/\alpha} - (t-\tau)_+^{H-1/\alpha}]$$

$$+b'[(t+h-\tau)_-^{H-1/\alpha} - (t-\tau)_-^{H-1/\alpha}]\Big] \text{ a.e. } t.$$

Since both sides of (2.4) represent functions continuous everywhere except at the four points $t=-h,\ 0,\tau-h,\tau$, relation (2.4) can be extended by continuity to all t not equal to these four points.

We claim that this implies

$$\left\{ \frac{a}{K_{\alpha,H}(a,b)} = \epsilon \frac{a'}{K_{\alpha,H}(a',b')}\, , \; \frac{b}{K_{\alpha,H}(a,b)} = \epsilon \frac{b'}{K_{\alpha,H}(a',b')} \right\}. \tag{2.5}$$

Indeed, differentiating (2.4) with respect to $t \neq -h, 0, \tau - h, \tau$, we get (2.6)

$$K^{-1}_{\alpha,H}(a,b)\Big[a[(t+h)_+^{H-1/\alpha-1} - t_+^{H-1/\alpha-1}] - b[(t+h)_-^{H-1/\alpha-1} - t_-^{H-1/\alpha-1}]\Big]$$

$$= \epsilon K^{-1}_{\alpha,H}(a',b')\Big[a'[(t+h-\tau)_+^{H-1/\alpha-1} - (t-\tau)_+^{H-1/\alpha-1}]$$

$$-b'[(t+h-\tau)_-^{H-1/\alpha-1} - (t-\tau)_-^{H-1/\alpha-1}]\Big]$$

for all t except $t = -h, 0, \tau - h, \tau$. Suppose W.L.O.G. that $a \neq 0$ and let $t \downarrow 0$. Then the L.H.S. of (2.6) tends to $\pm\infty$. For the R.H.S. to do the same it is necessary that $\tau = 0$ or $\tau = h$. To see that $\tau \neq h$, set $\tau = h$ and let $t \downarrow h$. This makes the L.H.S. of (2.6) tend to a finite number and the R.H.S. tend to $\pm\infty$ which is not possible. Hence $\tau = 0$ and therefore (2.5) holds.

Relation (2.5) implies that either
(i) $a = a' = 0$
(ii) $b = b' = 0$
(iii) a, a', b, b' are non-zero and

$$\epsilon \frac{K_{\alpha,H}(a,b)}{K_{\alpha,H}(a',b')} = \frac{a}{a'} = \frac{b}{b'}.$$

∎

REFERENCES

Cambanis, S. and Maejima, M. (1988), Two classes of self-similar stable processes with stationary increments, *Technical Report* **220**, Center for Stochastic Processes, University of North Carolina, Chapel Hill.

Hardin, C.D. Jr. (1984), Skewed stable variables and processes, *Technical Report* **79**, Center for Stochastic Processes, University of North Carolina, Chapel Hill.

Hunt, G.A. (1951), Random Fourier transform, *Transactions of the American Mathematical Society* **71**, 38-69.

Kanter, M. (1973), The L^p norm of sums of translates of a function, *Transactions of the American Mathematical Society*, **179**, 35-47.

Kolmogorov, A.N. (1940), Wienersche Spiralen und einige andere interessante Kurven in Hilbertschen Raum, *Comptes Rendus (Doklady) de l'Académie des Sciences de l' URSS (N.S.)*, **26**, 115-118.

Mandelbrot, B.B. and Van Ness, J.W. (1968), Fractional Brownian motion, fractional noises and applications, *SIAM Review*, **10**, 422-437.

Taqqu, M.S. (1986), A bibliographical guide to self-similar processes and long-range dependence, in *Dependence in Probability and Statistics, Progress in Probability and Statistics Series* (E. Eberlein and M.S. Taqqu, eds.), Birkhäuser, Boston, 137-162.

Taqqu, M.S. (1987), Random processes with long-range dependence and high variability, *Journal of Geophysical Research*, **92**, D8, 9683-9686.

Bonferroni-Type Probability Bounds as an Application of the Theory of Tchebycheff Systems

Stephen M. Samuels and William J. Studden*
Purdue University

1. INTRODUCTION

There is a sizeable literature on universal upper and lower bounds for the probability that at least m of N arbitrary events will occur (or that exactly m will occur) and related problems, subject to certain moment conditions on the random variable $X =$ the number of events which occur. The most famous of these are the classical Bonferroni (1937) inequalities:

$$(1) \qquad S_1 - S_2 + \cdots - S_{2m} \leq P(X \geq 1) \leq S_1 - S_2 + \cdots + S_{2m+1}$$

where S_k is the binomial moment, $E\binom{X}{k}$. Such bounds are of considerable theoretical interest and seem to be especially important for systems of events where N is large and only the first few values of the S_k's are given.

AMS 1980 Subject Classifications: Primary 60E15, Secondary 41A50.
Key words and phrases: Probability inequalities, moment inequalities, Bonferroni, Tchebycheff systems.
*Research supported by NSF Grants DMS 8503771 and 8802535.

PROBABILITY, STATISTICS, AND MATHEMATICS
Papers in Honor of Samuel Karlin

Copyright © 1989 by Academic Press, Inc.
All rights of reproduction in any form reserved.
ISBN-0-12-058470-0

Another important body of literature concerns moment inequalities in general, namely problems of the form:

> Find max(min)$E\psi(X)$ among all random variables, X, with support in Ω and prescribed moments
>
> $$E\phi_i(X) = s_i, \qquad i = 0, 1, \ldots, n,$$
>
> where $\phi_0, \ldots, \phi_n$ and ψ are a given set of functions with $\phi_0 \equiv 1$ and $s_0 = 1$.

If we take Ω to be the integers $0, 1, \ldots, N$, then any probability distribution on Ω, say $q(\cdot)$, can be the distribution of the number of events, say $A_1, A_2, \ldots, A_N$, which occur. (Indeed there are many ways of consistently assigning probabilities to the elements of the field generated by the A_i's. For example, we could take the set of N-digit binary integers as our sample space and define A_i to consist of those elements whose i-th digit is a one. If we assign probability $q(j)/\binom{N}{j}$ to each of the $\binom{N}{j}$ integers with j ones, then, not only is

$$P(\sum_{i=1}^{N} I_{A_i} = j) = q(j),$$

but, in addition, the A_i's are exchangeable.) Thus Bonferroni-type problems are just a special case of this general problem. So it is rather surprising that the two bodies of literature are, as far as we can tell, disjoint. Our chief aim, in this paper, is to remedy that situation by showing how the Bonferroni-type theory and proofs can be illuminated and simplified using the general theory of Tchebycheff systems.

Here is a widely known simple approach: For any random variables, $X \leq Y \Rightarrow EX \leq EY$; hence it is elementary that

$$\sum_{i=0}^{n} c_i\phi_i(x) \geq (\leq)\psi(x) \;\; \forall x \epsilon \Omega \quad \Rightarrow \quad \sum_{i=0}^{n} c_i s_i \geq (\leq) E\psi(X) \tag{2}$$

The bounds in (2) are sharp (best possible) if there are measures satisfying the moment conditions, and supported entirely on

$$\{x \epsilon \Omega : \sum_{i=0}^{n} c_i\phi_i(x) = \psi(x)\}. \tag{3}$$

For example, the well-known identity,

$$\sum_{r=0}^{k} \binom{m}{r}(-1)^r = \binom{m-1}{k}(-1)^k$$

for positive integers $k \leq m$, implies that, for each $m = 1, 2, \ldots$,

$$(4) \qquad \binom{x}{1} - \binom{x}{2} + \cdots - \binom{x}{2m} \leq$$
$$I_{\{x \geq 1\}} \leq \binom{x}{1} - \binom{x}{2} + \cdots + \binom{x}{2m+1} \qquad \forall x \epsilon \Omega$$

from which (1) follows by taking expectations. Notice that equality holds on the left side of (4) just for $x = 0, 1, \ldots, 2m$, and on the right side for $x = 0, 1, \ldots, 2m + 1$. Hence, only for certain specifications of the moments will the bounds in (1) be sharp. (Naive use of (1) can have unpleasant consequences; see Schwager (1984) for a nice example.) In particular, the lower bound $S_1 - S_2$ can be achieved if and only if $2S_2 \leq S_1$ (which implies $S_1 \leq 2$), in which case probabilities $1 - S_1 + S_2$, $S_1 - 2S_2$, and S_2 can be put at 0, 1, and 2, respectively. And the upper bound $S_1 - S_2 + S_3$ is achieved, with probabilities $1 - (S_1 - S_2 + S_3)$, $S_1 - 2S_2 + 3S_3$, $S_2 - 3S_3$, and S_3, at $0, 1, 2$, and 3, respectively, if and only if $3S_3 \leq S_2$ and $S_1 - S_2 + S_3 \leq 1$.

For other specifications of the moments there are other linear combinations of the $\phi_i(x) = \binom{x}{i}$'s for which the left side of (2) holds, and the right side is an equality (see Examples 5 and 6 in Section 3). This is guaranteed by a duality theorem (see Karlin and Studden (1966), Theorem 2.1 of Chapter 12, plus discussion of the origins of the theorem) which says that in general (i.e. for quite arbitrary Ω) whenever there is *any* linear combination of the ϕ_i's for which the left side of (2) holds, then there is essentially *always* a linear combination for which equality holds in the right side, and a corresponding measure for which the bound is attained. (This must be slightly modified in case the ϕ_i's are linearly dependent or the prescribed vector of moments is on the boundary of the space of all possible moment vectors.)

For the special case $\Omega = \{0, 1, \ldots, N\}$ which we are considering here, the above theorem is well-known as the duality theorem of linear programming. See Platz (1985) and Hadley (1962).

The key question, then, is how to find *which* linear combination of the ϕ_i's gives a sharp bound. Typically there are only a finite number of choices to consider. Various approaches, including linear programming, have been proposed. In the next section we will show how the theory of Tchebycheff systems provides a powerful and unified approach; and, in Section 3, we will illustrate it with numerous examples.

2. T-SYSTEMS ON $\{\Omega = 0, 1, \ldots, N\}$

By definition, a system of functions $\{\phi_i : i = 0, 1, \ldots, n\}$ defined on a subset Ω of the reals is a Tchebycheff system (abbreviated T-system) if the determinants $\det \|\phi_i(x_j)\|$ are strictly positive for all choices of x_i's in Ω with $x_0 < x_1 < \ldots < x_n$. If these determinants are merely non-negative for all choices of ordered x_i values, and are not linearly dependent on Ω, then the system is called a Weak-Tchebycheff or WT-system.

The simplest example of a T-system are the powers, $\phi_i(x) = x^i$, for which the determinants are the well-known Vandermonde determinants. A closely related example of a T-system, which we shall discuss in some detail, are the functions $\phi_i(x) = \binom{x}{i}$. This latter system is simply a linear transformation of the original powers. An interesting example of a WT-system is obtained by considering the subsequence $\phi_i(x) = \binom{x}{k_i}$, $0 = k_0 < k_1 < \ldots < k_n$. See Karlin and Studden (1966).

The majority of the properties of T-systems described below also hold for WT-systems. This can be demonstrated by perturbing the functions slightly so that the determinant inequalities become strict, and then applying limiting arguments.

All T-systems have been shown to share many of the basic properties of the ordinary powers. For example, linear combinations of $\{\phi_i(x) : i = 0, 1, \ldots, n\}$ can have at most n zeroes and Gauss quadrature formulas are available. Further, the Kreĭn-Markov-Stieltjes Theorem, described below, holds.

Interest centers on the set of all possible *moment* vectors:

$$\mathcal{M} = \{\mathbf{s} = (s_0, s_1, \ldots, s_n) : s_i = \int_\Omega \phi_i \, d\sigma, \quad i = 0, 1, \ldots, n\} \tag{5}$$

where σ varies over the set of non-negative measures on Ω. For any given

$s \epsilon \mathcal{M}$, we have the set

$$\Sigma(\mathrm{s}) = \{\sigma : \int_{\Omega} \phi_i \, d\sigma = s_i \quad i = 0, 1, \ldots, n\}$$

of all *representations* of s. A highlight of the theory of T-systems is that there are certain special representations which simultaneously achieve the upper or lower bounds

$$\max(\min)\{\int_{\Omega} \psi \, d\sigma : \sigma \epsilon \Sigma(\mathrm{s})\} \tag{6}$$

for a whole class of ψ's. To describe these special representations, we need the concept of the *index* of a finite subset of Ω (which henceforth will always be taken to be of the form $\Omega = 0, 1, \ldots, N$) and of a representation supported on that subset.

2.1. Index of a Representation

First of all, we define the index of a block of consecutive integers, $x, x + 1, \ldots, x + p - 1$, in Ω, to be simply p, the number of elements in the block, *except* if p is odd *and* the block is interior ($0 < x$; $x + p - 1 < N$), in which case the index is defined to be $p + 1$. So, for example, both a single interior element and a pair of adjacent interior elements have index two.

An arbitrary subset of Ω is uniquely expressible as a disjoint union of separated blocks; its index is defined to be the sum of the indices of those blocks.

The index of a representation is defined to be the index of its support set.

2.2. Principal and Canonical Representations

The general theory, as presented in Karlin and Studden (1966) and Kreĭn and Nudel'man (1977), can be described briefly as follows:

A point s is on the boundary of $\mathcal{M}$ if and only if it has a unique representation of index at most n. If so, this representation is the unique member of $\Sigma(\mathrm{s})$.

For each s in the interior of $\mathcal{M}$, there are precisely two representations, $\overline{\sigma} \equiv \overline{\sigma}(\mathrm{s})$ and $\underline{\sigma} \equiv \underline{\sigma}(\mathrm{s})$, of index $n + 1$; these are called, respectively, the *upper* and *lower principal representations.* One of these, $\overline{\sigma}$, always has a support set which includes N in a block with an *odd* number of elements.

For each s in the interior of $\mathcal{M}$, and each $m\epsilon\Omega$, there is a unique *canonical* representation $\sigma_m \equiv \sigma_m(\mathrm{s}) \epsilon \Sigma(\mathrm{s})$ whose mass at m, $\sigma_m\{m\}$, is *maximal.* (Every other representation in $\Sigma(\mathrm{s})$ has strictly smaller mass at m.) These *canonical* representations have index $n+1$ or $n+2$, and are mixtures of a measure concentrated on m and a measure (necessarily of index at most n) whose moment vector is on the boundary of $\mathcal{M}$. (Intuitively, think of extending a line from $(\phi_0(m), \phi_1(m), \ldots, \phi_n(m))$ through the point s until it hits the boundary of $\mathcal{M}$.)

In all cases, σ_N is the upper principal representation. If n is *even*, then σ_0 is the lower principal representation; for n odd, this may not be the case.

For example, suppose $n = 2$, $\phi_i(x) = x^i$, $i = 0, 1, 2$, and $s_0 = 1$. Then $\{(s_1, s_2) : (1, s_1, s_2)\epsilon\mathcal{M}\}$ is just the convex hull of $\{(x, x^2) : x = 0, 1, \ldots, N\}$. A subset of Ω of index one must be either $\{0\}$ or $\{N\}$; a subset of index two consists of a pair of adjacent points or a single interior point. Any σ with support on a set of index one or two is clearly a boundary point of $\mathcal{M}$. Any interior point of $\mathcal{M}$ has its lower principal representation on a set of the form $\{0, i, i+1\}$, where i is the integer part of s_2/s_1, and its upper principal representation on $\{j, j+1, N\}$, where j is the integer part of $(Ns_1 - s_2)/(N - s_1)$.

2.3. T-Systems and the Kreĭn-Markov-Stieltjes Theorem

Suppose we adjoin to the T-system $\phi_0, \phi_1, \ldots, \phi_n$ another function, ψ, such that $\phi_0, \phi_1, \ldots, \phi_n, \psi$ is also a T-system. Then a result of considerable importance is the following:

$$\int \psi \, d\underline{\sigma} \leq \int \psi \, d\sigma \leq \int \psi \, d\overline{\sigma} \quad \forall\sigma\epsilon\Sigma(\mathrm{s}). \tag{7}$$

Thus the upper and lower principal representations achieve the bounds (6) for all such ψ's.

A second important result, a special case of the Kreĭn-Markov-Stieltjes Theorem, says that, for each $m\epsilon\Omega$ with $m > 0$,

$$\int_{[m,N]} \psi \, d\sigma_{m-1} \leq \int_{[m,N]} \psi \, d\sigma \leq \int_{[m,N]} \psi \, d\sigma_m \quad \forall\sigma\epsilon\Sigma(\mathrm{s}). \tag{8}$$

Suppose now that $\phi_0, \phi_1, \ldots, \phi_n, \psi$ is a T-system. It is elementary that we may replace ψ by $\phi_0 + \varepsilon\psi$, where ε is an arbitrary positive constant, and still have a T-system. Substituting this into (8), letting $\varepsilon \to 0$, and using the fact that the canonical measures σ_m and σ_{m-1} do not depend on ε, we obtain the following corollary:

$$\int_{[m,N]} \phi_0 \, d\sigma_{m-1} \leq \int_{[m,N]} \phi_0 \, d\sigma \leq \int_{[m,N]} \phi_0 \, d\sigma_m \quad \forall \sigma \epsilon \Sigma(\mathbf{s}). \tag{9}$$

In the case of primary interest: $\phi_0 \equiv 1$ and $s_0 = 1$, this becomes:

$$\sum_{x=m}^{N} \sigma_{m-1}(\{x\}) \leq P(X \geq m) \leq \sum_{x=m}^{N} \sigma_m(\{x\}) \quad \forall X \ni E\phi_i(X) = s_i, \quad i = 1, \ldots, n. \tag{10}$$

Bounds for the individual terms, $P(X = m)$, can readily be obtained. Indeed, the general description of σ_m in Section 2.2 above indicates that the upper bound is $\sigma_m(\{m\})$. Kreĭn and Nudel'man (1977) show that the lower bound is given by either $\sigma_{m-1}(\{m\})$ or $\sigma_{m+1}(\{m\})$. These are always equal. Thus we have

$$\sigma_{m-1}(\{m\}) \leq P(X = m) \leq \sigma_m(\{m\}) \tag{11}$$

3. EXAMPLES

All of the examples to follow are concerned with the functions $\phi_i(x) = \binom{x}{i}, i = 0, 1, \ldots, n$ or subsets of these. The variable x ranges over $\Omega = \{0, 1, \ldots, N\}$ and $S_i = E\binom{X}{i}$. Throughout these examples we take $S_0 = 1$ as a prescribed moment condition; i.e. we limit ourselves to probability distributions. Hence we may, for convenience, refer to a random variable, X, which has the prescribed moments. We shall proceed as if the vector of prescribed moments is in the interior of the moment space defined in (5). If the system of functions is a T-system and the moment vector is a boundary point, there is a unique distribution with the prescribed moments and any tight upper and lower bounds that are given are in fact equal.

The first three examples can be found in Móri and Székely (1985). Notice how each is instrumental in solving the next. Example 4 is in Platz

(1985) and Prekopa (1987). Example 5 is in Dawson and Sankoff (1967), Kwerel (1975), Galambos (1977), Sathe, Pradhan, and Shah (1980), and Prekopa (1987); a weaker result appears in Chung and Erdös (1952). Example 6 is in Prekopa (1987).

Results in Kounias (1968) and Worsley (1982) lie outside the scope of this paper because their constraints are not on moments of functions of X. But see the comment in Section 3.5.

3.1. Bounds on S_ℓ in terms of S_k: $k < \ell$

Both $\{1, \binom{x}{k}\}$ and $\{1, \binom{x}{k}, \binom{x}{\ell}\}$ are T-systems or WT-systems. So, by (7), the upper and lower bounds on S_ℓ are achieved by the upper and lower principal representations of $(1, S_k)$, respectively; these have index $n+1 = 2$. From Section 2.2, the upper principal representation has support at 0 and N, while the lower one must have support on the unique pair $i-1$, i for which the prescribed moment condition, S_k, can be satisfied, namely $i \ni$

$$\binom{i-1}{k} < S_k \leq \binom{i}{k}. \tag{12}$$

The bounds can now be derived by first computing the uniquely determined probabilities on these support sets. By equation (7) we obtain

$$(1-\beta)\binom{i-1}{\ell} + \beta\binom{i}{\ell} \leq S_\ell \leq \alpha\binom{N}{\ell} \tag{13}$$

where i satisfies (12), $\alpha = S_k/\binom{N}{k}$, and $\beta = [S_k - \binom{i-1}{k}]/[\binom{i}{k} - \binom{i-1}{k}]$.

Alternatively, we may appeal directly to (2). We want inequalities of the form

$$c_0 + c_1\binom{x}{k} \geq (\leq)\binom{x}{\ell} \quad \forall x \epsilon \Omega = \{0, 1, \ldots, N\}$$

with equality holding at 0 and N (alt: at $i-1$ and i). Since $\{1, \binom{x}{k}, \binom{x}{\ell}\}$ is a WT-system, these are easy to construct; we have, for each $x\epsilon\Omega$, and for each $i = 1, 2, \ldots, N$,

$$\begin{vmatrix} 1 & 1 & 1 \\ 0 & \binom{x}{k} & \binom{N}{k} \\ 0 & \binom{x}{\ell} & \binom{N}{\ell} \end{vmatrix} \geq 0 \quad \text{and} \quad \begin{vmatrix} 1 & 1 & 1 \\ \binom{x}{k} & \binom{i-1}{k} & \binom{i}{k} \\ \binom{x}{\ell} & \binom{i-1}{\ell} & \binom{i}{\ell} \end{vmatrix} \geq 0. \tag{14}$$

Taking expectations, as in (2), we have, immediately, the upper bound:

$$\begin{vmatrix} 1 & 1 & 1 \\ 0 & S_k & \binom{N}{k} \\ 0 & S_\ell & \binom{N}{\ell} \end{vmatrix} \geq 0 \tag{15}$$

which can be rewritten as $S_\ell \leq \left[\binom{N}{\ell}/\binom{N}{k}\right] S_k$, and a collection of lower bounds:

$$\begin{vmatrix} 1 & 1 & 1 \\ S_k & \binom{i-1}{k} & \binom{i}{k} \\ S_\ell & \binom{i-1}{\ell} & \binom{i}{\ell} \end{vmatrix} \geq 0 \tag{16}$$

which, for $i > k$, can be rewritten as the left side of (13) or

$$S_\ell \geq \frac{\binom{i-1}{\ell-1}}{\binom{i-1}{k-1}} S_k - \left[\frac{\binom{i-1}{k}\binom{i}{\ell} - \binom{i}{k}\binom{i-1}{\ell}}{\binom{i-1}{k-1}}\right].$$

The general theory assures us that the right side is maximized when i satisfies (12).

3.2. Bounds on S_m in terms of S_k and S_ℓ : $k < \ell$

We shall proceed by assuming that $m > \ell$. If $m < \ell$, the same bounds hold. For $k < m < \ell$ the upper and lower bounds are interchanged. Since S_k and S_ℓ are prescribed moment values, necessarily, $0 \leq S_k \leq \binom{N}{k}$, and S_ℓ must satisfy (15), as well as (16). Both $\{1, \binom{x}{k}, \binom{x}{\ell}\}$ and $\{1, \binom{x}{k}, \binom{x}{\ell}, \binom{x}{m}\}$ are WT-systems; so, by (7), the upper and lower bounds on S_m are achieved by the upper and lower principal representations, respectively: these have index 3. From Section 2.2, the lower principal representation has support on a set of the form $\{0, i-1, i\}$, while the upper one has support on $\{j-1, j, N\}$ for some j. By the general theory, there are $i = i_0$ and $j = j_0$ for which the prescribed moment conditions can be satisfied. And the theory can also help us find these values, as follows:

Let

$$A_i = \begin{pmatrix} 1 & 1 & 1 \\ 0 & \binom{i-1}{k} & \binom{i}{k} \\ 0 & \binom{i-1}{\ell} & \binom{i}{\ell} \end{pmatrix} \quad \text{and} \quad B_j = \begin{pmatrix} 1 & 1 & 1 \\ \binom{j-1}{k} & \binom{j}{k} & \binom{N}{k} \\ \binom{j-1}{\ell} & \binom{j}{\ell} & \binom{N}{\ell} \end{pmatrix}.$$

For each $i \geq \ell$ and $j \geq k$, $\det(A_i) > 0$ and $\det(B_j) > 0$, so the equations

$$A_i \begin{pmatrix} p \\ q \\ r \end{pmatrix} = \begin{pmatrix} 1 \\ S_k \\ S_\ell \end{pmatrix} \quad \text{and} \quad B_j \begin{pmatrix} t \\ u \\ v \end{pmatrix} = \begin{pmatrix} 1 \\ S_k \\ S_\ell \end{pmatrix}$$

have solutions, say $p(i), q(i), r(i)$, and $t(j), u(j), v(j)$ where

$$|A_i|\,p(i) = \begin{vmatrix} 1 & 1 & 1 \\ S_k & \binom{i-1}{k} & \binom{i}{k} \\ S_\ell & \binom{i-1}{\ell} & \binom{i}{\ell} \end{vmatrix}$$

$$|A_i|\,q(i) = \begin{vmatrix} 1 & 1 & 1 \\ 0 & S_k & \binom{i}{k} \\ 0 & S_\ell & \binom{i}{\ell} \end{vmatrix}$$

$$\begin{aligned} |A_i|\,r(i) &= \begin{vmatrix} 1 & 1 & 1 \\ 0 & \binom{i-1}{k} & S_k \\ 0 & \binom{i-1}{\ell} & S_\ell \end{vmatrix} \\ &= -|A_{i-1}|\,q(i-1); \end{aligned} \tag{17}$$

and

$$|B_j|\,v(j) = \begin{vmatrix} 1 & 1 & 1 \\ \binom{j-1}{k} & \binom{j}{k} & S_k \\ \binom{j-1}{\ell} & \binom{j}{\ell} & S_\ell \end{vmatrix}$$

$$|B_j|\,t(j) = \begin{vmatrix} 1 & 1 & 1 \\ S_k & \binom{j}{k} & \binom{N}{k} \\ S_\ell & \binom{j}{\ell} & \binom{N}{\ell} \end{vmatrix}$$

$$\begin{aligned} |B_j|\,u(j) &= \begin{vmatrix} 1 & 1 & 1 \\ \binom{j-1}{k} & S_k & \binom{N}{k} \\ \binom{j-1}{\ell} & S_\ell & \binom{N}{\ell} \end{vmatrix} \\ &= -|B_{j-1}|\,t(j-1). \end{aligned} \tag{18}$$

($|A_{\ell-1}|\,q(\ell-1)|$ and $|B_{k-1}|\,t(k-1)$ are well-defined.)

For the lower and upper principal representations, we want to find the i_0 and j_0 for which the solutions are non-negative. This is always true of $p(i)$ and $v(j)$, by (16). By (17) and (18) non-negativity occurs at any $i \geq \ell$ and $j \geq k$ for which $q(i-1) < 0 \leq q(i)$ and $t(j-1) \leq 0 < t(j)$. This is clearly the case because $q(N) \geq 0$ by (15), $t(N-1) > 0$ by (16), $|B_{k-1}|\,t(k-1) = -|A_N|\,q(N)$, and $|A_{\ell-1}|\,q(\ell-1) < 0$ by inspection. Thus $\ell \leq i_0 \leq N$ and $k \leq j_0 < N$, i_0 is the minimal i such that

$$\begin{vmatrix} S_k & \binom{i}{k} \\ S_\ell & \binom{i}{\ell} \end{vmatrix} \geq 0 \tag{19}$$

and j_0 is the minimal j such that

$$\begin{vmatrix} 1 & 1 & 1 \\ S_k & \binom{j}{k} & \binom{N}{k} \\ S_\ell & \binom{j}{\ell} & \binom{N}{\ell} \end{vmatrix} \geq 0. \tag{20}$$

Solving for $(p(i_0), q(i_0), r(i_0))$ and $(t(j_0), u(j_0), v(j_0))$ we obtain $\underline{\sigma}$ and $\overline{\sigma}$. By (7),

$$E_{\underline{\sigma}}\binom{X}{m} \leq S_m \leq E_{\overline{\sigma}}\binom{X}{m} \tag{21}$$

As in the previous example, a convenient expression for (21) follows directly from (2). We want inequalities of the form

$$c_0 + c_1\binom{x}{k} + c_2\binom{x}{\ell} \geq (\leq)\binom{x}{m} \quad \forall x \epsilon \Omega = \{0, 1, \ldots, N\}$$

with equality holding at $j-1$, j, and N (alt: at 0, $i-1$, and i). The required inequalities are

$$\begin{vmatrix} 1 & 1 & 1 & 1 \\ \binom{x}{k} & \binom{j-1}{k} & \binom{j}{k} & \binom{N}{k} \\ \binom{x}{\ell} & \binom{j-1}{\ell} & \binom{j}{\ell} & \binom{N}{\ell} \\ \binom{x}{m} & \binom{j-1}{m} & \binom{j}{m} & \binom{N}{m} \end{vmatrix} \geq 0 \quad \text{and} \quad \begin{vmatrix} 1 & 1 & 1 & 1 \\ 0 & \binom{i-1}{k} & \binom{i}{k} & \binom{x}{k} \\ 0 & \binom{i-1}{\ell} & \binom{i}{\ell} & \binom{x}{\ell} \\ 0 & \binom{i-1}{m} & \binom{i}{m} & \binom{x}{m} \end{vmatrix} \geq 0,$$

Taking expectations, as in (2), yields the lower and upper bounds

$$\begin{vmatrix} 1 & 1 & 1 & 1 \\ S_k & \binom{j-1}{k} & \binom{j}{k} & \binom{N}{k} \\ S_\ell & \binom{j-1}{\ell} & \binom{j}{\ell} & \binom{N}{\ell} \\ S_m & \binom{j-1}{m} & \binom{j}{m} & \binom{N}{m} \end{vmatrix} \geq 0 \quad \text{and} \quad \begin{vmatrix} \binom{i-1}{k} & \binom{i}{k} & S_k \\ \binom{i-1}{\ell} & \binom{i}{\ell} & S_\ell \\ \binom{i-1}{m} & \binom{i}{m} & S_m \end{vmatrix} \geq 0,$$

respectively. These bounds are sharp for the i_0 and j_0 described above.

3.3. Bounds on $P(X = m)$ in terms of S_k and S_ℓ : $k < \ell$

The upper bound is achieved by the canonical representation, σ_m. Its possible support sets are $\{j-1, j, m\}$, or $\{m, i-1, i\}$, or $\{0, m, N\}$. The key to determining which applies is to start with the principal representations from the previous example. Let the support sets of the upper and lower principal representations be denoted by $\{j_0 - 1, j_0, N\}$, and $\{0, i_0 - 1, i_0\}$, respectively. Necessarily, $j_0 \leq i_0$ since both measures have the same pair of moments. We consider three cases:

Case 1: $m \geq i_0$. The existence of the lower principal representation is *prima facia* evidence that the moments, S_k and S_ℓ are achievable by probability measures on $\Omega' = \{0, 1, \ldots, m\}$. Now exactly the same argument used in Example 3.2 (18) goes through, with N replaced by m, to establish existence of a probability measure on $\{j-1, j, m\}$ where j is the minimal value such that

$$\begin{vmatrix} 1 & 1 & 1 \\ S_k & \binom{j}{k} & \binom{m}{k} \\ S_\ell & \binom{j}{\ell} & \binom{m}{\ell} \end{vmatrix} \geq 0. \tag{22}$$

Solving for the three weights on $j-1, j$, and m which give the prescribed moments, we find that

$$P(X = m) \leq \sigma_m(\{m\}) \tag{23}$$

where

$$\sigma_m(\{m\}) = \begin{vmatrix} 1 & 1 & 1 \\ \binom{j-1}{k} & \binom{j}{k} & S_k \\ \binom{j-1}{\ell} & \binom{j}{\ell} & S_\ell \end{vmatrix} \times \begin{vmatrix} 1 & 1 & 1 \\ \binom{j-1}{k} & \binom{j}{k} & \binom{m}{k} \\ \binom{j-1}{\ell} & \binom{j}{\ell} & \binom{m}{\ell} \end{vmatrix}^{-1}.$$

Case 2: $m \leq j_0 - 1$. Here we know that the prescribed moments are achievable by probability measures on $\Omega'' = \{m, m+1, \ldots, N\}$. The required measure is the lower principal representation relative to Ω''. The argument used in (17) of Example 3.2 goes through, with the "first column," where $x = 0$, replaced by $(1 \binom{m}{k} \binom{m}{\ell})'$, to establish the existence of a probability measure on $\{m, i-1, i\}$ where $i \geq i_0$. The resulting analysis gives the same bound as in Case 1.

Case 3: $j_0 \leq m < i_0$. Here we can see that there is always a probability measure on $\{0, m, N\}$. Indeed, the solution to

$$\begin{pmatrix} 1 & 1 & 1 \\ 0 & \binom{m}{k} & \binom{N}{k} \\ 0 & \binom{m}{\ell} & \binom{N}{\ell} \end{pmatrix} \begin{pmatrix} p \\ q \\ r \end{pmatrix} = \begin{pmatrix} 1 \\ S_k \\ S_\ell \end{pmatrix}$$

is

$$qD = \begin{vmatrix} 1 & 1 & 1 \\ 0 & S_k & \binom{N}{k} \\ 0 & S_\ell & \binom{N}{\ell} \end{vmatrix}, \quad pD = \begin{vmatrix} 1 & 1 & 1 \\ S_k & \binom{m}{k} & \binom{N}{k} \\ S_\ell & \binom{m}{\ell} & \binom{N}{\ell} \end{vmatrix}, \quad rD = -\begin{vmatrix} 1 & 1 & 1 \\ 0 & S_k & \binom{m}{k} \\ 0 & S_\ell & \binom{m}{\ell} \end{vmatrix}$$

where

$$D = \begin{vmatrix} \binom{m}{k} & \binom{N}{k} \\ \binom{m}{\ell} & \binom{N}{\ell} \end{vmatrix}.$$

Since $D > 0$, q is positive by (15); p is positive because $pD = |B_m|t(m)$ in (18) which is positive for $m \geq j_0$ by definition of j_0; and, likewise, $rD = -|A_m|q(m)$ in (17) which is positive for $m < i_0$. The resulting bound is

$$P(X = m) \leq q = \sigma_m(\{m\}).$$

For the lower bound, we may use $\sigma_{m-1}(\{m\})$ or $\sigma_{m+1}(\{m\})$ as described in Section 2.3. Alternately, we may use (2) directly. In order for the lower bound to be greater than zero, the inequalities

$$c_0 + c_1\binom{x}{k} + c_2\binom{x}{\ell} \leq \left\{ \begin{array}{ll} 0 & \text{if } x \neq m \\ 1 & \text{if } x = m \end{array} \right\}$$

must be satisfied with equality holding at m. This is impossible if $0 < m < k$, or if $m = 0$ and $k > 1$, because the above polynomials are identically c_0 for $x < k$.

In each of the remaining cases to be considered, there is only one polynomial with equality holding at m and as many as two other values. In each case the corresponding bound may or not be positive. Hence the lower bound we seek is its positive part. For $m = 0$ and $k = 1$, equality holds at $0, 1$, and n for the polynomial

$$1 - x + \frac{N-1}{\binom{N}{\ell}}\binom{x}{\ell}$$

which yields the bound

$$P(X = m) \geq \left(1 - S_1 + \frac{N-1}{\binom{N}{\ell}} S_\ell\right)^+ . \tag{24}$$

For $k \leq m < n$, equality holds at $m-1, m$, and $m+1$ for the polynomial

$$\begin{vmatrix} 1 & 1 & 1 \\ \binom{m-1}{k} & \binom{x}{k} & \binom{m+1}{k} \\ \binom{m-1}{\ell} & \binom{x}{\ell} & \binom{m+1}{\ell} \end{vmatrix} \times \begin{vmatrix} 1 & 1 & 1 \\ \binom{m-1}{k} & \binom{m}{k} & \binom{m+1}{k} \\ \binom{m-1}{\ell} & \binom{m}{\ell} & \binom{m+1}{\ell} \end{vmatrix}^{-1}$$

which yields the lower bound

$$\begin{vmatrix} 1 & 1 & 1 \\ \binom{m-1}{k} & S_k & \binom{m+1}{k} \\ \binom{m-1}{\ell} & S_\ell & \binom{m+1}{\ell} \end{vmatrix}^+ \times \begin{vmatrix} 1 & 1 & 1 \\ \binom{m-1}{k} & \binom{m}{k} & \binom{m+1}{k} \\ \binom{m-1}{\ell} & \binom{m}{\ell} & \binom{m+1}{\ell} \end{vmatrix}^{-1} .$$

Finally, for the case $m = N$, equality holds at $0, N-1$, and N for the polynomial

$$\begin{vmatrix} 1 & 1 & 1 \\ 0 & \binom{N-1}{k} & \binom{x}{k} \\ 0 & \binom{N-1}{\ell} & \binom{x}{\ell} \end{vmatrix} \times \begin{vmatrix} 1 & 1 & 1 \\ 0 & \binom{N-1}{k} & \binom{N}{k} \\ 0 & \binom{N-1}{\ell} & \binom{N}{\ell} \end{vmatrix}^{-1}$$

which yields the lower bound

$$\begin{vmatrix} 1 & 1 & 1 \\ 0 & \binom{N-1}{k} & S_k \\ 0 & \binom{N-1}{\ell} & S_\ell \end{vmatrix}^+ \times \begin{vmatrix} 1 & 1 & 1 \\ 0 & \binom{N-1}{k} & \binom{N}{k} \\ 0 & \binom{N-1}{\ell} & \binom{N}{\ell} \end{vmatrix}^{-1} .$$

3.4. Upper bound on $P(X \geq m)$ in terms of S_1 and S_2

By the Kreĭn-Markov-Stieltjes Theorem, the upper bound is attained by the canonical measure, σ_m, for which the implicit form was derived in the previous example. In this special case, we can give the bound explicitly, as follows.

We first obtain the lower and upper principal representations from Example 3.2. These have support sets $\{0, j_0 - 1, j_0\}$ and $\{i_0 - 1, i_0, N\}$, where $i_0 \geq j_0$. By (20), j_0 is the minimal j such that

$$\begin{vmatrix} 1 & 1 & 1 \\ S_1 & j & N \\ S_2 & \binom{j}{2} & \binom{N}{2} \end{vmatrix} \geq 0. \quad \text{or} \quad j_0 - 1 = \left[\frac{(N-1)S_1 - S_2}{N - S_1}\right].$$

By (19), i_0 is the minimal i such that

$$\begin{vmatrix} S_1 & i \\ S_2 & \binom{i}{2} \end{vmatrix} \geq 0 \quad \text{or} \quad i_0 - 1 = 1 + \left[\frac{2S_2}{S_1}\right].$$

If $m \leq j_0 - 1$, then, as shown in Example 3.3, σ_m has support on a set of the form $\{m, i-1, i\}$, where $i \geq i_0 \geq j_0 \geq m+1$. Thus, σ_m has mass one on $[m, N]$, so the best upper bound on $P(X \geq m)$ is one.

If $m \geq i_0$, then σ_m has support on $\{j-1, j, m\}$, where $j \leq j_0 \leq i_0 \leq m$. The upper bound is then $\sigma_m(\{m\})$. We find the pair, $j-1, j$ just as we did $j_0 - 1, j_0$ above, except that N is replaced by m. Thus

$$i - 1 = \left[\frac{(m-1)S_1 - S_2}{m - S_1}\right]$$

and the upper bound is

$$\begin{aligned} \sigma_m(\{m\}) &= \begin{vmatrix} 1 & 1 & 1 \\ i-1 & i & S_1 \\ \binom{i-1}{2} & \binom{i}{2} & S_2 \end{vmatrix} \times \begin{vmatrix} 1 & 1 & 1 \\ i-1 & i & m \\ \binom{i-1}{2} & \binom{i}{2} & \binom{m}{2} \end{vmatrix}^{-1} \\ &= \frac{i(i+1) - 2iS_1 + 2S_2}{(m-i)(m-i-1)}. \end{aligned}$$

For $j_0 \leq m \leq i_0 - 1$ the support set is on $\{0, m, N\}$, and the upper bound is $\sigma_m(\{m, N\})$. Alternately, the extremal polynomial is zero at $x = 0$ and one at $x = m$ and $x = N$. This polynomial is

$$1 - \begin{vmatrix} 1 & 1 & 1 \\ x & m & N \\ \binom{x}{2} & \binom{m}{2} & \binom{N}{2} \end{vmatrix} \times \begin{vmatrix} m & N \\ \binom{m}{2} & \binom{N}{2} \end{vmatrix}^{-1}.$$

The required upper bound is then

$$1 - \begin{vmatrix} 1 & 1 & 1 \\ S_1 & m & N \\ S_2 & \binom{m}{2} & \binom{N}{2} \end{vmatrix} \times \begin{vmatrix} m & N \\ \binom{m}{2} & \binom{N}{2} \end{vmatrix}^{-1}.$$

which reduces to

$$\frac{1}{mN}\left[(m+N-1)S_1 - 2S_2\right]. \tag{25}$$

Lower bounds for $P(X \geq m)$ can be obtained using $\sigma_{m-1}([m, N])$. The different cases can be handled as in the analysis above. We omit the details.

3.5. Bounds on $P(X \geq 1)$ in terms of S_1 and S_2

This is a special case of Example 3.3 by using the bounds for $P(X = 0)$. The upper bound is also a special case of the previous example. The bounds are

$$\frac{2}{i-1}S_1 - \frac{2}{i(i-1)}S_2 \leq P(X \geq 1) \leq \min\{1, S_1 - \frac{2}{N}S_2\}, \tag{26}$$

where $i - 1 = 1 + [2S_2/S_1]$.

From Example 3.4, the upper bound is clearly one unless $j_0 - 1 = 0$; i.e. unless S_1, S_2 has support set $\{0, 1, N\}$. In this case, $j_0 = m = 1$ and the upper bound is $S_1 - 2S_2/N$.

For the lower bound we use (2) in conjunction with the canonical representation, σ_0, with support on $\{0, i-1, i\}$. The extremal polynomial is

$$1 - \begin{vmatrix} 1 & 1 & 1 \\ x & i-1 & i \\ \binom{x}{2} & \binom{i-1}{2} & \binom{i}{2} \end{vmatrix} \times \begin{vmatrix} i-1 & i \\ \binom{i-1}{2} & \binom{i}{2} \end{vmatrix}^{-1}$$

which gives the lower bound. It is interesting to note that the bound in Worsley (1982),

$$P(\cup_{i=1}^{N} A_i) \leq \sum_{i=1}^{N} P(A_i) - \sum_{i=1}^{N-1} P(A_i A_{i+1}),$$

reduces to the right side of (26) when the A_i's are exchangeable.

3.6. Bounds on $P(X \geq 1)$ in terms of S_1, S_2, and S_3

The principal representations of (S_1, S_2, S_3) have index $n+1=4$ and must necessarily be of the form $\{0, i, i+1, N\}$ for some i, or of the form $\{i, i+1, j, j+1\}$ for some $i < j-1$. As noted in Section 2, the upper principal representation must be of the first type. Also, all "polynomials" satisfying

$$c_0 + c_1\binom{x}{1} + c_2\binom{x}{2} + c_3\binom{x}{3} \geq (\leq) I_{\{x \geq 1\}} \quad \forall x \epsilon \Omega = \{0, 1, \ldots, N\}$$

have equality holding on a set of index at most four, because they change direction at most twice. Thus, the lower bound is achieved by the upper principal representation, and the upper bound is strictly less than one if and only if the lower principal representation is supported on a set of the form $\{0, 1, i, i+1\}$.

For the upper bound, the polynomials to consider are:

$$1 + \frac{(x-1)(x-i)(x-(i+1)}{i(i+1)} =$$

$$1 - \begin{vmatrix} 1 & 1 & 1 & 1 \\ 1 & \binom{x}{1} & \binom{i}{1} & \binom{i+1}{1} \\ 0 & \binom{x}{2} & \binom{i}{2} & \binom{i+1}{2} \\ 0 & \binom{x}{3} & \binom{i}{3} & \binom{i+1}{3} \end{vmatrix} \times \begin{vmatrix} 1 & 1 & 1 & 1 \\ 1 & 0 & \binom{i}{1} & \binom{i+1}{1} \\ 0 & 0 & \binom{i}{2} & \binom{i+1}{2} \\ 0 & 0 & \binom{i}{3} & \binom{i+1}{3} \end{vmatrix}^{-1}.$$

Taking expectations and simplifying, we get the upper bounds

$$P(X \geq 1) \leq S_1 - \frac{2(2i-1)}{i(i+1)}S_2 + \frac{6}{i(i+1)}S_3. \tag{27}$$

The minimum of these bounds is achieved at

$$i = 2 + \left[\frac{3S_3}{S_2}\right]. \tag{28}$$

If the minimal bound is less than one, it must, by the theory, be the least upper bound and attainable by the lower principal representaion. Otherwise, the least upper bound is one.

Alternatively, solving for the non-negative solution, $\mathbf{p}$, to

$$\begin{pmatrix} 1 & 1 & 1 & 1 \\ 0 & 1 & i & i+1 \\ 0 & 0 & \binom{i}{2} & \binom{i+1}{2} \\ 0 & 0 & \binom{i}{3} & \binom{i+1}{3} \end{pmatrix} \mathbf{p} = \begin{pmatrix} 1 \\ S_1 \\ S_2 \\ S_3 \end{pmatrix}, \tag{29}$$

leads to the same result, namely (28).

For the lower bound, we work first with the ordinary moments, $\nu_i = EX^i$, and then convert the results to the binomial moments,

$$\begin{aligned} S_1 &= \nu_1 & \nu_1 &= S_1 \\ S_2 &= (\nu_2 - \nu_1)/2 & \nu_2 &= 2S_2 + S_1 \\ S_3 &= (\nu_3 - 3\nu_2 + 2\nu_1)/6 & \nu_3 &= 6S_3 + 6S_2 + S_1. \end{aligned}$$

The polynomials to consider are

$$h_i(x) \equiv 1 + \frac{(x-i)(x-(i+1))(x-N)}{i(i+1)N} =$$

$$\frac{1}{i(i+1)N}x^3 - \left(\frac{1}{i(i+1)} + \frac{1}{iN} + \frac{1}{(i+1)N}\right)x^2 + \left(\frac{1}{i} + \frac{1}{i+1} + \frac{1}{n}\right)x$$

Taking expectations, we get lower bounds $Eh_i(X)$. After simplifying, these satisfy

$$Eh_i(X) - Eh_{i-1}(X) = -\frac{2}{(i-1)i(i+1)N}[i(N\nu_1 - \nu_2) - (N\nu_2 - \nu_3)].$$

Hence we conclude that the greatest lower bound is achieved at

$$i = \left[\frac{N\nu_2 - \nu_3}{N\nu_1 - \nu_2}\right] = 1 + \left[2\frac{(N-2)S_2 - 3S_3}{(N-1)S_1 - 2S_2}\right],$$

necessarily by the upper principal representation.

REFERENCES

Bonferroni, C. E. (1937), Teoria statistica delle classi e calcolo delle probabilità, *Volume in onore di Riccardo Dalla Volta*, Università di Firenze, 1-62.

Chung, K. L. and Erdös, P. (1952), On the application of the Borel-Cantelli lemma, *Transactions of the American Mathematical Society*, **72**, 179-186.

Dawson, D. A. and Sankoff, D. (1967), An inequality for probabilities, *Proceedings of the American Mathematical Society*, **18**, 504-507.

Galambos, J. (1977), Bonferroni inequalities, *Annals of Probability*, **5**, 577-581.

Galambos, J. and Mucci, R. (1980), Inequalities for linear combinations of binomial moments, *Publicationes Math*, **27**, 263-269.

Hadley, G. (1962), *Linear Programming*, Addison-Wesley, Reading, MA.

Karlin, S. J. and Studden, W. J. (1966), *Tchebycheff Systems*, Wiley Interscience, New York.

Kounias, Eustratios G. (1968), Bounds for the probability of a union, with applications, *Annals of Mathematical Statistics*, **39**, 2154-2158.

Kreĭn, M. G. and Nudel'man, A. A. (1977), *The Markov Moment Problem and Extremal Problems*, Translations of Mathematical Monographs, No. 50, American Mathematical Society, Providence R. I.

Kwerel, S. M. (1975), Most stringent bounds on aggregated probabilities of partially specified dependent probability systems, *Journal of the American Statistical Association* **70**, 472-479.

Móri, T. F. and Székely, G. J. (1985), A note on the background of several Bonferroni-Galambos-type inequalities, *Journal of Applied Probability*, **22**, 836-843.

Platz, O. (1985), A sharp upper probability bound for the occurrence of at least m out of n events, *Journal of Applied Probability*, **22**, 978-981.

Prékopa, A. (1987), Boole-Bonferroni inequalities and linear programming, *Operations Research*,

Sathe, Y. S., Pradhan, M., and Shah, S. P. (1980), Inequalities for the probability of the occurrence of at least m out of n events, *Journal of Applied Probability*, **17**, 1127-1132.

Schwager, Steven J. (1984), Bonferroni sometimes loses, *The American Statistician*, **38**, 192-197.

Worsley, K. J. (1982), An improved Bonferroni inequality and applications, *Biometrika*, **69**, 297-302.

The $\sqrt{N}$ Law and Repeated Risktaking

Paul A. Samuelson*
Massachusetts Institute of Technology

1. INTRODUCTION AND REVIEW

P.A. Samuelson (1963) pointed out a widespread fallacy concerning how insurance works: it is not by *adding* independent risks that risk exposure is eased, but by *subdividing* through pooling the added risks. Insuring 100 independent ships increases risk 10-fold over insuring 1 ship; by pooling risk coverage of the 100 over 100 insurers, we reduce the risk on each to effectively one-tenth.

This topic evoked considerable discussion: e.g., L.L. Lopes (1981), A. Tversky and M. Bar-Hille (1983), S.H. Chew and L. Epstein (1988), J.W. Pratt and R.J. Zeckhauser (1987), P.A. Samuelson (1988); but since it appeared in a limited-circulation journal, some recapitulation will be useful.

AMS Subject Classification: Primary 90A09.
Key words and phrases: Proper utility, insurance, risk aversion.

* I owe thanks to the MIT Center for Real Estate Development for partial support; to Aase Huggins, Eva Hakala, and Martha Adams for editorial assistance; to Gerad Gennotte for needed mathematical auditing; and to discussions with Joram Mayshar of Jerusalem University, Joseph Stiglitz of Stanford, and to Richard Zeckhauser and John Pratt of Harvard. John Pratt's companion piece provides an interesting alternative proof of my analysis and ingenious findings in its own right.

Copyright © 1989 by Academic Press, Inc.
All rights of reproduction in any form reserved.
ISBN-0-12-058470-0

The exposition began with Stan Ulam's whimsy that, *A coward is one who won't bet either way on a favorable gamble.* Actually, it can be rational for a maximizer of expected utility, of $E\{U(W)\}$ with $U(W)$ strictly concave and in this sense risk averse, to refuse a sizeable bet favorable in money if dollars of gain bring less marginal utility than dollars of loss take away. For $U(W)$ a *smooth* concave function, Ulam would be correct to require that a sufficiently-scaled-down favorable bet be accepted.

I quoted the discussion of Ulam's remark by an unnamed colleague, who asserted he would shun an even-odds bet with \$200 gain and \$100 loss—but would embrace a set of a large enough number of such independent bets. Since it will be seen that it was gratuitous for me to attribute to my colleague reliance on a fallacious version of the Law of Large Numbers, I can identify him as E. Cary Brown, my long-time MIT boss. The Brownian motive need not have been a belief that, when a favorable gamble is repeated a large enough number of times N, it is virtually certain to leave you ahead.

Stock market letters and baseball announcers do often assert what boils down to the following false version of the Law of Large Numbers: after 1000 or more plays of a gamble, you are virtually certain to be near to 1000 times the expected gain per play; or

> *If each play of gamble brings positive expected return,* $E\{X_j\} = \mu > 0$, *then*

$$\text{Prob}\,\{|(X_1 + \ldots + X_N) - N\mu| < \epsilon\} \to 1\ , \quad \text{for } 0 < \epsilon << 1\ , \quad \text{as } N \to \infty. \tag{1.1}$$

"In the long run, the odds dominate and chance errors cancel out."

Mathematicians know this to be monstrously misleading as stated, since for any $1/\epsilon$ however large, as N gets larger and larger it becomes virtually certain the $\sum_1^N X_j$ will diverge absolutely from $N\mu$ by more than $1/\varepsilon$. Riskiness does not cancel out for the longterm investor in a random-walk course, but rather grows like $\sqrt{N}$. This is so even though it is mathematically valid to assert that, for all positive ϵ,

$$\lim_{N\to\infty} \text{Prob}\left\{\left|\frac{\sum_1^N X_j}{N\mu} - 1\right| < \epsilon\right\} = 1 \tag{1.2}$$

$$\lim_{N\to\infty}\left\{a < \frac{\sum_1^N X_j - \mu N}{N^{1/2}\sigma}\right\} < b\Bigg\} = \int_a^b (2\pi)^{-\frac{1}{2}} e^{-\frac{1}{2}t^2}\,dt \;,$$

where $\sigma^2 = \text{Variance}\,\{X_j\} < \infty$; and also

$$\lim_{N\to\infty} \text{Prob}\left\{\sum_1^N X_j > 0\right\} = 1 \;.$$

These Law-of-Large-Number and Central-Limit-Law truths do not justify belief that, when a favorable bet is repeatable a large enough number of times, it must end you up ahead. Nor should you be misled into following decisions simply because, with limiting probability of unity, they will leave you ahead—as in the dubious longrun investing counsel of Latané, Kelley, Breiman, Markowitz, Hakannson, and others. How much disutility harm you suffer when you are behind must rationally be given its proper weight. (Samuelson (1972, 1977, 1986, Chs. 204, 207, 245, 246, 329) nags away on this.)

Despite (1.2), there are many probability gambles with positive money gain $E\{X_j\}$, such that for admissible strictly-concave $U(W)$ of risk averters, for *all* N it will be the case that

$$E\left\{U\left(W_0 + \sum_1^N X_j\right)\right\} < U(W_0) \;. \tag{1.3}$$

The possibility of (1.3) could be verified by inductive calculation for the numerical case

$$U(W) = -2^{-W} \;, \tag{1.4}$$
$$\text{Prob}\,\{X_j = -1\} = 1/2 = \text{Prob}\,\{X_j = 1+\epsilon\} \;, \qquad 0 < \epsilon << 1.$$

For (1.4) it can be shown that (1.3)'s left-hand side diminishes with each increment of N above $N = 0$.

Constructively, Samuelson (1963) stated the theorem that, a maximizer of $E\{U(W)\}$ who is risk averse enough to refuse a specified favorable bet *at every* W_0 level cannot ever rationally accept a set of $2, 3, \ldots,$ or $N, \ldots$ such favorable bets. (Remark: Recently Chew and Epstein (1988) showed that a risk-averse investor, who satisfies the axioms of Mark Machina (1982) that are weaker than those of the Expected $\{U\}$ dogma, is *not* bound by the 1963 theorem.)

2. PRESENT PURPOSES

The following problem naturally suggests itself as one aspect of the property Pratt and Zeckhauser (1987) call "proper" utility: *For what* $U(W)$ *functions is it the case that, whenever you are indifferent to 1 of a specified favorable gamble, you will also be indifferent to* $2, 3, \ldots, N, \ldots$ *of such gambles as a set?* And vice versa, in the sense that being indifferent to K such gambles means you will be indifferent to any J such gambles, $J \gtreqless K$? Once we are given a $U(W)$ function that meets this repeated-indifference test, we are assured that it will also reject (or accept) 2 repeated independent gambles whenever it rejects (or accepts) one such gamble.

The resulting restriction on $U(W)$ is the rather natural one adepts at portfolio theory would guess it to be—namely constancy of absolute risk aversion. Of greater interest are the functional equations derived here to tackle the questions. In half a century of theorizing about economics, I have rarely encountered any importance in the third derivative of a function, as with $U^{(3)}(W)$, much less encountered importance in such higher derivatives as the fifth and sixth derivatives of relevant economic functions.

Also, there is independent interest in the alternative approach to the posed problem used by John Pratt (1989) in a companion paper evoked by our private discussions. I have no doubt that still other approaches could be useful.

3. AMBIGUOUS CASES AND REASONS TO SELF-INSURE

Before analyzing the posed problem, I should mention that the cited 1963 theorem is of limited economic application. Thus for investors with constant relative risk aversion, a popular textbook case, at large W_0 values the specified favorable gamble is effectively scaled down enough in size to make it acceptable; therefore, the premise of the 1963 theorem is not met. Indeed, the ruling empirical dogma—as in K.J. Arrow (1964)—hypothesizes that people typically have diminishing absolute risk aversion, so that $-U''(W)/U'(W)$ drops indefinitely with W. Hence, at high enough W, practically everyone will come to accept the original *single* favorable gamble. So my theorem may be in the Pirandello role of a syllogism looking for an application. Who therefore is to say that Brown, contemplating the odds that he will end up at such high W levels after undertaking a *large* number of favorable gambles, is irrational? In the original version circulated of this paper, I worked out such a valid defense for him by using the pioneering E.D. Domar and R.A. Musgrave (1944) $U(W)$, which consists of straight lines meeting at a corner so that each dollar of loss hurts more than each dollar of gain by a constant multiple. (Below, between Equations (5.7) and (5.8), I describe another defense for Brown along lines suggested by John Pratt.)

Although constant-relative-risk-aversion $U(W)$ functions elude the grasp of the 1963 theorem, we must not infer from this that Brown with such a utility may rationally accept a repeated number of a gamble that is singly unacceptable. That is not possible since such utilities are not, in the pejorative terminology of Pratt-Zeckhauser (1984), so "improper."

Thus, suppose I have Bernoulli logarithmic utility, $U = \log W$. Suppose I have W_0 so low that I will reject the favorable gamble

$$\text{(3.1)} \quad \text{Prob}\,\{X_1 = H + h\} = \frac{1}{2} = \text{Prob}\,\{X_1 = H - h\}\ , \qquad 0 < H < h\ .$$

Then I can never, at such W_0, accept the package of 2 independent such gambles.

Thus,

$$\frac{1}{2}\log(W_0 + H + h) + \frac{1}{2}\log(W_0 + H - h) < \log(W_0)\ , \qquad 0 < H < h\ ,$$

implies

$$\frac{1}{4}\log(W_0 + 2H + 2h) + \frac{1}{2}\log(W_0 + 2H + h - h) + \frac{1}{4}\log(W_0 + 2H - 2h)$$

$$\text{(3.2)} \qquad\qquad < \log(W_0)\ .$$

Similarly, Pratt-Zeckhauser (1987) proved for any function $W^{1-C}/(1-C)$, $0 < C \neq 1$, function that refusal of one favorable bet implies refusing any integral number of such independent bets. The whole family is in this Pratt-Zeckhauser sense "proper."

The present problems have practical concern beyond that for gamblers or portfolio investors. Since insurance companies charge a premium that more than covers their loss risk, when I desist from buying insurance on an item, I am embracing a favorable gamble. When I do insure the item, I am refusing a favorable gamble.

Most individuals with 1 or 2 autos prefer to buy insurance. Many corporations with fleets of trucks or taxis are willing to self-insure. Is this rational?

It may be. Thus, the firm with trucks may already have many shareholders. What is called self-insuring may then really be a case of subdividing risk by sharing it among many different risk bearers. In terms of Samuelson (1963), we are then dealing with the $1/\sqrt{N}$ law—not the $\sqrt{N}$ law. "Self" insurance is then only a hidden form of pooling insurance.

Here is a different argument. The accounting reports of a 1-car taxi company will oscillate wildly in the absence of insurance. The decision whether to buy into the company will be contaminated by noise in the data due to stochastic risk. While it is true that the company with 100 taxis will have 10 times the oscillations in accident payments, their *percentage* importance in per share earnings of the large company will be only one-tenth as great. So there is less accounting objection to self insurance that comes with size. So goes this second line of argument. On examination, it also appears to involve the $1/\sqrt{N}$ rather than the $\sqrt{N}$ law. (To be sure, the $1/\sqrt{N}$ law would not be valid in its sphere if it were not the case that the $\sqrt{N}$ law is valid in its: really, they are two aspects of the same truth.)

Here is a third consideration. Suppose 1 person owns the huge company. It is her utility function that is alone relevant. And, almost by definition, the W in her $U(W)$ function is a large one. Insuring a single taxi, looked at from her perspective, has been scaled down to a degree that makes this favorable gamble attractive. And if she is rich enough, the same logic will make it desirable to self-insure for a great number of taxis. (Rather than self-insure on M of N taxis, insuring the other N-M completely, it will no doubt be better to make more complicated reinsurance contracts that involve various deductibility clauses. Details about avoidable transaction and information costs become critical in arriving at the optimal arrangement.)

Clearly, $\sqrt{N}$ laws cannot be alone decisive. If she is rich as Croesus, but has $(-e^{-CW})$ utility, it will be seen that the size of her fleet will not affect her self-insurance decisions.

4. DEFINING $U(W)$ NEUTRAL TOWARD REPETITION OF RISKS

$U(W)$'s come in several general categories: i) Those on the razor's edge where, if you will accept (reject, or be indifferent to) 1 favorable gamble, you will also accept (reject, or be indifferent to) 2, 3, ..., of such independent gambles. ii) Those in which you will take some maximum number M of such gambles, or less, but not more than M, where $M = 1, 2, \ldots$. iii) those in which you will take some minimum number M or more, but *will not take less than* M. iv) Those with more idiosyncratic properties. Depending upon the W level in question, one may of course generally go from one category to another. And of course, gambles of different compositions may affect one differently. However, for gambles sufficiently "compact" or " small," the composition of the gamble will presumably become immaterial.

Category i) is in a sense the razor's-edge border between the next two categories and our problem is to characterize its $U(W)$s.

To explore where i)'s broad border of indifference falls, I first proceed heuristically, asking the following question: For what local properties of the $U(W)$ function will it be the case that, when I am broadly indifferent about taking either one of the two specified (independent, compact, identical) gambles, I am also broadly indifferent to taking both of them together?

Suppose my $U(W)$ is such that at W_0 I opt to just take 1 of a sufficiently-favorable gamble that gives me with even odds a gain of $H + h$ or a loss of $h - H$. With what new utility function (of W near W_0) do I subsequently contemplate whether to take a second such independent gamble? My new ("post-one-gamble") utility function will be $\overline{U}(W)$ defined as follows.

$$\overline{U}(W) = \frac{1}{2}U(W + H_1(h; W_0) + h) + \frac{1}{2}U(W + H_1(h; W_0) - h) \ , \tag{4.1}$$

where

$$H_1(h; W_0) = 0h + (-U''(W_0)/U'(W_0))h^2 + \text{higher powers of } h \ , \tag{4.2}$$

and $H_1(h; W_0)$ is the unique root of

$$\frac{1}{2}U(W_0 + H_1(h; W_0) + h) + \frac{1}{2}U(W_0 + H_1(h; W_0) - h) = U(W_0) \ . \tag{4.3}$$

Within Category i), one expects the risk coefficient of $\overline{U}(W)$ to equal that of $U(W)$:

$$-\frac{U''(W)}{U'(W)} = -\frac{\overline{U}''(W)}{\overline{U}'(W)} \tag{4.4}$$

$$= -\frac{U''(W+H_1+h)+U''(W+H_1-h)}{U'(W+H_1+h)+U'(W+H_1-h)} .$$

Now, for h small, and h^3 negligible compared to h^2, we apply Taylor's expansions to the numerator and denominator, to get

$$-\frac{U''(W)}{U'(W)} = -\frac{U''(W)+(-1U^{(3)}U''(-U')^{-1}+U^{(4)})h^2+\cdots}{U'(W)+(-1U''U''(-U')^{-1}+U^{(3)})h^2+\cdots} , \tag{4.5}$$

where third, fourth, and higher derivatives of U are written as $U^{(3)}$, $U^{(4)}$, ... For (4.5)'s equivalence to hold approximately near W, we must have to a good approximation,

$$\frac{U^{(4)}}{U'} = 2\left(\frac{U^{(3)}}{U'}\right)\left(\frac{U''}{U'}\right) - \left(\frac{U''}{U'}\right)^3 . \tag{4.6a}$$

In terms of the absolute risk-aversion function, $A = -U''/U'$, this 4th order equation in U becomes the 2nd order differential equation:

$$A''(W) = A'(W)A(W) . \tag{4.6b}$$

This is one necessary condition on admissible $A(W)$. An important subclass of (4.6b)'s solutions is verified to be the well-known case of constant-absolute-risk aversion:

$$A(W) \equiv a > 0 , \tag{4.7a}$$

$$U(W) = b + d\left(-e^{-aW}\right) , \qquad (d,a) > 0 , -\infty < b < +\infty . \tag{4.7b}$$

There is a more general 2-parameter infinity of other solutions to (4.6b), each uniquely defined for

$$\tilde{a} = A(W_0) > 0 , \qquad A'(W_0) = \overline{a} \gtreqless 0 , \tag{4.8a}$$

$$A(W) = f(W; \tilde{a}, \overline{a}) \ . \tag{4.8b}$$

With $\tilde{a}$ prescribed positive, $A(W)$ will be positive and risk averse in some interval around specified W_0. For $\overline{a}$ prescribed to vanish, we have (4.7)'s special case; and, as will be seen, this is the only admissible case when necessary conditions additional to (4.6b) are deduced to hold.

If all we require is a rough approximation to indifference, then any member of the $f(W; \tilde{a}, \overline{a})$ family will be very nearly indifferent about repeating just-indifferent bets that are small in scale.

As we introduce higher-order terms in $(h^3, h^4, h^5, h^6, \ldots)$ in the Taylor's expansions of (4.4)'s numerators and denominators, and eschew approximations, we deduce further necessary conditions involving $A(W)$'s higher derivatives. We soon discover that having to satisfy additional necessary conditions beyond (4.6b) narrows the admissible $A(W)$ functions to the case of *constant*-absolute-risk aversion (including of course the risk-neutral case where $A(W)$ vanishes and U is linear in W).

5. EXACT DERIVATION

I shall now show that strict constancy of $A(W)$ is a necessary as well as sufficient condition for our defined neutrality toward repetition of just-indifferent gamble.

First, define the respective premium that you must be given as a bribe to induce you to tolerate an even-odds chance of gaining $+h$ or losing it. In (3.1) this was called H_1 or $H_1(h; W)$ when a single-gamble indifference is involved.

To deal with 2 repetitions of a gamble and your indifference, define H_2 or $H_2(h; W)$, the premium you must be bribed to be just willing to chance 2 independent bets that each yield $+h$ and $-h$ with even odds.

Clearly, H_1 and H_2 are defined as roots of the following respective equivalences:

$$\frac{1}{2}U(W + H_1 + h) + \frac{1}{2}U(W + H_1 - h) = U(W) \ , \tag{5.1}$$

$$0 < H_1 = H_1(h; W) < h \ ,$$

$$(5.2)\quad \frac{1}{4}U(W+H_2+2h)+\frac{1}{2}U(W+H_2+0)+\frac{1}{4}U(W+H_2-2h)=U(W)\,,$$

$$0 < H_2 = H_2(h;W) < 2h\ .$$

Category i) consists then of the $U(W)$s such that it will be exactly the case that the bribe to just take 2 bets must be exactly twice the bribe to just take 1 bet.

For what $U(W)$s will it be the case that

$$(5.3)\qquad H_2(h;W) \equiv 2H_1(h;W)$$

In terms of formal Taylor's expansions

$$(5.4a)\quad H_1 = 0+0h+A(W)h^2+(1/3!)H_1^{(3)}(0;W)h^3+(1/4!)H_1^{(4)}(O;W)$$

$$+(1/5!)H_1^{(5)}(O;W)+(1/6!)H_1^{(6)}(0;W)+R_7\ ,$$

$$(5.4b)\quad H_2 = 0+0h+(2)A(W)h^2+\sum_{j=3}^{6}(1/j!)H_2^{(j)}(0;W)h^j + \text{Remainder}\ .$$

The requisite derivatives of H_2 and $2H_1$ in (5.4a) and (5.4b) must be made to match in accordance with (5.3)'s requirement. It can be verified that each jth order derivative of an H depends on all U derivatives up to that same order, and therefore depends on all A derivatives up to order $j-2$.

$$(5.5)\qquad 2H_1^{(j)}(0;W) = g^j(A(W),A'(W),\ldots,A^{(j-2)}(W))\ ,$$

$$H_2^{(j)}(0;W) = G^j(A(W),A'(W),\ldots,A^{(j-2)}(W))\ .$$

Thus, equating coefficients we verify relations binding on admissible $A(W)$'s such as

$$(5.6)\qquad \begin{aligned} 0 &= G^4(A,A',A'')-g^4(A,A',A'') \\ &= A''(W)-A'(W)A(W)\ , \end{aligned}$$

as in (4.6b). Because the odd fifth derivatives vanish for all U and A functions, we need only adjoin to (5.6) the matching of sixth derivatives of $2H_1$ and H_2. When Professor Gerard Gennotte of UC (Berkeley) was an MIT graduate student, he kindly slogged through the monstrous calculations to deduce finally that we must in the end adjoin to (5.6) the further necessary condition

$$0 = A(W)A'(W)^2 \ . \tag{5.7}$$

Joseph Stiglitz, doodling during my Stanford lecture on this subject, arrived at this same conclusion.

Together (5.6) and (5.7) do entail the constancy of $A(W)$ and of absolute risk aversion.

This completes the demonstration that the only risk-averse $U(W)$ that leads to indifference toward repetition of indifferent gambles is (4.7)'s $-e^{-aW}$.

What holds for indifference holds also for strong preference. Thus, iff my $A(W)$ is constant, when I accept 1 gamble, I will accept $2, 3, \ldots$ gambles. Likewise for rejection decisions.

Now Pratt's suggested defense for Brown can be sketched. For all W less than a critical W_1, let Brown have constant risk aversion—say at $A(W) \equiv 1$ as for $-e^{-W}$. Let him face the single even-odds bet $(-1+H, 1+H)$ where e^H is epsilon below $\frac{1}{2}(e + e^{-1})$. Then for all W_0 below $W_1 - 1 - H$, Brown will definitely refuse a single bet, albeit just barely refuse it. Above $W = W_1$, let Brown's $A(W)$ fall off smoothly from unity, to level off at $1/2$ for all W above $W_3 > W_1$. Since Brown has been made to be more appreciative of gains to any W above W_1 than he was before, and since he would in any case have almost been willing to have taken a single gamble, we deduce that there is an identifiable initial wealth level, call it W_2, $W_1 - 1 - H < W_2 < W_3$, at which he will just be indifferent to taking a single gamble (and, for initial W_0 above W_2 he will gladly accept a single gamble). Now we realize that for all initial W_0's on the interval $(W_1 - H - 1, W_2)$, Brown will reject the single gamble but definitely want to take 2 gambles—because the attractiveness of winning 2 in a row has been made finitely large, which is enough to outweigh the arbitrarily small unattractiveness contrived for the single gamble. QED, namely such a Brown is proved to be rational to refuse a single gamble while accepting a multiple repetition of it.

Let me record a remark about mathematical heuristics. The present demonstration of constancy of absolute risk aversion involved only special probability gambles—2 outcomes at even-odds. Yet practitioners in this field can have an easy mind that the result extends to *any* (non-degenerate) probability distribution.

Thus, replace the even-odds case by a probability-density gamble:

$$\int_{-\infty}^{\infty} p(h)hdh = \mu > 0 . \tag{5.8}$$

For it, neutrality toward repetition of a just-acceptable gamble replaces the few-term sums of (5.1) and (5.2) by the following integral equations involving $U(W)$:

If $H_1(W)$ is the unique root of

$$\int_{-\infty}^{\infty} dh\, U(W + H_1 + h)p(h) = U(W) , \tag{5.9}$$

and $H_2(W)$ the root of

$$\int_{-\infty}^{\infty} dh\, U(W + H_2 + h) \int_{-\infty}^{\infty} p(h - h')p(h')dh' = U(W) , \tag{5.10}$$

then only for

$$U(W) = b - de^{-aW} , \quad d > 0 , a > 0 , \quad -\infty < b < +\infty , \tag{5.11}$$

is

$$H_2(W) \equiv 2H_1(W) . \tag{5.12}$$

This remarkable simplicity is encountered again and again in this branch of probability. If at every wealth level I invest the same amount in one specified kind of equity, then for any other specified kind of equity we know my $U(W)$ is such that my absolute investment is independent of initial wealth level.

Thus, define $x = f_j(W)$ as the root of

$$0 = \int_{-\infty}^{\infty} dhU'(W - x + xh)(h - 1)p_j(h) , \qquad j = 1, 2 . \tag{5.13}$$

Then the $U'(W)$ that assures

$$f_1(W) \equiv f_1(1) ; \tag{5.14}$$

namely $U' = be^{-aW}$, also assures

$$f_2(W) = f_2(1) \ , \tag{5.15}$$

even though $p_1(h)$ and $p_2(h)$ are very different densities. We should count our blessing that such simplicities do obtain.

Depending on the algebraic sign at any particular W of (4.6b)'s $A''(W) - A'(W)A(W)$, and of $A'(W)$ itself, one expects to be able to state various strong *local* theorems: thus, for the former of a specified sign, if you will reject 1 of a particular sufficiently small gamble, you must reject 2 of it; or, for alternative specified sign, even if you reject 1 sufficiently small gamble, you will want to accept some number of them. Even more intricate local uniformities can be deduced such as, If you just barely reject 1 sufficiently small gamble, and $A'' - a'A$ is of a specified sign, you will accept some multiple of those gambles. John Pratt (1989), in his companion paper, states and proves a number of such delicate relationships.

6. FINAL REFLECTIONS

One can win a logical victory and lose an actual war. Academics have long warned practical money managers against a naive belief that investors for the long run can prudently invest more heavily in risky equities because time diversification will tend toward cancelling out risk when repeated plays are involved. For three decades those same academics have pointed out that rational portfolio managers, confronted by random-walk stock prices and possessing constant relative risk aversion, will select the same equity exposure when old as when young.

In the last few years statisticians have begun to believe that there is negative autocorrelation in price changes over periods measured in years rather than in days or weeks. It would be wrong to think that this new element of enhanced cancellation of risk over time would provide a rational basis for all long-horizon investors to hold relatively more equities than short-horizon investors would hold. No such sweeping result is valid: thus if you act to maximize expected value of log (Terminal Wealth), à la Daniel Bernoulli, you will choose the same equity fraction when old or young; if you act to maximize expected value of $\sqrt{W}$, à la Gabriel Cramer, you will be *more venturesome when old* than when young. But if you are like most investors, and seem to act as if more risk averse than Bernoulli, I was able to announce at the May 1988 Tobin *Fest* at Yale the following confirmation of practical investors' folk wisdom: *Even when long-run riskiness still grows eventually in proportional to* $\sqrt{\text{Time Horizon}}$, *you as a youngster with many periods ahead of you should hold more of risky equities than your older uncles should.*

There is a second question. Would empirical findings of significant long-period negative autocorrelation in equity price changes, and/or new findings concerning attitudes toward repeated gambles, weaken or strengthen the criticisms of Latané-Kelly-Breiman strategy of maximizing "long-term growth?" I believe that a rational person who is truly more risk averse than a Daniel Bernoulli will remain adamant in eschewing this irrelevant counsel and will perceive that any notion of long-run utility dominance is illusory.

Thirdly, what theory of the distribution of wealth and income can we fabricate from the assumption that all of us are confronted in life with quasi-independent favorable gambles that are repeatable? If very poor and very rich people are hypothesized to have lower risk tolerance than those in between (because the poor are desperate to avoid utter poverty, and the rich are jaded about becoming super-rich), then the middle classes will constantly be relatively more venturesome, and will accordingly be better rewarded (per dollar of wealth). Many of them tend to rise in affluence; some by chance sink into poverty. So long as some of the poor, like some of the rich, will move to the adjacent class, we can indeed deduce an evolution to an ergodic wealth-distribution steady state.

The following Markov transition matrix, M, captures our hypotheses:

$$(6.1) \quad (\text{Prob}\{\text{going from class } i \text{ at time } t-1 \text{ to class } j \text{ at } t\}) = (M_{ij})$$

$$= \begin{pmatrix} 0.5 & 0.5 & 0 \\ 0.2 & 0.5 & 0.3 \\ 0 & 0.5 & 0.5 \end{pmatrix}, \qquad \begin{cases} j=1 & \text{means being poor} \\ j=2 & \text{means middle class} \\ j=3 & \text{means being rich} \end{cases}.$$

Half of each class stand to remain in the same class. Half of the rich descend to the middle class. Half of the poor rise to the middle class. Whereas three-tenths rise to affluence from the middle class, only two-tenths of them fall into poverty.

The limiting ergodic state, given by the rows of powers of M, M^t as $t \to \infty$, puts half the people ultimately in the middle class, twenty percent in poverty, and thirty percent in the upper crust.

Schumpeter was right. Capitalism is like a hotel. Its upper rooms are always full, but the occupants are constantly changing according to the fortunes of competition.

REFERENCES

Arrow, K.J. (1964), Aspects of the Theory of Risk-Bearing, *Yrjo Jahnsson Lectures,* Yrjo Jahnsson Foundation, Helsinki.

Bernoulli, D. (1738), Specimen theoriae navae de mensura sortis, *Commentarii,* St. Petersburg Academy. Translated in *Econometrica,* **22** (1954), 23–36; reprinted in *Precursors in mathematical Economics: An Anthology,* Baumol, W. and Goldfield, S., eds., Series of Reprints of Scarce Works on Political Economy, No. 19, London: School of Economics and Political Science, 1968, 15–26.

Chew, S.H. and Epstein, L. (1988), The law of large numbers and the attractiveness of compound gambles, *Journal of Risk and Uncertainty,* **1**, 125–132.

Domar, E.D. and R.A. Musgrave (1944), Proportional income taxation and risk-taking, *Quarterly Journal of Economics,* **58,** 388–422.

Lopes, L.L. (1981), Decision making in the short run, *Journal of Experimental Psychology: Learning, Memory and Cognition,* **7,** 377–385.

Machina, M. (1982), 'Expected utility' analysis without the independence axiom, *Econometrica,* **50,** 277–323.

Markowitz, H.L. (1959), *Portfolio Selection: Efficient Diversification of Investments,* John Wiley and Sons, Inc., New York.

Pratt, J.W. (1988), Proof of a proposition of Paul Samuelson on exponential utility, to appear in 1989 Karlin *Festschrift.*

Pratt, J.W. and R.J. Zeckhauser (1987), Proper risk aversion, *Econometrica,* **55**,, No. 1, 143–154.

Samuelson, P.A. (1963), Risk and uncertainty: a fallacy of large numbers, *Scientia,* 6th Series, April-May. Reprinted in *Collected Scientific Papers of Paul A. Samuelson,* Volume 1, 1985, MIT Press, Cambridge.

Samuelson, P.A. (1972, 1977, 1986), *Collected Scientific Papers of Paul A. Samuelson,* MIT Press, Cambridge, Volumes 3, 4, and 5.

Samuelson, P.A. (1988), How a certain 'internal consistency' entails the expected utility dogma, *Journal of Risk and Uncertainty,* **1,** 339–343.

Tversky, A. and M. Bar-Hillel (1983), Risk: the long and the short, *Journal of Experimental Psychology: Learning, Memory and Cognition* **9**, 713–717.

A Theorem in Search of a Simple Proof

Herbert E. Scarf*
Yale University

I offer these remarks to Samuel Karlin, on the occasion of his sixty-fifth birthday, with a deep sense of gratitude. Sam played an extremely important role in my life. In the mid-1950s, he invited me to spend several years at Stanford University, thereby providing a professional path which otherwise would not have been available to me. I collaborated with Sam and Kenneth Arrow in a series of papers on inventory theory, which eventually appeared in a monograph; this was a heady experience for a young person at the very beginning of his career. I learned the rudiments of mathematical programming and economic theory from this pair of awesome mentors, and in particular, I was exposed, in our investigation of inventory problems with set-up costs, to an important and natural occurrence of economies of scale in production. This topic, and its generalization to discrete programming, is one to which I have devoted a sizable fraction of my professional attention.

In 1985, I published a paper entitled "Integral Polyhedra in Three Space," which contained a theorem about certain families of tetrahedra arising in the study of integer programming problems with three variables and three inequalities. As we shall see, the theorem is relatively easy to state, but

AMS 1980 Subject Classifications: Primary 90C10, Secondary 90C27.

Key words and phrases: Integer programming, neighbors, Geometry of Numbers, integral Polyhedra

*The research reported in this paper was supported by a grant from the National Science Foundation.

Copyright © 1989 by Academic Press, Inc.
All rights of reproduction in any form reserved.
ISBN-0-12-058470-0

In 1985, I published a paper entitled "Integral Polyhedra in Three Space," which contained a theorem about certain families of tetrahedra arising in the study of integer programming problems with three variables and three inequalities. As we shall see, the theorem is relatively easy to state, but its validity is far from obvious. The proof given in the paper is based on a tedious examination of a large number of special cases and is extraordinarily untidy. It is quite likely that the paper has had only one serious reader; after a certain point in his report, the extremely conscientious referee threw up his hands in frustration and took the remainder of the argument on faith.

I would like to take the present opportunity to describe this theorem and to make some suggestions that might lead to a more transparent proof. Let us begin by considering a surprising result, first demonstrated by Roger Howe, which forms the basis for the subsequent discussion.

THEOREM (HOWE). *Let T be a tetrahedron in three space, with positive volume, and with integer vertices v^0, v^1, v^2, v^3. Suppose, in addition, that the tetrahedron contains no other lattice points in its interior or boundary. Then there is a unimodular transformation and a translation by a lattice vector which carries the vertices of T into*

$$\begin{aligned} v^0 &= (1,0,0) \\ v^1 &= (0,1,0) \\ v^2 &= (0,0,1) \\ v^3 &= (1,p,q) \end{aligned}$$

with p and q relatively prime nonnegative integers.

An argument for Howe's theorem is given in the paper cited above and will not be repeated here. I may say, however, that it is quite easy to construct a unimodular transformation and a translation by a lattice vector so that three of the vertices of T become unit vectors and the remaining vertex a nonnegative lattice point. The more difficult part of the proof is to demonstrate that one of the three coordinates of the fourth vertex must be unity. Several arguments are available; the reader may enjoy constructing his own version.

The *lattice width* of the tetrahedron T is defined to be the minimum of

$$w(v,T) = \max\{vx | x \varepsilon T\} - \min\{vx | x \varepsilon T\}$$

as v varies over all non-zero integral vectors in Z^3. The lattice width of a non degenerate tetrahedron with integral vertices is clearly a positive integer. Howe's theorem asserts that if the tetrahedron is free of additional lattice points, then its lattice width is unity: the four vertices lie on two adjacent translates of a lattice plane, which we shall term a *characteristic plane* associated with the tetrahedron T. The corresponding theorem is false in higher dimensions: a simplex in $n > 3$ dimensions, with integral vertices and which contains no other lattice points, may have a lattice width greater than one.

In order to introduce our *family of tetradehra*, let A be a real matrix, with four rows and three columns, which is fixed through the remaining discussion. We make the following assumption:

ASSUMPTION.

1. There is a positive vector π, unique aside from scale, such that $\pi A = 0$.
2. Each row of A is independent over the integers in the sense that if for some particular $i, a_i h = 0$ for an integral h, then $h = 0$.

Let us select a vector c such that the interior of $K_c = \{x : Ax \geq c\}$ is free of lattice points and such that any relaxation obtained by lowering some or all of the coordinates of c contains at least one lattice point. It follows that there is a single lattice point, which we denote by v^i, on each planar facet $a_i x = c_i$. The inscribed tetrahedron T_c with vertices v^i will have integer vertices and contain no additional lattice points. It will therefore possess a characteristic plane yielding a lattice width of unity.

This procedure determines a set of tetrahedra, intrinsically associated with the matrix A, which I have called *primitive sets* in a previous publication [Scarf (1981)]. These tetrahedra are closely related to the family of integer programming problems

$$\begin{aligned} \max \quad & a_0 h \quad \text{subject to} \\ & a_i h \geq b_i \quad \text{for } i = 1, 2, 3, \text{ and } h \text{ integer.} \end{aligned}$$

In order to see one such connection, let us label each lattice point h in Z^3 with an integer label $l(h)$ according to the following rule:

1. $l(h) = 0$ if all of the inequalities $a_i h \geq b_i$, for $i = 1, 2, 3$, are satisfied;
2. $l(h) = i$ if i is the largest subscript for which the ith inequality is violated.

Given this particular labeling, it can be demonstrated [Scarf (1981)] that there is a unique primitive set all of whose vertices are differently labeled; moreover the vertex with label zero in such a primitive set is the optimal solution to the above integer program.

It may also be shown that there is a *replacement operation* for the vertices of primitive sets, in the following sense. Let v^0, v^1, v^2, v^3 be the vertices of a particular primitive set. Then there is a unique replacement for any given vertex which yields another primitive set in conjunction with the remaining three vertices. If a catalogue of the distinct primitive sets were known, the techniques used in approximating fixed points of a continuous mapping could therefore be applied directly to the determination of a completely labeled primitive set and consequently to the solution of integer programming problems.

Two lattice points are defined to be *neighbors* of each other if they are members of a common primitive set. The neighborhood relation is symmetric and invariant under translation by a common lattice point. It may be shown that the particular neighborhood relation deriving from primitive sets is sufficient for a *local maximum* of the above integer program to be a *global maximum*, in the following sense: let h be a lattice point which satisfies the constraints of the integer program and suppose that all neighbors of h are either infeasible or yield a lower value of the objective function. Then h is indeed the global maximum. Moreover this particular definition of a neighborhood relation is the unique, minimal neighborhood system associated with the matrix A, which is symmetric and translation invariant and which also possesses the property that a local maximum is a global maximum for all right-hand sides b.

The number of distinct primitive sets associated with a particular matrix A is finite if two primitive sets which are translates of each other by a lattice point are identified. There are particular cases in which the number of distinct primitive sets is small. For example, if A has the sign pattern of a Leontief matrix:

$$\begin{bmatrix} - & - & - \\ + & - & - \\ - & + & - \\ - & - & + \end{bmatrix}$$

with positive row sums for rows 1, 2 and 3, then there are 6 distinct primitive sets, each obtained by taking $v^o = 0$ and $v^{i+1} - v^i$ equal to the three unit vectors in some order. In this case, primitive sets give rise to the conventional simplicial subdivision of the unit cube used frequently in fixed point computations. In general, however, the number of distinct primitive sets may be arbitrarily large as A varies over the set of matrices satisfying our assumption 2. In order to appreciate the complexity that may arise, let me present the following 16 tetrahedra which are the distinct primitive sets arising from a small perturbation of the matrix

$$\begin{vmatrix} 1 & 1 & 1 \\ 6 & -1 & -1 \\ -5 & 2 & -5 \\ -2 & -2 & 5 \end{vmatrix}$$

$$\begin{bmatrix} 0 & 0 & 0 & 0 \\ 0 & 1 & 0 & 1 \\ 0 & 1 & 1 & 0 \end{bmatrix} \quad \begin{bmatrix} 0 & 0 & 0 & 0 \\ 0 & 2 & 1 & 1 \\ 0 & 1 & 1 & 0 \end{bmatrix} \quad \begin{bmatrix} 0 & 0 & 0 & 0 \\ 0 & 3 & 2 & 1 \\ 0 & 1 & 1 & 0 \end{bmatrix} \quad \begin{bmatrix} 0 & 0 & 0 & 0 \\ 0 & 5 & 2 & 3 \\ 0 & 2 & 1 & 1 \end{bmatrix}$$
$$\det = 0 \qquad \det = 0 \qquad \det = 0 \qquad \det = 0$$

$$\begin{bmatrix} 0 & 1 & 1 & 1 \\ 0 & 5 & 2 & 0 \\ 1 & 2 & 1 & 0 \end{bmatrix} \quad \begin{bmatrix} 0 & 1 & 1 & 1 \\ 1 & 5 & 0 & 3 \\ 0 & 2 & 0 & 1 \end{bmatrix} \quad \begin{bmatrix} 0 & 0 & 1 & 0 \\ 0 & 1 & 5 & 1 \\ 1 & 1 & 2 & 0 \end{bmatrix} \quad \begin{bmatrix} 0 & 0 & 1 & 0 \\ 0 & 0 & 0 & 1 \\ 0 & 1 & 0 & 0 \end{bmatrix}$$
$$\det = 1 \qquad \det = 1 \qquad \det = 1 \qquad \det = 1$$

$$\begin{bmatrix} 0 & 0 & 1 & 1 \\ 0 & 1 & 2 & 5 \\ 0 & 0 & 0 & 1 \end{bmatrix} \quad \begin{bmatrix} 0 & 0 & 1 & 1 \\ 0 & 1 & 1 & 4 \\ 0 & 0 & 0 & 1 \end{bmatrix} \quad \begin{bmatrix} 0 & 0 & 1 & 1 \\ 0 & 1 & 4 & 2 \\ 0 & 0 & 1 & 0 \end{bmatrix} \quad \begin{bmatrix} 0 & 0 & 1 & 1 \\ 0 & 1 & 0 & 3 \\ 0 & 0 & 0 & 1 \end{bmatrix}$$
$$\det = -1 \qquad \det = -1 \qquad \det = 1 \qquad \det = -1$$

$$\begin{bmatrix} 0 & 0 & 1 & 1 \\ 0 & 1 & 3 & 1 \\ 0 & 0 & 1 & 0 \end{bmatrix} \quad \begin{bmatrix} 0 & 0 & 1 & 1 \\ 0 & 0 & 2 & 0 \\ 0 & 1 & 1 & 0 \end{bmatrix} \quad \begin{bmatrix} 0 & 1 & 1 & 0 \\ 0 & 5 & 0 & 1 \\ 1 & 2 & 0 & 0 \end{bmatrix} \quad \begin{bmatrix} 0 & 0 & 1 & 1 \\ 0 & 0 & 4 & 2 \\ 0 & 1 & 2 & 1 \end{bmatrix}$$
$$\det = 1 \qquad \det = -2 \qquad \det = 7 \qquad \det = -2$$

This example illustrates a number of properties of the family of primitive sets associated with a general 4×3 matrix A. The columns of each tableau are, of course, the vertices of a primitive set, ordered so that $a_i v^i = c_i$ for $i = 0, 1, 2, 3$. Underneath each primitive set is the determinant of

$$\begin{bmatrix} 1 & 1 & 1 & 1 \\ v^0 & v^1 & v^2 & v^3 \end{bmatrix}.$$

The reader may notice that the sum of the determinants over all of the distinct primitive sets is equal to 6; the algebraic sum of the volumes of the distinct primitive sets is unity. This is a general property which is obtained first by ordering the rows of A such that $\det(\pi, A) > 0$ and then associating an index of $-1, 0, +1$ with each primitive set, given by the sign of its determinant. It may be shown that the algebraic sum of the indices of those primitive sets which contain a generic vector x in their convex hulls is unity. If we then integrate over all x in a large cube, we obtain the statement that the algebraic sum of the determinants is itself one. This particular result is demonstrated for a general matrix with $n+1$ rows and n columns satisfying assumption 2 in [White (1983)]. The first of these results may be interpreted geometrically as stating that the family of primitive sets gives rise to a *piecewise linear manifold*, whose linear pieces are simplices with integer vertices, which is topologically equivalent to R^n but embedded in R^n in a nontrivial way, and which is also identical to itself under translation by an arbitrary lattice point.

Notice also that in this example there are four degenerate primitive sets lying on the lattice plane $h_1 = 0$. Each of these degenerate primitive sets is a parallelogram of unit area. Moreover, the sequence of degenerate primitive sets can be ordered linearly in the following sense: consider any primitive set in the middle of the sequence. If any of its four vertices are replaced, we move to another degenerate primitive set; two of these replacements will take us to a lattice translate of the degenerate primitive set immediately to the left, and the other two replacements to the degenerate primitive set immediately to the right. Two of the replacements in the first degenerate primitive set will lead to nondegenerate tetrahedra, with three vertices on the lattice plane $h_1 = 0$ and a single vertex on $h_1 = -1$. Similarly two of the replacements in the last degenerate primitive set will lead to nondegenerate tetrahedra with three vertices on the lattice plane $h_1 = 0$ and a single vertex on $h_1 = 1$. The remaining tetrahedra have two vertices on this lattice plane and two on an adjacent plane. In order to move to another pair of adjacent planes, by a sequence of replacement steps, it is necessary to pass through a full set of degenerate tetrahedra.

It is also important to notice that this lattice plane is the characteristic lattice plane for *all* of the nondegenerate primitive sets arising from this matrix. The above structure of primitive sets, with a large linearly ordered sequence of degenerate primitive sets lying on a lattice plane which is the common characteristic lattice plane for all nondegenerate primitive sets, is valid for the general 4×3 matrix satisfying assumption 2. This is the essence of the theorem whose extremely complicated proof appears in [Scarf (1985)],

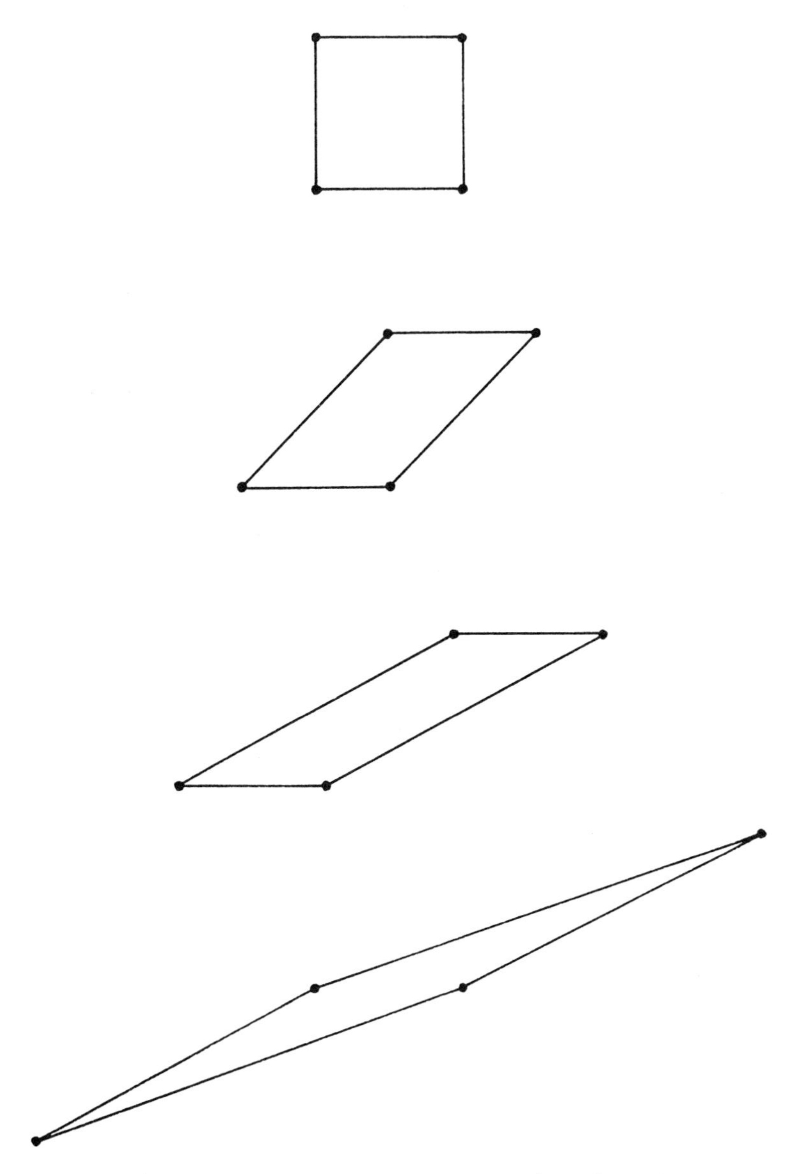

Degenerate Primitive Sets on the Plane $h_1 = 0$.

and for which a simpler and more intuitive argument would be of great interest.

THEOREM 1. *All primitive sets associated with the same matrix share a common characteristic lattice plane.*

Let me make some remarks about a strategy that might be followed in demonstrating Theorem 1. It is relatively easy to give a geometrical condition which is necessary and sufficient for a particular lattice plane, $w \cdot h =$ integer, to be the common characteristic lattice plane for all of the primitive sets determined by A. Take any vector c for which the set $\{Ax \geq c\}$ is free of lattice points *on the particular lattice plane*. If none of these sets contains lattice points on both sides of the lattice plane, then the plane is the common characteristic lattice plane. The difficulty that I experienced in proving Theorem 1 was in demonstrating the existence of a lattice plane with this particular property. After a good deal of frustrating experimentation, I adopted the following rather desperate line of argument: I began with a matrix A which was selected so as to have a lattice plane with this property, and then perturbed the entries in A so that one of the lattice free regions on the plane admitted lattice points on both sides. I was then able to show that at just this moment, another lattice plane with the correct property became available, thereby allowing a continuous deformation from the initial matrix to an arbitrary matrix satisfying the assumption. There is, however, an extraordinarily large number of ways in which this property can be lost under perturbation, and each one of them required a separate argument yielding a new lattice plane.

Of course, it is not necessary to examine *all* lattice free regions of the form $\{Ax \geq c\}$ on the prospective lattice plane; it is sufficient to examine those regions on the plane, determined by the four inequalities, whose interior contains no lattice points on the plane, and such that each inequality is satisfied with equality by a single lattice point on the plane. If none of these regions—which are primitive sets determined by four inequalities in a lattice plane—admit lattice points on both sides of the plane, then neither will any other lattice free region, and the particular lattice plane will be the characteristic plane for all of the three-dimensional primitive sets.

It is elementary to argue that each such maximal lattice free region is defined either by four or three lattice points. In the former case, the convex hull of the four lattice points is a parallelogram of unit area; in the latter case, the convex hull is a triangle of area 1/2. Precisely as in the example given above, the sequence of parallelograms has a natural linear order, with the additional property that two of the replacements of the leftmost parallelogram will lead to a pair of triangles and similarly for the rightmost parallelogram. The following important property can be demonstrated by an elementary argument.

THEOREM 2. *Consider the linear sequence of primitive sets on a lattice plane. Those primitive sets which admit a lattice point on one side of the lattice plane begin with the pair of triangles on the left and continue up to a certain parallelogram. Those which admit a lattice point on the other side of the lattice plane begin with the pair of triangles on the right and continue backwards to a possibly different parallelogram.*

Theorem 4 had an immediate consequence: if there is a degenerate primitive set, then the lattice plane containing this primitive set is a common characteristic lattice plane for the entire family of primitive sets. In other words, Theorem 1 has an elementary, geometrical argument if there is at least one degenerate primitive set. Moreover, this is a property which does, in fact, occur for the vast majority of matrices A. It is the absence of this property for the occasional matrix which forced me to embark on the elaborate proof of Theorem 1.

In order to provide a simple proof of Theorem 1, it would be sufficient to restrict our attention to those matrices A all of whose primitive sets are nondegenerate. Is there a geometrical argument for this special case? Given the fact that the family of primitive sets is a piecewise linear manifold, topologically equivalent to R^3, and invariant under translations by lattice points, the absence of degenerate primitive sets suggests, to me, that the manifold is embedded in R^3 without singularities, and should therefore have a structure which is capable of an elementary description. I would suspect that in this special case, the manifold is a simplicial decomposition of some parallelepiped of unit volume whose faces are lattice plane, but I am not enough of a geometer to provide a direct argument. If this were correct, Theorem 1 would have an appealing proof.

There is an additional observation, interesting in its own right, and possibly useful in an alternative argument. The bodies $K_c = \{x : Ax \geq c\}$ are tetrahedra, which are all scaled and translated versions of a single one of them. It follows that the difference bodies $K_c - K_c$ are convex bodies, symmetric about the origin and differing only in scale. There is, therefore, a distance function $F(x)$ such that the family of difference bodies is given by $F(x) \leq t$, as t varies. Consider the concept of *successive minima* given by Minkowski: let b^1, b^2, b^3 be three lattice points such that b^1 is the smallest nonzero lattice point according to F, b^2 is the shortest nonzero lattice point linearly independent of b^1, and b^3 the shortest lattice point linearly independent of both b^1 and b^2. As a special case of a more general result presented in Kannan, Lovász and Scarf (1988), it may be argued that all three of these vectors are neighbors of the origin for the matrix A. Given the validity of Theorem 1, it follows that if the normal vector to the common characteristic lattice plane is w, then $w \cdot b^i = c_i$ with $c_i = -1, 0$ or 1. This provides us with a very limited set of possibilities for the common characteristic plane, which could possibly be exploited in constructing an alternative argument for its existence.

REFERENCES

Kannan, R., L. Lovász and H. E Scarf (1988), *The Shapes of Polyhedra*, in preparation.

Scarf, H. E. (1981), Production sets with indivisibilities, *Econometrica*, **49**, 1-32.

Scarf, H. E. (1985), Integral polyhedra in three space, *Mathematics of Operations Research*, **10**, 403–438.

White, Philip M. (1983), Discrete activity analysis, Ph.D. Thesis, Yale University.

Grade of Membership Representations: Concepts and Problems

Burton Singer*
Yale University

1. INTRODUCTION

Heterogeneous populations defined by a large number of discrete conditions, no combination of which occurs with high frequency, are especially difficult to model in a meaningful way. Mixtures of distributions are the most common form of mathematical structure utilized for this purpose. However, a fundamental weakness of mixture models is the presumption that the underlying population can be decomposed into distinct well-defined categories at some level of refinement. Two contexts in which such decompositions have not been empirically defensible are medical diagnosis [Woodbury and Manton (1982) and Swartz et al. (1987)] and health status assessment of communities of elderly persons [Berkman et al. (1988)].

In the context of medical diagnosis, clinical textbooks designate extensive lists of symptoms and/or measurement outcomes which characterize particular diseases. However, most people who actually have a particular

AMS 1980 Subject Classification: Primary 62H99, Secondary 62620.
Key words and phrases: Heterogeneity, profiles, contingency tables.
*This research was supported by NIA Contract No. N01-AG-02105.

PROBABILITY, STATISTICS,
AND MATHEMATICS
Papers in Honor of Samuel Karlin

Copyright © 1989 by Academic Press, Inc.
All rights of reproduction in any form reserved.
ISBN-0-12-058470-0

disease do not exhibit the full text book list of symptoms. Instead they may be thought of as being some degree of similarity to one disease and varying degrees of similarity to other disorders based on partial and overlapping symptomatologies for several diseases. Similarly in attempting to characterize health status in communities of elderly persons, one is confronted with the fact that many individuals have multiple disabilities, no combination of which occurs with high frequency. Thus crisply defined health status categories cannot be meaningfully constructed.

An important response to these analytical difficulties are the Grade of Membership (GOM) models introduced by Woodbury et al. (1978) and subsequently developed and applied in a variety of medical, epidemiological and health policy settings [see e.g. Manton et al. (1987), Manton and Vertrees, (1988) and Swartz et al. (1987)]. The purpose of this paper is to make precise many of the intuitive ideas about GOM models in the extant literature. This necessitates the introduction of new concepts for both the static and stochastic process versions of GOM. A diverse array of open mathematical and statistical problems is exposed in our discussion, thereby revealing a challenging research domain for the future.

In Section 2 we discuss a 2-variable GOM model in detail in order to illustrate the principal ideas in the simplest possible setting. Section 3 contains a formal specification of GOM representations for multidimensional contingency tables and introduces the concepts of admissible and extreme admissible profiles which are implicit in the empirical applications of this class of models. In Section 4 we discuss estimation and goodness-of-fit assessments for static GOM models, emphasizing the centrality of vector goodness-of-fit criteria. Section 5 contains several formulations of GOM for stochastic processes, emphasizing admissible pure-type processes appropriate for characterizing the health status dynamics of heterogeneous communities. We conclude in Section 6 with the beginnings of a research agenda exposed by the formulations in Sections 2-5.

2. AN ELEMENTARY SPECIAL CASE

We begin with a 2-variable GOM model as the simplest prototypical situation where most of the subtleties of the general case reveal themselves, but in a very transparent form. To this end, consider an artificial data set $\mathbf{X}^{(i)} = (X_1^{(i)}, X_2^{(i)})$ for $1 \leq i \leq I$, $I =$ (number of individuals) where the component responses are health status measures interpreted as

$$X_j^{(i)} = \begin{cases} 1 & \text{if individual } i \text{ has deleterious condition } j \\ 0 & \text{otherwise} \end{cases}$$

Individuals with response vector $(1,1)$ will be called 'severely impaired' while those with response vector $(0,0)$ will be called 'healthy'. These extreme response vectors may be viewed as medically and biologically meaningful profiles of conditions (or their absence) that occur together. They are each defined by logical AND statements–e.g. the 'severely impaired' profile, $(1,1)$, means condition 1 AND condition 2 occur together. Our goal here, and in the general formulation in Sections 3-5, is to be able to interpret each individual's response vector as being some 'degrees-of- similarity' to scientifically interpretable special profiles.

If we identify the pair $(0,0)$ and $(1,1)$ as a set of scientifically interpretable profiles – call them profiles 1 and 2, respectively – then individuals with response vectors $(0,1)$ or $(1,0)$ should be thought of as intermediate between them. We quantify this idea by assigning to each individual, i, the grade of membership (GOM) scores $\gamma_i = (\gamma_{i;1}, \gamma_{i;2})$ where $\gamma_{i;k}$ means 'degree of similarity of individual i to profile k'. We insist that $\gamma_{i;k} \geq 0$ and $\gamma_{i;1} + \gamma_{i;2} = 1$. In the present example it is intuitively reasonable to assign scores according to the rule:

$$\gamma_i = \begin{cases} (1,0) & \text{if the } i^{th} \text{ individual has response vector } (0,0) \\ (0,1) & \text{if the } i^{th} \text{ individual has response vector } (1,1) \\ (1/2,1/2) & \text{if the } i^{th} \text{ individual has response vector } (0,1) \text{ or } (1,0) \end{cases}$$

Although these assignments are intuitively obvious for our 2-variable health status example, such assignments, and the definition of 'interpretable' profiles, are by no means obvious for community health status studies where response vectors may contain 40-50 components, [see in this regard Berkman et al. (1988)]. Furthermore, each component has a biological defense for its inclusion in the investigation. Thus one hopes that a formal GOM model is sufficiently sensitive to identify meaningful GOM scores relative to what an investigator regards as 'interpretable' profiles of conditions. To see that this is indeed the case in our present 2-variable formulation, consider modeling response probabilities $P(\mathbf{X} = \ell) = P(X_1 = \ell_1, X_2 = \ell_2)$, where $\ell_j = 0$ or 1, by

$$
\begin{aligned}
(1) \qquad P(\mathbf{X}^{(\mathbf{g})} = \ell) &= \int_{\mathbf{S}_K} P(\mathbf{X}^{(g)} = \ell \mid \mathbf{g} = \gamma) d\mu(\gamma) \\
&= \int_{\mathbf{S}_K} \sum_{j=1}^{2} P(X_j^{(g)} = \ell_j \mid \mathbf{g} = \gamma) d\mu(\gamma) \\
&= \int_{\mathbf{S}_K} \sum_{j=1}^{2} [\sum_{k=1}^{K} \gamma_k P(Y_j^{(k)} = \ell_j)] \, d\mu(\gamma)
\end{aligned}
$$

$\mathbf{S}_K = \{\gamma = (\gamma_1, \ldots, \gamma_K) : \gamma_k \geq 0, \sum_{k=1}^{K} \gamma_k = 1\}$ and $\mu(\cdot)$ is a probability measure on $\mathbf{S}_K$.

The GOM scores, $\mathbf{g}$, may be viewed as random parameters in the population, and a basic model assumption is that the response variables are independent conditional on values, γ, of $\mathbf{g}$. The 'interpretable' profiles described above are a special case of a more general notion of profile that must be introduced to clarify the full meaning of the representation, (1). In particular, we define a set of K extreme admissible profiles to be a collection of level vectors $\ell^{(1)}, \ldots, \ell^{(K)}$, where $\ell^{(k)}$ occurs with probability 1 for individuals identified with the characteristics of profile k. We define the random variables $\{Y_j^{(k)}\}$ to represent the possible responses on variable j for an individual in profile k. These variables satisfy $P(Y^{(k)} = \ell_j^{(k)})$, which automatically implies that $P(\mathbf{Y}^{(k)} = \ell^{(k)}) = 1$.

Although we singled out the set of extreme admissible profiles $\ell^{(1)} = (0, 0)$ and $\ell^{(2)} = (1, 1)$ a priori – i.e. $K = 2$ – in the context of a community study of health status, it is important to notice that GOM models may also be used in an exploratory mode. In this situation we would ask for estimates of sets of extreme admissible profiles that are best supported by the data. We would then assess whether or not the 'best fitting' set of profiles is indeed substantively interpretable. For 2-variable GOM models the possible sets of extreme admissible profiles are:

$K=2$	$K=3$	$K=4$
(0,0); (0,1)	(0,0); (0,1); (1,0)	(0,0); (0,1); (1,0); (1,1)
(0,0); (1,0)	(0,0); (0,1); (1,1)	
(0,0); (1,1)	(0,1); (1,0); (1,1)	
(0,1); (1,0)	(0,0); (1,0); (1,1)	
(0,1); (1,1)		
(1,0); (1,1)		

With this list at hand we determine a 'best fitting' GOM model by maximizing the conditional likelihood

$$\mathcal{L} = \prod_{i=1}^{I}[\prod_{j=1}^{2}\sum_{k=1}^{K}\gamma_{i;k}P(Y_j^{(k)} = \ell_j)]^{\delta_{i;\ell}} \tag{2}$$

$$\text{where } \delta_{i,\ell} = \begin{cases} 1 & \text{if individual } i \text{ has response vector } \ell = (\ell_1, \ell_2) \\ 0 & \text{otherwise} \end{cases}$$

For each value of K, maximization is over the set of $\binom{4}{K}$ possible extreme admissible profiles and over all possible GOM scores $(\gamma_{i;1}, \ldots, \gamma_{i;K})$ for $1 \leq i \leq I$. When $K = 2$ the maximum is attained for the pair of profiles $(\ell^{(1)}, \ell^{(2)})$ which define the level vectors *complementary* to those – call them ℓ^- and ℓ^{--} – for which $\min_{\ell,\ell'}(n_\ell + n_{\ell'})$ is attained. Here $n_\ell = \sum_{i=1}^{I}\delta_{i;\ell}$ = (number of individuals with response vector ℓ). The GOM scores for which (2) is maximized are given by

$$\gamma_i = (\gamma_{i;1}, \gamma_{i;2}) = \begin{cases} (1,0) & \text{if } \delta_{i;\ell^{(1)}} = 1 \\ (0,1) & \text{if } \delta_{i;\ell}^{(2)} = 1 \\ (1/2, 1/2) & \text{if } \delta_{i,\ell}^{-} = 1 \text{ or } \delta_{i,\ell}^{--} = 1 \end{cases}$$

The value of $\ell n\mathcal{L}$ at which the maximum is attained is $-\min_{\ell,\ell'}(n_\ell + n_{\ell'})\ell n\, 4$. Observe that what we designated, a priori, as 'interpretable' profiles in the context of health status – i.e. $\ell^{(1)} = (0,0)$ and $\ell^{(2)} = (1,1)$ – would arise as the unconstrained maximum likelihood profiles only if $\min_{\ell,\ell'}(n_\ell + n_{\ell'})$ was attained for ℓ^- and ℓ^{--} given by $(0,1)$ and $(1,0)$. Our intuitively prescribed GOM scores, $(1/2, 1/2)$, for persons with 'intermediate' levels of disability – i.e. with response vectors $(0,1)$ or $(1,0)$

– are also the maximum likelihood estimates when $(0,0)$ and $(1,1)$ are the estimated extreme admissible profiles.

When $K = 3$, let ℓ^- be the level vector for which $\min_\ell n_\ell$ is attained. Then $\mathcal{L}$ is maximized when $(\ell^{(1)}, \ell^{(2)}, \ell^{(3)})$ are the profiles complementary to ℓ^- and

$$\gamma = \begin{cases} (1,0,0) & \text{if } \delta_{i;\ell^{(1)}} = 1 \\ (0,1,0) & \text{if } \delta_{i;\ell^{(2)}} = 1 \\ (0,0,1) & \text{if } \delta_{i;\ell^{(3)}} = 1 \\ (1/2,1/2,0) & \text{if } \delta_{i;\ell^-} = 1 \text{ and } \prod_{j=1}^2 \sum_{k=1}^3 \gamma_k P(Y_j^{(k)} = \ell_j^-) = \gamma_1\gamma_2 \\ (1/2,0,1/2) & \text{if } \delta_{i;\ell^-} = 1 \text{ and } \prod_{j=1}^2 \sum_{k=1}^3 \gamma_k P(Y_j^{(k)} = \ell_j^-) = \gamma_1\gamma_3 \\ (0,1/2,1/2) & \text{if } \delta_{i;\ell^-} = 1 \text{ and } \prod_{j=1}^2 \sum_{k=1}^3 \gamma_k P(Y_j^{(k)} = \ell_j^-) = \gamma_2\gamma_3 \end{cases}$$

The value of $\ell n\ \mathcal{L}$ at which the maximum is attained is $-\min_\ell(n_\ell)\ell\ 4$

Finally for $K = 4$, all possible level vectors form the set of extreme admissible profiles. Each individual is assigned a GOM score equal to 1 for the profile corresponding to his/her response vector. The value of $\ell n\ \mathcal{L}$ in this situation is 0 and, obviously, represents a perfect fit to the original data.

The maximum likelihood estimate is unique for $K = 2$ provided $\min_{\ell,\ell'}(n_\ell + n_{\ell'})$ is attained for a unique pair of responses. If, for example, $n_{0,0} = n_{0,1} = n_{1,0} < n_{1,1}$ then $\min_{\ell,\ell'}(n_\ell + n_{\ell'}) = n_{0,0} + n_{0,1} = n_{0,0} + n_{1,0} = n_{0,1} + n_{1,0}$ and the indistinguishable sets of maximum likelihood extreme admissible profiles are $(1,1); (1,0), (1,1); (0,1)$, and $(0,0); (1,1)$. When $K = 3$ the GOM model estimate is unique provided $\min_\ell(n_\ell)$ is attained for a unique response vector. Thus we have a complete characterization of the maxima for the 2-variable binary response GOM model.

With the intuition provided by this special case we now consider GOM models for general multidimensional contingency tables and stochastic processes.

3. MULTIDIMENSIONAL CONTINGENCY TABLES

Let $\mathbf{X} = (X_1, \ldots, X_J)$ be a vector whose components are discrete variables, each of which can only assume a finite number of possible values.

Thus any continuous variable will be assumed to have been approximated by an ordinal categorical variable having a similar distribution. We denote the distribution of $\mathbf{X}$ by $P(\mathbf{X} = \ell)$ where $\ell = (\ell_1, \ldots, \ell_J)$ is a vector whose coordinates are possible levels of the variables $X_1, \ldots, X_J$. The basic idea of a GOM representation for $P(\mathbf{X} = \ell)$ is the transferring of the stochastic dependence among the original variables into degree of similarity (or GOM) OBscores for individuals and then conditional on these scores, assuming that the original variables are independent.

More formally we associate with each individual a set of scores $\mathbf{g} = (g_1, \ldots, g_K)$ such that $g_k \geq 0$ and $\sum_k g_k = 1$. We label an individual's response vector as $\mathbf{X}^{(\mathbf{g})} = (X_1^{(\mathbf{g})}, \ldots, X_J^{(\mathbf{g})})$ and, henceforth, denote the distribution of $\mathbf{X}$ by $P(\mathbf{X}^{(\mathbf{g})} = \ell)$. Now we *assume* that conditional on the values of $\mathbf{g}$, the original variables are independent. This leads immediately to the representation

$$P(\mathbf{X}^{(\mathbf{g})} = \ell) = \int_{\mathbf{S}_K} \prod_{j=1}^{J} P(X_j^{(\mathbf{g})} = \ell_j \mid \mathbf{g} = \gamma) d\mu(\gamma) \tag{3}$$

where $\mu(\gamma)$ is the distribution of GOM scores and $\mathbf{S}_K = \{\gamma = (\gamma_1, \ldots, \gamma_K) : \gamma_k \geq 0, \sum_k \gamma_k = 1\}$ is the unit simplex with K vertices. We further *assume* that the conditional probabilities in (3) are given by

$$\begin{aligned} P(X_j^{(\mathbf{g})} = \ell_j \mid \mathbf{g} = \gamma) &= \sum_{k=1}^{K} \gamma_k P(Y_j^{(k)} = \ell_j) \\ &\equiv \sum_{\ell=1}^{K} \gamma_k \lambda_{k,j,\ell_j} \end{aligned} \tag{4}$$

where $\sum_{\ell_j \in L_j} \lambda_{k,j,\ell_j} = 1$ for $1 \leq k \leq K$; $1 \leq j \leq J$ and $L_j = \{$ possible levels of jth variable$\}$.

The random variables $Y_j^{(k)}$ for $1 \leq k \leq K$ are interpreted as the jth variable, X_j, but for individuals identified with what we will refer to as the kth pure-type (or ideal profile). Their distributions $\{\lambda_{k,j,\ell_j} = P(Y_j^{(k)} = \ell_j)\}$ are the basis for defining what we will mean by *extreme admissible profiles* and a weaker formulation which will be referred to as simply *admissible profiles*.

The essential idea behind an extreme admissible profile is that distinguished levels of particular variables – i.e. substantively important levels such as 'presence of a physical disability' in a health status study – occur with probability 1 in a pure-type. Especially, $P(Y_j^{(k)} = d_j) = 1$ for $j \in J' \subset \{1, 2, \ldots, J\}$ where d_j denotes a distinguished level for variable j. This implies that $P(\bigcap_{j \in J'} (Y_j^{(k)} = d_j)) = 1$, which in turn implies a logical AND statement defining a conjunction of conditions – $\{d_j\}_{j \in J}$ – which occur together. The conjunction should represent – in the context of medical diagnosis, for example – an ideal symptomatology for a disease or class of diseases.

More formally we have

DEFINITION 1. A family of probability distributions $\{\lambda_{k,j,\ell_j}\}_{\ell_j \in L_j}$, $1 \leq j \leq J$, $1 \leq k \leq K$ will be said to define a set of K extreme admissible profiles if:

(i) There is at most one profile, call it k_0, such that $\lambda_{k_0,j,\ell_j} = 0$ for $1 \leq j \leq J$ and all *distinguished* levels $\ell_j = d_j \in L_j$.

REMARK. The profile, k_0, is identified as one in which no distinguished characteristic occurs. In the context of health status assessments this would correspond to healthy people – i.e. they do not have any disabilities.

(ii) For all profiles other than k_0 there is at least one variable for which $\lambda_{k,j,d_j} = 1$, where d_j = (set of distinguished levels on variable j). For all variables not satisfying this condition, $\lambda_{k,j,d_j} = 0$.

(iii) For each pair of profiles (k, k') not including k_0 there is at least one variable for which $\lambda_{k,j,k_j} = 1$ and $\lambda_{k',j,d_j} = 0$.

REMARK. This means that there is at least one condition that distinguishes each profile from all of the others.

With this definition at hand we can interpret (3) and (4) with $\{\lambda_{k,j,\ell_j}\}_{\ell_j \in Lj}$ defining an extreme admissible profile as inducing a point mapping from the J-coordinate space $X_1 \times \cdots \times X_J$ (associated with the variables $X_1, \ldots, X_J$) to the simplex $\mathbf{S}_K$ whose vertices are identified with extreme admissible profiles. Each individual's GOM score $\mathbf{g} = (g_1, \ldots, g_K)$ defines a location in barycentric coordinates for that individual in $\mathbf{S}_K$.

For some applications criteria (i)-(iii) prove to be excessively stringent. A somewhat more general formulation which is implicit in many empirical analyses [e.g. Manton and Vertrees (1988)] is to replace (ii) and (iii) by the requirement that a distinguished level enters an admissible profile if λ_{k,j,d_j} is substantially larger than the marginal frequency, $P(X_j = d_j)$, for the same variable. Condition (i) is weakened to the requirement that λ_{k_0,j,d_j} is roughly of the same magnitude or less than the marginal frequency $P(X_j = d_j)$ for *all* variables.

Thus far we have assumed that GOM scores and admissible profiles are based on all variables for a given individual. However, to ensure an interpretable set of profiles it is important to distinguish between those variables that are viewed as central to a classification system and those that should not be included in the production of admissible profiles. Variables which, on substantive grounds, are considered to be potentially important components of admissible profiles, will be referred to as *internal* variables. All other variables will be called *external*. For example, when characterizing the health status of elderly populations, measures of physical and cognitive impairment, presence or absence of a variety of chronic diseases, and direct physiological measurements (e.g. systolic and diastolic blood pressure) are designated to be internal variables. Demographic characteristics such as age, sex, geographic location, and income are treated as external variables. This distinction is reflected in (3) and (4) if we rewrite them as

$$P(\mathbf{X}^{(\mathbf{g})} = \ell) = \prod_{j \in J_E} \prod_{j \in J_I} \sum_{k=1}^{K} \gamma_k P(Y_j = \ell_j^{(k)}) \tag{5}$$

where

$$J_E = \{j : X_j \text{ is an external variable}\}$$
$$J_I = \{j : X_j \text{ is an internal variable}\}.$$

Then $P(Y_j^{(k)} = \ell_j) = \lambda_{k,j,\ell_j}$ is only subject to admissibility or extreme admissibility restrictions for $j \in J_I$. The pure-type probabilities $P(Y_J^{(k)} = \ell_j)$ for $j \in J_E$ are not subject to a priori profile restrictions.

4. ESTIMATION

There are two distinct statistical estimation problems associated with GOM representations. These are:

(i) Determine a set of extreme or just admissible profiles with the smallest value of K and a distribution μ on the simplex $\mathbf{S}_K$ which best fits – according to some pre specified optimality criteria – the empirical frequencies $P(\mathbf{X} = \ell)$.

(ii) Determine a set of extreme or just admissible profiles with the smallest value of K and a vector, $\mathbf{g} = (g_1, \ldots, g_K)$, for each individual which best fits – according to some optimality criteria – $P(\mathbf{X} = \ell)$.

It is well known – see e.g. Tukey (1974) – that these are not equivalent problems. In the present paper we focus attention on (ii). Problem (i) is entirely uncharted territory in the context of GOM.

Estimation strategies to-date for problem (ii) are all based on maximization of a conditional likelihood, where the conditioning is on the GOM score, $\mathbf{g}$. [See e.g. Manton and Stallard (1988); Woodbury et al. (1978)]. In particular for individual i, we identify a vector $\mathbf{X}^{(i)} = (X_1^{(i)}, \ldots, X_J^{(i)})$ of observations and set

$$\delta_{i,\ell j} = \begin{cases} 1 & \text{if } X_j^{(i)} = \ell_j \\ 0 & \text{otherwise} \end{cases}$$

Assuming that individuals generate response vectors independent of each other, the likelihood conditional on GOM scores is given by

$$\mathcal{L} = \prod_{i=1}^{I} \prod_{j \in J_E} \prod_{j \in J_I} \prod_{\ell_j \in L_j} [\sum_{k=1}^{K} \gamma_{i,k} \lambda_{k,j,\ell_j}]^{\delta_{i,\ell_j}} \tag{6}$$

where $\{\gamma_{i,k}\}$, $1 \leq k \leq K$ are the GOM scores for individual i and I = total number of individuals. Now let A_K = collection of all sets of K extreme or just admissible profiles. Then estimation of GOM scores and pure type probabilities reduces to the optimization problem of calculating $\gamma^{(1)}, \ldots, \gamma^{(I)}$ and $\{\lambda_{k,j,\ell}\}$ such that

$$\sup_{\{\lambda_{k,j,\ell_j}\}, j \in J_E} \left[\sup_{\substack{\gamma^{(1)}, \ldots, \gamma^{(I)} \in \mathbf{S}_K \\ \{\lambda_{k,j,\ell_j}\} \in A_K}} \mathcal{L} \right] \tag{7}$$

is attained. This calculation presumes that K = number of profiles is fixed a priori. When fitting GOM representations to data, we begin with $K = 2$

and increase K until the value of $\mathcal{L}$ at which the optimum is attained does not increase appreciably for two successive values of K. Under most circumstances we insist on $K \leq 9$, which seems, based on experience to-date, to be the limit of interpretability across a diverse range of problems.

When using GOM to explore high dimensional data for interpretable admissible profiles and individual GOM scores, it is useful to calculate $\gamma^{(1)}, \ldots, \gamma^{(I)}$ and $\{\gamma_{k,j,\ell_j}\}$ as solutions of the optimization:

$$(8) \qquad \sup_{\{\lambda_{k,j,\ell_j}\}, j \in J_E} \left[\sup_{\substack{\gamma^{(1)},\ldots\gamma^{(I)} \in \mathbf{S}_K \\ \{\lambda_{k,j,\ell_j}\} j \in J_I}} \mathcal{L} \right]$$

The λ's are, of course, constrained to be probabilities for each variable and each pure-type, k. Then we check the λ's for all $j \in J_I$ for admissibility or approximate extreme admissibility. This strategy is useful in that it allows the data to suggest for which one or two values of K admissible profiles are defined.

Having obtained a best fitting – in the sense of constrained conditional likelihood maximization – GOM representation, we further impose the following more fine-grained criteria for asserting that a given representation is adequate.

1. *All* individuals with GOM scores $\gamma_{i,k}$ satisfying $.9 \leq \gamma_{i,k} \leq 1$ for some profile k should have response vector $\mathbf{X}^{(i)}$ with the same distinguished levels – or absence of them in the case of k_0 – as appear in the kth profile.

2. Individuals with $\gamma_{i,k} + \gamma_{i,k'} = 1$ for some pair (k, k') of profiles and whose GOM scores satisfy $.2 \leq \gamma_{i,k} \leq .8$ should have response vectors with distinguished levels only from profiles k and k'.

3. Individuals for whom $\min_{1 \leq k \leq K} \gamma_{i,k} \geq .15$ must have response $\mathbf{X}^{(i)}$ that contain at least one distinguished level from each profile that has one or more distinguished levels.

Criteria 1-3 represent stringent tests of a GOM representation, since they require that *all* individuals whose GOM scores are in the designated ranges have response vectors of a prescribed character. A slight weakening of these conditions – particularly to take account of the inevitable missing data on selected variables – would be to allow at most 1% of persons satisfying given GOM score criteria to violate the corresponding response vector restriction(s). Although more stringent tests than 1-3 above could be imposed – tailored to specific contexts – we view this list as a minimum set of requirements to support an assertion that a given GOM representation 'fits the data'. These criteria focus on the ability of GOM models to both decompose high dimensional contingency tables *and* indicate how a vector of diverse responses for an individual can now be interpreted as the individual being different degrees of similarity away from admissible profiles.

5. STOCHASTIC PROCESS

Let $\mathbf{X}(t)$, $t = 0, 1, 2, \ldots$ be a vector stochastic process all coordinates of which assume a finite set of values. Associate a grade of membership process $\mathbf{g}(t)$ with $\mathbf{X}(t)$ such that the finite dimensional joint distributions satisfy

(9)

$$P[\bigcap_{t=0}^{T}(\mathbf{X}(t) = \ell_t] \equiv P[\bigcap_{t=0}^{T}(\mathbf{X}^{\mathbf{g}(t)}(t) = \ell_t)]$$

$$= \int_{\mathbf{S}_0^T} P(\bigcap_{t=0}^{T}(\mathbf{X}^{\mathbf{g}(t)}(t) = \ell_t) \mid \mathbf{g}(s) = \gamma_s, s \leq T) d\mu(\gamma_0, \ldots, \gamma_T)$$

$$= \int_{\mathbf{S}_0^T} \prod_{j=1}^{J} P(\bigcap_{t=0}^{T} X_j^{\mathbf{g}(t)}(t) = \ell_{t;j} \mid \mathbf{g}(s) = \gamma_s, s \leq T) d\mu(\gamma_0, \ldots, \gamma_T)$$

$$= \int_{\mathbf{S}_0^T} \prod_{j=1}^{J} \prod_{t=1}^{T} [\sum_{k_t=1}^{K_t} \gamma_{k_t,t} P(Y_j^{(k_t)}(t) = \ell_{t;k_tj} \mid Y_j^{(k_s)}(s) = \ell_{s;k_0,j}, s \leq t-1)]$$

$$\times \sum_{k_0=1}^{K_0} \gamma_{k_0,0} P(Y_j^{(k_0)}(0) = \ell_{0;k_tj}) d\mu(\gamma_0, \ldots, \gamma_T)$$

Here $\mathbf{S}_0^T = \mathbf{S}_{K_0} \times \cdots \times \mathbf{S}_{K_T}$ is a Cartesian product of simplices containing $K_0, K_1, \ldots,$ vertices respectively and $\gamma(t)$, $t \geq 0$ is a sequence of GOM scores such that $\gamma(t)$ represents the position of an individual in the simplex, $\mathbf{S}_{K_t}$. This specification assumes that the coordinate processes are independent conditional on the GOM process $\gamma(t)$ $t \leq T$. The pure-type process $Y_j^{(k_t)}(t)$, $t \geq 0$ must be further constrained but, nevertheless, allow for a variety of formulations of 'admissible profile processes'. The time-dependent indices k_t reflect the fact that the number of admissible profiles and the distinguished levels defining them can vary with time.

Two examples of admissible profile processes will illustrate the myriad of possibilities depending on scientific context.

EXAMPLE 1. ONSET AND TERMINATION PROCESS.

In the context of health status dynamics of elderly populations, considerable interest centers around the age of onset of chronic conditions and their consequential disabilities. Complementary to age of onset is age of termination of chronic conditions due to mortality selection in a given cohort. Formalization of these notions in the setting of evolving extreme admissible profiles can be introduced via the following

DEFINITION. $Y_j^{(k_t)}(t)$, $t \geq 0$ is an extreme admissible onset process

if $\exists$ a sequence of pure-type indices k_t, $t \geq 0$ such that

$$P(Y_j^{(k_t)}(t) = d_j) = \begin{cases} 0 \text{ if } t < t_j^\star \text{ for some fixed } T_j^\star > 0 \\ 1 \text{ if } t \geq T_j^\star \end{cases}$$

Correlatively, $Y_j^{(k_t)}(t)$, $t \geq 0$ is an extreme admissible termination process

if $\exists$ a sequence of pure-type indices k_t, $t \geq 0$ such that

$$P(Y_j^{(k_t)}(t) = d_j) = \begin{cases} 1 & \text{if } t < t_j^{\star\star} \\ 0 & \text{if } t \geq T_j^{\star\star} \text{ for some fixed } T_j^{\star\star} > 0 \end{cases}$$

REMARK. Observe that the age of onset formulation implies that

$$P(Y_j^{(k_t)}(t) = d_j \mid Y_j^{(k_s)}, s < t) = \begin{cases} 1 & \text{if } t \geq T^\star \\ 0 & \text{if } t < T^\star \end{cases}$$

Thus a family of profiles is defined to be extreme admissible with selected age of onset coordinates if:

(ii) The extreme admissibility criteria are satisfied for all marginal frequencies for $t = 0, 1, 2, \ldots, T$

(ii) The extreme onset restrictions hold for some subset of the internal variables.

EXAMPLE 2. ADMISSIBLE MARKOV CHAINS.

We will call $Y_j^{(k_t)}(t)$, $t \geq 0$ an admissible time-homogeneous Markov profile process if:

(i) $\exists$ a sequence of pure-type indices k_t, $t \geq 0$ such that

$$P(Y_j^{(k_t)}(t) = \ell_{t;k_t,j} \mid Y_j^{(k_s)}(s) = \ell_{s;k_s,j}, \; s < t)$$
$$= P(Y_j^{(k_t)}(t) = \ell_{t;k_t,j} \mid Y_j^{(k_{t-1})}(t-1) = \ell_{t-1;k_{t-1},j}))$$

and the probability only depends on the state labels and not on time;

(ii) letting $t_j = \#(t : Y_j^{(k_t)} = d_j, \; t \leq T)$ and

$$U_j = \#(t : X_j(t) = d_j, \; t \leq T),$$

the occupation times of $Y_j^{(k_t)}(t)$ and $X_j(t)$, respectively, at distinguished level, d_j, we have

$$\frac{t_j - U_j}{U_j} > .20$$

REMARK. This 20% difference in occupation time at the distinguished level is an ad-hoc choice at the discretion of the investigator. Indeed, requiring a 50% difference could be appropriate depending on the scientific context.

In addition to the internal-variable based profile processes, external processes – i.e. those not used to define admissible profiles – are of central importance for applications of GOM. We illustrate the role of external processes by considering survival analysis conditional on GOM scores. This methodology is an important alternative to the widely used hazard regression technology but suited to a setting – many statistically dependent covariates – where regression models would be both numerically unstable and substantively uninterpretable.

To clarify the issues, let

$$Z^{(\mathbf{g})}(t) = \begin{cases} 1 & \text{if an individual with GOM vector } \mathbf{g} \\ & \text{is alive at the end of time period } t \\ 0 & \text{otherwise} \end{cases}$$

and identify $\mathbf{g}$ with $\mathbf{g}(0)$. All internal variables will be evaluated at $t = 0$. Hence the only dynamics to be considered is the mortality process $Z^{(\mathbf{g})}(t)$. The original contingency table data are represented – in terms of GOM scores and profile probabilities – by

$$P(\mathbf{X}^{(\mathbf{g})} = \ell, \bigcap_{t=0}^{T}(Z^{(\mathbf{g})}(t) = i_t))$$
$$= \int_{\mathbf{S}_k} \prod_{j=1}^{J} [\sum_{k=1}^{K} \gamma_k \lambda_{k,j,\ell_j}] \cdot P(\bigcap_{t=0}^{T} Z^{(\gamma)}(t) = i_t) d\mu(\gamma)$$

where i_t can assume the values 0 or 1 and $T =$ (last observation period for a given study). In terms of the process $Z^{(\mathbf{g})}(t)$, $t = 0, 1, \ldots, T$ we define

$$L = \max(t : Z^{(\mathbf{g})}(t) = 1) + 1$$
$$= \text{time period in which death occurs for an individual with GOM vector } \mathbf{g}.$$

Then the survivor functions conditional on GOM scores are given by $P(L^{(\mathbf{g})} > t \mid \mathbf{g} = \gamma)$ for $t \geq 0$. In mortality studies [e.g. Berkman et al. (1988)] one usually isolates subsets of GOM scores, $G \subset \mathbf{S}_K$ and then computes Kaplan-Meier estimates of $P(L^{(\mathbf{g})} > \tau \mid \mathbf{g} \in G)$. This is survivor analysis conditional on a system of admissible or extreme admissible profiles and a restricted family of GOM scores.

6. DISCUSSION

The formal specification of GOM models with the accompanying notions of admissible profiles makes explicit many intuitive ideas from the empirical literature that attempts to characterize heterogeneous populations when crisp classification into meaningful categories is not possible. The distinguishing feature of GOM representations relative to mixture models is that they are convex in parameters rather than in distributions. To clarify this distinction it is useful to compare GOM with P. Lazarsfeld's latent structure models [see e.g. Lazarsfeld and Henry, (1968)]. For this we begin with the GOM formalism

$$P(X^{(\mathbf{g})} = \ell) = \int_{S_{K^\star}} P(\mathbf{X}^{(\mathbf{g})} = \ell \mid \mathbf{g} = \gamma) d\mu(\gamma)$$

and then restrict μ to be a point mass $d\mu(\gamma) = \begin{cases} 1 & \text{if } \gamma = \gamma^\star \\ 0 & \text{otherwise} \end{cases}$. This yields

$$P(\mathbf{X}^{(\mathbf{g})} = \ell) = P(\mathbf{X}^{(g)} = \ell \mid \mathbf{g} = \gamma^\star).$$

Now we assume that

$$\begin{aligned} P(\mathbf{X}^{(\mathbf{g})} = \ell \mid \mathbf{g} = \gamma^\star) &= \sum_{k=1}^{K^\star} \gamma_k^\star P(\mathbf{Z}^{(k)} = \ell) \quad \text{(a \textit{mixture} of distributions)} \\ &= \sum_{k=1}^{K^\star} \gamma_k^\star \prod_{j=1}^{J} P(Z_j^{(k)} = \ell_j) \end{aligned}$$

Thus our original multidimensional contingency table is a mixture of 'independence' tables, and this is Lazarsfeld's latent structure decomposition. If we restricted μ in the GOM representation to being a point mass at $\gamma^{\star\star}$ we would have

$$P(\mathbf{X}^{(\mathbf{g})} = \ell) = \prod_{j=1}^{J} \sum_{k=1}^{K} \gamma_k^{\star\star} P(Y_j^{(k)} = \ell_j)$$

Observe that K and $K^\star$ are not necessarily equal. The key point, however, is that the representations are *not* equivalent since addition and multiplication are not commutative operations.

With this comparison at hand we conclude by listing several open problems which are suggested by our formalization.

The notion of an acceptable GOM model is based on multiple test criteria which are designed to put any such model in jeopardy from several directions. The general philosophy about goodness-of-fit tests that we are adopting – [see Bush and Mosteller (1959) and Heckman and Walker (1987)] – is that acceptability of a model in a particular analysis should be assessed by introducing a list of features of empirical data that the model will be required to reproduce with some fidelity and then checking whether *all* or at least a major portion of the features are indeed reproduced. The measures of success of a GOM model rest on: (i) its ability to reveal meaningful interpretations for what otherwise might look like a morass of conditions on an individual record; (ii) its ability to define interpretable admissible profiles and distributions of GOM scores. Formalizing a multicriteria assessment-of-fit process is a major challenge for the future. Tukey (1987) contains some ideas in this direction; however, this appears to be largely unexplored territory.

GOM parameter estimates are increasingly 'stable' as the number of internal variables increases in the sense that omission of one or a few variables does not qualitatively change the estimated pure-type probabilities for the remaining variables if the 2-stage optimization defined by (8) is utilized for estimation purposes. Empirical experience indicates that this stability phenomena occurs with a diverse collection of data sets. However, a rigorous explanation lies in the future.

The above mentioned stabilizing effect is somehow related to the ability of GOM models to satisfy the stringent acceptability criteria in which an individual's response vector – often based on 40 or more responses – identified by his/her GOM score, $\mathbf{g}$, is often a morass which becomes interpretable within the GOM representation. This is particularly perplexing in that GOM is underidentified in the usual senses of identifiability. Only equivalence classes of models defined by regions of the simplex, $\mathbf{S}_K$ and moments of the pure-type distributions should be identifiable. However, a precise formulation of equivalence classes of identifiable GOM models is a major challenge for the future.

It is our hope that these problems will be resolved in the near future and thereby help to establish a quite novel methodology on a more rigorous basis.

REFERENCES

Berkman, L.F., B. Singer and K. Manton (1988), *Black/White Differences in Health Status and Mortality Among the Elderly*, Demography, (in press).

Bush, R.R. and F. Mosteller (1959), A comparison of eight models (R.R. Bush and W.K. Estes, eds.), *Studies in Mathematical Learning Theory*, Stanford University Press, 292–307.

Heckman, J.J. and J.R. Walker (1987), Using goodness-of-fit and other criteria to choose among competing duration models: a case study of Hutterite data (C. Clogg, ed.), *Sociological Methodology*, American Sociological Association, Washington, D.C., 247–307.

Lazarsfeld, P.F. and N. Henry (1968), *Latent Structure Analysis*, Houghton-Mifflin, Boston.

Manton, K.G., E. Stallard, M.A. Woodbury, H.D. Tolley and A.I. Yashin (1987), Grade of membership techniques for studying complex event history processes with unobserved covariates (C. Clogg, ed.), *Sociological Methodology*, American Sociological Association, Washington, D.C., 309-345.

Manton, K.G. and E. Stallard (1988), *Chronic Disease Modelling: Measurement and Evaluation of the Risks of Chronic Disease Processes*, Charles Griffin, Ltd., London.

Manton, K.G. and J. Vertrees (1988), *A Population–Based Evaluation of the Quality of Care Provided by a Medicaid Health Insuring Organization in Philadelphia*, Medical Care, in press.

Swartz, M., D. Blazer, M.A. Woodbury, L. George and K.G. Manton (1987), Study of somatization disorder in a community population utilizing grade of membership analysis, *Psychiatric Developments*, **3**, 219-237.

Tukey, J.W. (1974), Named and faceless values: An initial exploration in memory of Prasanta C. Mahalanobis, *Sankhyā, Series A*, **36**, Part 2, 125-176.

Tukey, J.W. (1987), Configural polysampling, *SIAM Review*, **29**, 1-20.

Woodbury, M.A., J. Clive and A. Garson (1978), Mathematical typology: A grade of membership technique for obtaining disease definition, *Computers and Biomedical Research*, **11**, 277-298.

Woodbury, M.A. and K.G. Manton (1982), A new procedure for analysis of medical classification, *Methods of Information in Medicine*, **21**, 210-220.

Uniform Error Bounds Involving Logspline Models

Charles J. Stone*
University of California, Berkeley

1. INTRODUCTION

Splines are of increasing importance in statistical theory and methodology. In particular, Stone and Koo (1986) and Stone (1988) considered exponential families of densities in which the logarithm of the density is a spline. Such exponential families are the subject of the present paper, as are corresponding exponential response models. In each context we use an extension of a key result of de Boor (1976) to obtain a bound on the L_∞ norm of the approximation error associated with maximizing the associated expected log-likelihood.

Let Y be a real-valued random variable ranging over a compact interval $\mathcal{I}$; without loss of generality, let $\mathcal{I} = [0, 1]$. Suppose that Y has a density f that is continuous and positive on $\mathcal{I}$.

AMS 1980 Subject Classifications: Primary 62G05, Secondary 62F12.

Key words and phrases: B-splines, uniform approximation, exponential family, density estimate.

*Research supported in part by National Science Foundation Grant DMS-8600409.

PROBABILITY, STATISTICS,
AND MATHEMATICS
Papers in Honor of Samuel Karlin

Copyright © 1989 by Academic Press, Inc.
All rights of reproduction in any form reserved.
ISBN-0-12-058470-0

Let S be a standard vector space of spline functions of a given order $q \geq 1$ on $\mathcal{I}$ (piecewise polynomials of degree $q-1$ or less that are right-continuous on $\mathcal{I}$ and continuous at 1) having finite dimension $K \geq 2$. Let $B_1, \ldots, B_K$ be a B-spline basis of S (see de Boor, 1978). Then $B_1, \ldots, B_K$ are nonnegative and sum to 1 on $\mathcal{I}$. Let $\theta_1, \ldots, \theta_K$ be real constants. Set

$$c(\theta_1, \ldots, \theta_K) = \log\left(\int \exp\left(\sum_k \theta_k B_k(y)\right) dy\right)$$

and

$$f(y; \theta_1, \ldots, \theta_K) = \exp\left(\sum_k \theta_k B_k(y) - c(\theta_1, \ldots, \theta_K)\right), \quad y \in \mathcal{I}.$$

This defines an exponential family of densities on $\mathcal{I}$. Observe that, for $a \in \mathbf{R}$,

$$c(\theta_1 + a, \ldots, \theta_K + a) = c(\theta_1, \ldots, \theta_K) + a$$

and hence

$$f(y; \theta_1 + a, \ldots, \theta_K + a) = f(y; \theta_1, \ldots, \theta_K), \quad y \in \mathbf{R}.$$

Consequently the exponential family fails to be identifiable. In order to make it identifiable, we require that $\theta_K = 0$.

Let Θ denote the collection of ordered $(K-1)$-tuples $\theta_1, \ldots, \theta_{K-1}$ of real numbers. For $\boldsymbol{\theta} = (\theta_1, \ldots, \theta_{K-1}) \in \Theta$, set

$$s(y; \boldsymbol{\theta}) = \theta_1 B_1(y) + \cdots + \theta_{K-1} B_{K-1}(y), \quad y \in \mathcal{I},$$

$$C(\boldsymbol{\theta}) = \log\left(\int \exp(s(y; \boldsymbol{\theta})) dy\right),$$

and

$$f(y; \boldsymbol{\theta}) = \exp(s(y; \boldsymbol{\theta}) - C(\boldsymbol{\theta})), \quad y \in \mathcal{I}.$$

This defines an identifiable exponential family; it is referred to as a *logspline model* since $\log(f(\cdot; \boldsymbol{\theta})) \in \mathcal{S}$.

Let $Y_1 \ldots Y_n$ be independent random variables having common density f, which is not necessarily a member of the logspline model. The log-likelihood function $l(\boldsymbol{\theta})$, $\boldsymbol{\theta} \in \Theta$, corresponding to the model is defined by

$$l(\boldsymbol{\theta}) = \sum_i \log(f(Y_i; \boldsymbol{\theta})) = \sum_i [s(Y_i; \theta) - C(\boldsymbol{\theta})], \quad \boldsymbol{\theta} \in \Theta.$$

Suppose that (for given values of $Y_1, \ldots, Y_n$) the log-likelihood function has a maximizing value $\hat{\theta} \in \Theta$. Then this maximizing value is unique and is called the maximum-likelihood estimate of $\boldsymbol{\theta}$; the corresponding density $\hat{f}$ defined by $\hat{f}(y) = f(y; \hat{\theta})$ for $y \in \mathcal{I}$, is referred to as the *logspline density estimate* corresponding to the given logspline model.

The expected log-likelihood function $\lambda(\boldsymbol{\theta})$, $\boldsymbol{\theta} \in \Theta$, is defined by

$$\lambda(\boldsymbol{\theta}) = El(\boldsymbol{\theta}) = n\left[\int s(y;\boldsymbol{\theta})f(y)dy - C(\boldsymbol{\theta})\right], \quad \boldsymbol{\theta} \in \Theta.$$

It follows by a convexity argument that the expected log-likelihood function has a unique maximizing value $\boldsymbol{\theta}^* \in \Theta$. (Recall that f is a positive density on $\mathcal{I}$ and that $s(\cdot;\boldsymbol{\theta})$ is a nonconstant function for $\boldsymbol{\theta} \neq 0$.) Consider the corresponding density $Q_S f$ on $\mathcal{I}$ defined by $Q_S f(y) = f(y; \boldsymbol{\theta}^*)$, $y \in \mathcal{I}$. The density f belongs to the logspline model if and only if $f = Q_S f$ on $\mathcal{I}$. When f does not belong to this model, the function $f - Q_S f$ plays an important role in the analysis of the asymptotic behavior of the logspline density estimate (see Stone, 1988); roughly speaking, it acts as a bias term.

Given a real-valued function g on $\mathcal{I}$, set $\| g \|_\infty = \sup_{\mathcal{I}} | g(y) |$. Let $\mathcal{F}$ denote a family of positive densities on $\mathcal{I}$ such that the family $\{\log(f) : f \in \mathcal{F}\}$ is an equicontinuous family. Set

$$\delta_{\mathcal{S}}(f) = \inf_{s \in \mathcal{S}} \| \log(f) - s \|_\infty, \quad f \in \mathcal{F}.$$

(For an upper bound to $\delta_{\mathcal{S}}(f)$ in terms of the smoothness of $\log(f)$, see Theorem XII.1 of de Boor, (1978.) In Section 4 we will obtain an inequality of the form

$$\| \log(f) - \log(Q_S f) \|_\infty \leq M\delta_{\mathcal{S}}(f), \quad f \in \mathcal{F}, \tag{1}$$

where the positive constant M depends only on $\mathcal{F}$, the order of $\mathcal{S}$, and a bound on a suitable "global mesh ratio" of $\mathcal{S}$. The main point of this result is that M does not depend on $K = \dim(\mathcal{S})$. It follows from (1) that

$$\| f - Q_S f \|_\infty \leq [\exp(M\delta_{\mathcal{S}}(f) - 1)] \| f \|_\infty, \quad f \in \mathcal{F}.$$

Suppose now that the distribution of Y depends on a real variable x that ranges over a compact interval $\mathcal{I}$; without loss of generality, let $\mathcal{I} = [0,1]$. Let $f(\cdot \mid x)$ denote the dependence of density of Y on x. It is supposed that $f(y \mid x)$, $x, y \in \mathcal{I}$, is a continuous and positive function.

Let $\mathcal{H}$ be a standard finite-dimensional vector space of spline functions of a given order on $\mathcal{I}$ having dimension $J \geq 1$, and let $H_1, \ldots, H_J$ be a B-spline basis of $\mathcal{H}$.

Let $\mathcal{B}$ denote the collection of $J \times (K-1)$ matrices $\boldsymbol{\beta} = (\beta_{jk})$ of real numbers β_{jk}, $1 \leq j \leq J$ and $1 \leq k \leq K-1$. Let $\boldsymbol{\beta} \in \mathcal{B}$. For $1 \leq k \leq K-1$, let $h_k(\cdot;\boldsymbol{\beta})$ be the real-valued function on $\mathcal{I}$ defined by

$$h_k(x;\boldsymbol{\beta}) = \sum_j \beta_{jk} H_j(x), \quad x \in \mathcal{I}.$$

Set

$$\mathbf{h}(x;\boldsymbol{\beta}) = (h_1(x;\boldsymbol{\beta}), \ldots, h_{K-1}(x;\boldsymbol{\beta})), \quad x \in \mathcal{I}.$$

Then $\mathbf{h}(\cdot;\boldsymbol{\beta})$ is an $\mathbf{R}^{K-1}$-valued function on $\mathcal{I}$.

The logspline response model corresponding to $\mathcal{H}$ and $\mathcal{S}$ is defined by

$$f(y \mid x;\boldsymbol{\beta}) = f(y;\mathbf{h}(x;\boldsymbol{\beta})) = \exp(s(y;\mathbf{h}(x;\boldsymbol{\beta})) - C(\mathbf{h}(x;\boldsymbol{\beta})))$$

for $\boldsymbol{\beta} \in \mathcal{B}$ and $x, y \in \mathcal{I}$. Observe that, for $\boldsymbol{\beta} \in \mathcal{B}$ and $x \in \mathcal{I}$, $f(\cdot \mid x;\boldsymbol{\beta})$ is a positive density on $\mathcal{I}$.

Let $x_1, \ldots, x_n \in \mathcal{I}$ and let $Y_1, \ldots, Y_n$ be independent random variables such that Y_i has density $f(\cdot \mid x_i)$. The corresponding log-likelihood function $l(\boldsymbol{\beta})$, $\boldsymbol{\beta} \in \mathcal{B}$, is defined by

$$l(\boldsymbol{\beta}) = \sum_i \log(f(Y_i \mid x_i;\boldsymbol{\beta})) = \sum_i (s(Y_i;\mathbf{h}(x_i;\boldsymbol{\beta})) - C(\mathbf{h}(x_i;\boldsymbol{\beta}))), \quad \boldsymbol{\beta} \in \mathcal{B}.$$

The expected log-likelihood function $\lambda(\boldsymbol{\beta})$, $\boldsymbol{\beta} \in \mathcal{B}$, is defined by

$$\lambda(\boldsymbol{\beta}) = El(\boldsymbol{\beta}) = \sum_i \left[\int s(y;\mathbf{h}(x_i;\boldsymbol{\beta})) f(y \mid x_i) dy - C(\mathbf{h}(x_i;\boldsymbol{\beta})) \right], \quad \boldsymbol{\beta} \in \mathcal{B}.$$

Suppose that $\mathcal{H}$ is identifiable from $x_1, \ldots, x_n$; that is, that if $h \in \mathcal{H}$ and $h(x_1) = \cdots = h(x_n) = 0$, then $h = 0$ on $\mathcal{I}$. Then, by a convexity argument, the expected log-likelihood function has a unique maximum $\boldsymbol{\beta}^* \in \mathcal{B}$. Consider the corresponding function $Q_{\mathcal{S}} f$ on $\mathcal{I} \times \mathcal{I}$ defined by

$$Q_{\mathcal{S}} f(y \mid x) = f(y \mid x;\boldsymbol{\beta}^*), \quad x, y \in \mathcal{I}.$$

Let $\mathcal{T}$ denote the tensor product of $\mathcal{H}$ and $\mathcal{S}$; that is, the vector space of real-valued functions on $\mathcal{I} \times \mathcal{I}$ spanned by functions of the form $h(x)s(y)$, $x, y \in \mathcal{I}$, as h and s range over $\mathcal{H}$ and $\mathcal{S}$ respectively. Then $\mathcal{T}$ has dimension

JK, and the functions $H_j(x)B_k(y)$, $x, y \in \mathcal{I}$, $1 \le j \le J$ and $1 \le k \le K$ form a basis of $\mathcal{T}$.

Given a real-valued function g on $\mathcal{I} \times \mathcal{I}$, set $\| g \|_\infty = \sup_{\mathcal{I}\times\mathcal{I}} g(x, y)$. Let $\mathcal{F}$ denote a family of continuous and positive functions f on $\mathcal{I} \times \mathcal{I}$ such that $f(\cdot \mid x)$ is a density on $\mathcal{I}$ for $x \in \mathcal{I}$ and $\{\log(f) : f \in \mathcal{F}\}$ is an equicontinuous family of functions on $\mathcal{I} \times \mathcal{I}$. Set

$$\delta_{\mathcal{T}}(f) = \inf_{t \in \mathcal{T}} \| \log(f) - t \|_\infty, \quad f \in \mathcal{F}.$$

(For an upper bound to $\delta_{\mathcal{T}}(f)$ in terms of the smoothness of $\log(f)$, see Theorem 12.8 of Schumaker, 1981.) In Section 5 we will obtain an inequality of the form

$$\text{(2)} \qquad \| \log(f) - \log(Q_{\mathcal{T}} f) \|_\infty \le M \delta_{\mathcal{T}}(f), \quad f \in \mathcal{F},$$

where the positive constant M depends on $\mathcal{F}$, the orders of $\mathcal{H}$ and $\mathcal{S}$, bounds on the global mesh ratios of $\mathcal{H}$ and $\mathcal{S}$, and a measure of regularity of $x_1, \ldots, x_n$ that depends on $\mathcal{H}$. The main point of this result is that M does not depend on $J = \dim(\mathcal{H})$ or $K = \dim(\mathcal{S})$.

2. PRELIMINARY INEQUALITIES

The bound on the global mesh ratio for $\mathcal{S}$ described in de Boor (1976) is equivalent to a bound of the form

$$\text{(3)} \qquad M^{-1}K^{-1} \le \int B_k(y)dy \le M_1 K^{-1}, \quad 1 \le k \le K,$$

where $M_1 > 1$ is a constant. Since the support of B_k is an interval having length $q \int B_k(y)dy$, where q is the order of $\mathcal{S}$, (3) can be written as a two-sided bound on this length. Under (3) there is a constant $M_2 > 1$ (depending on the order of $\mathcal{S}$) such that, for $\theta_1, \ldots, \theta_K \in \mathbf{R}$,

$$\text{(4)} \quad M_1^{-1} M_2^{-1} K^{-1} \sum_k \theta_k^2 \le \int \left(\sum_k \theta_k B_k(y) \right)^2 dy \le M_1 K^{-1} \sum_k \theta_k^2$$

(see (7) of de Boor, 1976).

Similarly, we assume that

$$(5) \qquad M_1^{-1}J^{-1} \le \int H_j(x)dx \le M_1 J^{-1}, \quad 1 \le j \le J.$$

Under (5) it can be assumed that, for $\beta_1, \ldots, \beta_J \in \mathbf{R}$,

$$(6) \quad M_1^{-1}M_2^{-1}J^{-1}\sum_j \beta_j^2 \le \int \left(\sum_j \beta_j H_j(x)\right)^2 dx \le M_1 J^{-1} \sum_j \beta_j^2.$$

For a given order q of $\mathcal{H}$, the functions in $\mathcal{H}$ are piecewise polynomials of degree $q-1$ or less. In light of (5), a natural regularity assumption on $x_1, \ldots, x_n$ is that

$$(7) \qquad M_3^{-1} n \int h^2(x)dx \le \sum_i h^2(x_i) \le M_3 n \int h^2(x)dx, \quad h \in \mathcal{H},$$

for some constant $M_3 > 1$. It follows from (7) that $\mathcal{H}$ is identifiable from $x_1, \ldots, x_n$. It also follows from (7), by choosing M_3 larger if necessary depending on the order of $\mathcal{H}$, that

$$(8) \qquad \sum_i H_j(x_i) \le M_3 J^{-1} n, \quad 1 \le j \le J.$$

(Let h denote the sum of the H_k's whose support overlaps with that of H_j; note that $H_j \le 1 = h = h^2$ on the support of H_j.)

Let ρ be a positive (Borel) function on $\mathcal{I}$ such that, for some constant $M_4 > 1$,

$$(9) \qquad M_4^{-1} \le \rho(y) \le M_4, \quad y \in \mathcal{I}.$$

For the real-valued function g on $\mathcal{I} \times \mathcal{I}$, let $\| g \|_2$ be the nonnegative square root of

$$\| g \|^2 = \sum_i \int g^2(x_i, y)\rho(y)dy.$$

For $1 \le j \le J$ and $1 \le k \le K$, define B_{jk} on $\mathcal{I} \times \mathcal{I}$ by

$$B_{jk}(x,y) = H_j(x)B_k(y), \quad x, y \in \mathcal{I}.$$

It follows from (4), (6), (7) and (9) that, for $\beta \in \mathcal{B}$,

$$\frac{n}{M_1^2 M_2^2 M_3 M_4 JK} \sum_j \sum_k \beta_{jk}^2 \le \left\| \sum_j \sum_k \beta_{jk} B_{jk} \right\|_2^2 \tag{10}$$
$$\le \frac{M_1^2 M_3 M_4 n}{JK} \sum_j \sum_k \beta_{jk}^2.$$

3. THE INVERSE GRAM MATRIX

Consider the $K \times K$ matrix M whose (k,l)th entry is $\int B_k(y)B_l(y)\rho(y)dy$. It follows from (4) that $\mathbf{M}$ is invertible. Let α_{kl} denote the (k,l)th entry of $\mathbf{M}^{-1}$. Then

$$\| \mathbf{M}^{-1} \|_\infty \le \max_k \sum_l | \alpha_{kl} | .$$

By a slight extension of a result in de Boor (1976), there is a constant $M_8 > 1$, depending on M_1, M_2 and M_4, such that

$$\| \mathbf{M}^{-1} \|_\infty \le M_8 K \tag{11}$$

(see the proof of (18) below). This has the following consequence.

LEMMA 1. *Set* $g = \sum_k \theta_k B_k$. *Then*

$$\max_k | \theta_k | \le M_8 K \max_k \left| \int g(y) B_k(y) \rho(y) dy \right| .$$

For real-valued functions g_1 and g_2 on $\mathcal{I} \times \mathcal{I}$ such that the norms $\| g_1 \|_2$ and $\| g_2 \|_2$ are finite, set

$$\langle g_1, g_2 \rangle = \sum_i \int g_1(x_i, y) g_2(x_i, y) \rho(y) dy.$$

Then $\| g \|_2^2 = \langle g, g \rangle$. Consider now the $JK \times JK$ matrix $\mathbf{M}$ whose $((j,k),(l,m))$th entry is the inner product $\langle B_{jk}, B_{lm} \rangle$ of B_{jk} and B_{lm}. It follows from (10) that $\mathbf{M}$ is invertible. Let α_{jklm} denote the $((j,k),(l,m))$th entry of $\mathbf{M}^{-1}$. Then

$$\| \mathbf{M}^{-1} \|_\infty = \max_{j,k} \sum_l \sum_m | \alpha_{jklm} | . \tag{12}$$

We will now imitate the elegant proof of (11) above in de Boor's paper (see also Descloux, 1972).

Set

$$f_{jk} = \sum_l \sum_m \alpha_{jklm} B_{lm}.$$

Then $\langle f_{jk}, B_{lm} \rangle$ equals 1 if $j = l$ and $k = m$ and it equals zero otherwise. Consequently,

$$0 < \| f_{jk} \|_2^2 = \alpha_{jkjk}.$$

Set $M_5 = M_1^2 M_2^2 M_3 M_4 > 1$. Then, by (10),

$$M_5^{-1} J^{-1} K^{-1} n \alpha_{jkjk}^2 \leq M_5^{-1} J^{-1} K^{-1} n \sum_l \sum_m \alpha_{jklm}^2 \leq \| f_{jk} \|_2^2 = \alpha_{jkjk}.$$

Therefore

$$\alpha_{jkjk} \leq M_5 JKn^{-1}$$

and

$$\sum_l \sum_m \alpha_{jklm}^2 \leq M_5 JKn^{-1} \alpha_{jkjk} \leq (M_5 JKn^{-1})^2. \tag{13}$$

Set $M_6 = M_1^2 M_2 M_3 M_4 > 1$.

LEMMA 2. *There is a constant $M_7 > 1$, depending on M_6, such that*

$$| \alpha_{jklm} | \leq M_5 M_6 M_7 JK M_7^{-(|j-l|+|k-m|)} n^{-1}.$$

PROOF. Let (j,k) be given and let $v, w \in \mathbf{R}$ with $v^2 + w^2 = 1$. For $c \in \mathbf{R}$, set

$$S_c = \{(l,m) : v(l-j) + w(m-k) \geq c\}$$

and

$$g_c = \sum_{S_c} \sum \alpha_{jklm} B_{lm}.$$

Let $c > 0$. Since f_{jk} is orthogonal to B_{lm} for $(l, m) \neq (j, k)$, g_c is orthogonal to f_{jk}. There is a positive constant u, depending only on the order of $\mathcal{H}$ and $\mathcal{S}$, such that if $(l, m) \in S_c$ and $(l_1, m_1) \neq S_{c-u}$, then B_{lm} and $B_{l_1 m_1}$ have disjoint support and hence are orthogonal to each other. Consequently, g_c is orthogonal to $f_{jk} - g_{c-u}$ and hence to g_{c-u}. Therefore,

$$\| g_{c-u} \|_2^2 + \| g_c \|_2^2 = \| g_{c-u} - g_c \|_2^2$$

and hence

$$\| g_{c-u} \|_2^2 \leq \| g_{c-u} - g_u \|_2^2 . \tag{14}$$

Now

$$g_{c-u} - g_u = \sum\sum_{S_{c-u,c}} \alpha_{jklm} B_{lm},$$

where

$$S_{c-u,c} = S_{c-u} \backslash S_c = \{(l, m) : c - u \leq v(l - j) + w(m - k) < c\}.$$

We conclude from (10) and (14) that

$$\sum\sum_{S_{c-u,c}} \alpha_{jklm}^2 \geq M_6^{-2} \sum\sum_{S_{c-u}} \alpha_{jklm}^2, \quad c > 0. \tag{15}$$

Set

$$a_\nu = \sum\sum_{S_{c+(\nu-1)u,c+\nu u}} \alpha_{jklm}^2, \quad \nu = 0, 1, 2, \ldots .$$

By (15),

$$| a_\nu | \geq M_6^{-2}(| a_\nu | + | a_{\nu+1} | + \cdots), \quad \nu = 0, 1, 2, \ldots . \tag{16}$$

According to Lemma 2 of de Boor (1976), (16) implies that

$$| a_\nu | \leq | a_0 | \, | M_6 |^2 \, (1 - M_6^{-2})^\nu, \quad \nu = 0, 1, 2, \ldots . \tag{17}$$

By (13) and (17),

$$| a_\nu | \leq (M_5 M_6 J K n^{-1})^2 \, (1 - M_6^{-2})^\nu, \quad \nu = 0, 1, 2, \ldots .$$

It follows by choosing v, w, and c appropriately that if

$$\nu \leq u^{-1}[(l - j)^2 + (m - k)^2]^{1/2},$$

then

$$| \alpha_{jklm} | \leq M_5 M_6 J K (1 - M_6^{-2})^{\nu/2} n^{-1}.$$

This yields the conclusion of the lemma. ∎

Set

$$M_8 = M_5 M_6 M_7 (M_7 + 1)^2 (M_7 - 1)^{-2} > 1.$$

It follows from (12) and Lemma 2 that

$$\| \mathbf{M}^{-1} \|_\infty \leq M_8 J K n^{-1}. \tag{18}$$

This inequality has the following implication.

LEMMA 3. *Set*

$$g = \sum_j \sum_k \beta_{jk} B_{jk}.$$

Then

$$\max_{j,k} | \beta_{jk} | \leq M_8 J K n^{-1} \max_{j,k} | \langle g, B_{jk} \rangle | .$$

4. LOGSPLINE MODELS

In this section, we obtain (1). For f a positive density on I and $0 \leq a < 1$, let f_a denote the density on $\mathcal{I}$ defined by

$$f_a(y) = \frac{(f(y))^a}{\int (f(y))^a dy}.$$

It can be assumed that $f_a \in \mathcal{F}$ for $f \in \mathcal{F}$ and $0 \leq a < 1$. (Extend $\mathcal{F}$ if necessary.)

Choose $s \in \mathcal{S}$ and define the real-valued function g on $\mathbf{R}$ by

$$\int \exp(ts(y) - g(t)) Q_{\mathcal{S}} f(y) dy = 1.$$

Then

$$g'(0) = \int s(y) Q_{\mathcal{S}} f(y) dy.$$

Also

$$\int [\log(Q_{\mathcal{S}} f(y)) + ts(y) - g(t)] f(y) dy$$

is maximized at $t = 0$; hence

$$g'(0) = \int s(y)f(y)dy.$$

Thus

$$\int s(y)[Q_S f(y) - f(y)]dy = 0.$$

Consequently,

$$\int B_k(y)[Q_S f(y) - f(y)]dy = 0, \quad 1 \le k \le K, \tag{19}$$

or, equivalently,

$$\int B_k(y)[Q_S f(y) - f(y)]dy = 0, \quad 1 \le k \le K-1. \tag{20}$$

Formula (20) can also be written as

$$\frac{\partial C}{\partial \theta_k}(\boldsymbol{\theta}^*) = \int B_k(y)f(y)dy, \quad 1 \le k \le K-1. \tag{21}$$

Let K be a fixed positive integer and let $\mathcal{S}$ otherwise vary subject to (3). Then $B_1 \dots B_K$ depend continuously (in the L_2 norm) on the knot sequence defining $\mathcal{S}$. Thus it follows from (21) and the properties of the Hessian matrix of $C(\cdot)$ (e.g., it is negative definite) that $\boldsymbol{\theta}^*$ depends continuously on $\int B_k(y)f(y)dy$, $1 \le k \le K-1$, and the knot sequence defining f.

Let $f \in \mathcal{F}$. There is an $s \in \mathcal{S}$ such that $\| \log(f) - s \|_\infty = \delta_{\mathcal{S}}(f)$. Since f is a density on $\mathcal{I}$, we conclude that

$$\left| \log \left(\int \exp(s(y))dy \right) \right| \le \delta_{\mathcal{S}}(f).$$

Consequently, there is a $\bar{\boldsymbol{\theta}} \in \Theta$ such that

$$\| \log(f) - \log(f(\cdot;\bar{\boldsymbol{\theta}}) \|_\infty \le 2\delta_{\mathcal{S}}(f). \tag{22}$$

Note that $Q_S \bar{f} = \bar{f}$, where $\bar{f} = \bar{f}(\cdot;\bar{\boldsymbol{\theta}})$. Thus it follows from (22) and the continuity properties of $\boldsymbol{\theta}^*$ described above that there is a positive constant M_{1K} (depending on M_1 and $\mathcal{F}$ as well as K) such that

$$\| \log(f(\cdot;\boldsymbol{\theta}^*)) - \log(f(\cdot;\bar{\boldsymbol{\theta}})) \|_\infty \le M_{1K}\delta_{\mathcal{S}}(f)$$

and hence

$$\| \log(f) - \log(Q_S f) \|_\infty \leq (M_{1K} + 2)\delta_S(f), \quad f \in \mathcal{F}. \tag{23}$$

Choose $\bar{\theta} \in \Theta$ such that (22) holds and set $\bar{f} = f(\cdot; \bar{\theta})$. Then

$$\| \log(f) - \log(\bar{f}) \|_\infty \leq 2\delta_S(f). \tag{24}$$

There are constants M_9, $M_{10} > 1$, depending on $\mathcal{F}$, such that

$$\| f - \bar{f} \|_\infty \leq M_9 \delta_S(\mathcal{F}) \tag{25}$$

and

$$M_{10}^{-1} \leq \bar{f}(y) \leq M_{10}, \quad y \in \mathcal{I}. \tag{26}$$

By (3), (19) and (25),

$$\left| \int B_k(y)[Q_S f(y) - \bar{f}(y)]dy \right| \leq M_1 M_9 K^{-1} \delta_S(f), \quad 1 \leq k \leq K. \tag{27}$$

Write

$$\log(Q_S f) - \log(\bar{f}) = \sum_k \theta_k B_k$$

and set $\epsilon = \max_k \mid \theta_k \mid$. Now $\| \log(Q_S f) - \log(\bar{f}) \|_\infty \leq \epsilon$ and hence

$$\| \log(f) - \log(Q_S f) \|_\infty \leq \epsilon + 2\delta_S(f). \tag{28}$$

It follows from (viii) on Page 155 of de Boor (1978) that there is a positive constant M_{11}, depending on the order of $\mathcal{S}$, such that

$$\epsilon \leq M_{11} \| \log(Q_S f) - \log(\bar{f}) \|_\infty . \tag{29}$$

Suppose that $\epsilon \leq 1$. Since $Q_S f = \bar{f} \exp(\sum_k \theta_k B_k)$, we conclude from (26) that

$$\left\| Q_S f - \bar{f} - \bar{f} \sum_k \theta_k B_k \right\|_\infty \leq M_{10} \epsilon^2$$

and hence from (3) and (27) that, for $1 \leq k \leq K$,

$$\left| \int B_k(y) \sum_l \theta_l B_l(y) \bar{f}(y) dy \right| \leq M_1 M_9 K^{-1} \delta_S(f) + M_1 M_{10} K^{-1} \epsilon^2. \tag{30}$$

According to (26), (30) and Lemma 1, there is a constant $M_{12} > 1$, depending on M_1, M_2 and M_{10}, such that

$$\epsilon \leq M_1 M_9 M_{12} \delta_{\mathcal{S}}(f) + M_1 M_{10} M_{12} \epsilon^2.$$

Suppose now that

$$M_1 M_{10} M_{12} \epsilon \leq \frac{1}{2}. \tag{31}$$

Then $\epsilon \leq 2M_1 M_9 M_{12} \delta_{\mathcal{S}}(f)$ and hence, by (28),

$$\| \log(f) - \log(Q_{\mathcal{S}} f) \|_\infty \leq M_{13} \delta_{\mathcal{S}}(f), \tag{32}$$

where $M_{13} = 2(M_1 M_9 M_{12} + 1)$. According to (29), a sufficient condition for (31) and hence for (32) is

$$\| \log(Q_{\mathcal{S}} f) - \log(\bar{f}) \|_\infty \leq M_{14}^{-1}, \tag{33}$$

where $M_{14} = 2M_1 M_{10} M_{11} M_{12}$.

Let

$$0 < \delta < 2^{-1} M_{13}^{-1} M_{14}^{-1}.$$

There is a positive integer K_0, depending on M_1 and the order of $\mathcal{S}$, such that

$$\delta_{\mathcal{S}}(f) \leq \delta, \quad K \geq K_0 \text{ and } f \in \mathcal{F} \tag{34}$$

(see Page 167 of de Boor, 1978). Let $K \geq K_0$. Suppose that

$$\| \log(f) - \log(Q_{\mathcal{S}} f) \|_\infty \leq 2^{-1} M_{14}^{-1}. \tag{35}$$

Then (33) follows from (24), so (32) holds.

We will now verify that (35) necessarily holds for $K \geq K_0$. Suppose not. Now

$$\| \log(f_a) - \log(Q_{\mathcal{S}} f_a) \|_\infty$$

is continuous in a for $0 \leq a < 1$ and it approaches 0 as $a \to 0$. (According to an earlier argument, $\boldsymbol{\theta}^*$ is continuous in a.) Thus there is a value of $a \in (0,1)$ such that

$$\| \log(f_a) - \log(Q_{\mathcal{S}} f_a) \|_\infty = 2^{-1} M_{14}^{-1}.$$

By the previous argument, (32) and (34) hold with f replaced by f_a; hence

$$\| \log(f_a) - \log(Q_{\mathcal{S}} f_a) \|_\infty \leq M_{13} \delta_{\mathcal{S}}(f_a) \leq M_{13} \delta < 2^{-1} M_{14}^{-1},$$

which yields a contradiction.

We have now shown that

$$\| \log(f) - \log(Q_S f) \|_\infty \le M_{13}\delta_S(f), \quad K \ge K_0 \text{ and } f \in \mathcal{F}. \tag{36}$$

The desired inequality (1) follows from (36) together with (23) for $1 \le K < K_0$. ∎

5. LOGSPLINE RESPONSE MODELS

In this section, we obtain (2). For f a positive function on $\mathcal{I} \times \mathcal{I}$ such that $f(\cdot \mid x)$ is a density on $\mathcal{I}$ for each $x \in \mathcal{I}$ and for $0 < a < 1$, let f_a be defined on $\mathcal{I} \times \mathcal{I}$ by

$$f_a(y \mid x) = \frac{(f(y \mid x))^a}{\int (f(y \mid x))^a dy}.$$

It can be assumed that $f_a \in \mathcal{F}$ for $f \in \mathcal{F}$. (Extend $\mathcal{F}$ if necessary.)

Let $1 \le k \le K - 1$. Choose $h \in \mathcal{H}$ and let $\mathbf{h}$ be the $\mathbf{R}^{K-1}$-valued function on $\mathcal{I}$ whose kth component is h and whose other components are zero. Define the real-valued function g on $\mathbf{R}$ by

$$g(t) = \sum_i \left[\int s(y; \mathbf{h}(x_i; \boldsymbol{\beta}^*) + t\mathbf{h}(x_i)) f(y \mid x_i) dy - C(\mathbf{h}(x_i; \boldsymbol{\beta}^*) + t\mathbf{h}(x_i)) \right].$$

Then

$$0 = g'(0) = \sum_i h(x_i) \left[\int B_k(y) f(y \mid x_i) dy - \frac{\partial C}{\partial \theta_k}(\mathbf{h}(x_i; \boldsymbol{\beta}^*)) \right].$$

Thus, for $1 \le j \le J$ and $1 \le k \le K - 1$,

$$\sum_i H_j(x_i) \frac{\partial C}{\partial \theta_k}(\mathbf{h}(x_i; \beta^*)) = \sum_i H_j(x_i) \int B_k(y) f(y \mid x_i) dy, \tag{37}$$

which can also be written as

$$\sum_i H_j(x_i) \int B_k(y)[f(y \mid x_i) - Q_T f(y \mid x_i)] dy = 0$$

or, equivalently, as

$$\sum_i H_j(x_i) \int B_k(y)[f(y \mid x_i) - Q_{\mathcal{T}} f(y \mid x_i)]dy = 0. \tag{38}$$

Let $f \in \mathcal{F}$. There is a $t \in \mathcal{T}$ such that $\| \log(f) - t \|_\infty = \delta_{\mathcal{T}}(f)$. Let $x \in \mathcal{I}$. Since $f(\cdot \mid x)$ is a density on $\mathcal{I}$, we conclude that

$$\left| \log\left(\int e^{t(x,y)} dy \right) \right| \leq \delta_{\mathcal{T}}(f), \quad x \in \mathcal{I}.$$

Consequently, there is a $\bar{\beta} \in \mathcal{B}$ such that

$$\| \log(f) - \log(f(\cdot \mid \cdot; \bar{\beta}) \|_\infty \leq 2\delta_{\mathcal{T}}(f). \tag{39}$$

Let J and K be fixed positive integers and let $\mathcal{H}, \mathcal{S}$ and $x_1 \dots x_n$ otherwise vary subject to (3), (5) and (7). It follows from (37) that there is a positive constant M_{JK} (depending on M_1, M_3 and $\mathcal{F}$ as well as J and K) such that

$$\| \log(\boldsymbol{f}(\cdot \mid \cdot; \boldsymbol{\beta}^*) - \log(\boldsymbol{f}(\cdot \mid \cdot; \bar{\boldsymbol{\beta}}) \|_\infty \leq \boldsymbol{M_{JK}\delta_{\mathcal{T}}(f)}. \tag{40}$$

We conclude from (39) and (40) that

$$\| \log(f) - \log(Q_{\mathcal{T}} f) \|_\infty \leq (M_{JK} + 2)\delta_{\mathcal{T}}(f), \quad f \in \mathcal{F}. \tag{41}$$

There are positive integers J_0 and K_0 and there is a positive constant M_9, depending on $\mathcal{F}, M_1 \dots M_4$ and the orders of $\mathcal{H}$ and $\mathcal{S}$ such that

$$\| \log(f) - \log(Q_{\mathcal{T}} f) \|_\infty \leq M_9 \delta_{\mathcal{T}}(f), \quad J \geq J_0, \ K \geq K_0 \text{ and } f \in \mathcal{F}. \tag{42}$$

The argument used to prove (42) is a refinement of that used to prove (36). To start off, choose $\bar{t} \in \mathcal{T}$ such that $\| \log(f) - \bar{t} \|_\infty = \delta_{\mathcal{T}}(f)$, set

$$\bar{c}(x) = \log\left(\int \exp(\bar{t}(x,y)) dy \right), \quad x \in \mathcal{I},$$

and note that

$$| \bar{c}(x) | \leq \delta_{\mathcal{T}}(f), \quad x \in \mathcal{I}.$$

Define $\bar{f}$ on $\mathcal{I} \times \mathcal{I}$ by $\bar{f}(y \mid x) = \exp(\bar{t}(x,y) - \bar{c}(x))$. Then

$$\| \log(f) - \log(\bar{f}) \|_\infty \leq 2\delta_{\mathcal{T}}(f).$$

There are constants M_{10}, $M_{11} > 1$, depending on $\mathcal{F}$, such that

$$\| f - \bar{f} \|_\infty \leq M_{10}\delta_{\mathcal{T}}(f) \tag{43}$$

and

$$M_{11}^{-1} \leq \bar{f}(y \mid x) \leq M_{11}, \quad x, y \in \mathcal{I}.$$

By (3), (8), (38) and (43),

$$\left| \sum_i H_j(x_i) \int B_k(y)[Q_{\mathcal{T}} f(y \mid x_i) - \bar{f}(y \mid x_i)]dy \right| \leq \frac{M_1 M_3 M_{10}}{JK} n\delta_{\mathcal{T}}(f)$$

for $1 \leq j \leq J$ and $1 \leq k \leq K$.

Write

$$\log(Q_{\mathcal{T}} f(y \mid x)) = t^*(x, y) - c^*(x), \quad x, y \in \mathcal{I},$$

where $t^* \in \mathcal{T}$, and set $t = t^* - \bar{t}$. Then

$$Q_{\mathcal{T}} f(y \mid x) = \exp(t(x, y) + \bar{c}(x) - c^*(x))\bar{f}(y \mid x), \quad x, y \in \mathcal{I},$$

$$\begin{aligned} c^*(x) &= \log\left(\int \exp(t(x, y) + \bar{c}(x))\bar{f}(y \mid x)dy\right) \\ &= \log\left((1 + \int [\exp(t(x, y) + \bar{c}(x)) - 1]\bar{f}(y \mid x)dy)\right) \end{aligned}$$

for $x \in \mathcal{I}$, and

$$Q_{\mathcal{T}} f(y \mid x) - \bar{f}(y \mid x) = [\exp(t(x, y) + \bar{c}(x) - c^*(x)) - 1]\bar{f}(y \mid x), \quad x, y \in \mathcal{I}.$$

Thus

$$c^*(x) - \bar{c}(x) \approx \int t(x, y)\bar{f}(y \mid x)dy, \quad x \in \mathcal{I},$$

and hence

$$Q_{\mathcal{T}} f(y \mid x) - \bar{f}(y \mid x) \approx \left[t(x, y) - \int t(x, y)\bar{f}(y \mid x)dy\right] \bar{f}(y \mid x) \tag{44}$$

for $x, y \in \mathcal{I}$.

Write

$$t(x, y) = \sum_j \sum_k \beta_{jk} H_j(x) B_k(y), \quad x, y \in \mathcal{I}.$$

It follows by a double application of (viii) on Page 155 of de Boor (1978) that there is a positive constant M_{12}, depending on the order of $\mathcal{H}$ and $\mathcal{S}$, such that

$$\max_{j,k} | \beta_{jk} | \leq M_{12} \| t \|_\infty .$$

Choose $\eta > 0$. Now

$$\int t(x,y)\bar{f}(y \mid x)dy = \sum_k \int B_k(y) \sum_j \beta_{jk} H_j(x) \bar{f}(y \mid x)dy.$$

Choose x_j in the support of H_j. Define $h \in \mathcal{H}$ by

$$\begin{aligned} h(x) &= \sum_k \int B_k(y) \sum_j \beta_{jk} H_j(x) \bar{f}(y \mid x_j)dy \\ &= \sum_j H_j(x) \sum_k \beta_{jk} \int B_k(y) \bar{f}(y \mid x_j)dy. \end{aligned}$$

There is a positive integer J_0, depending on M_1, M_{12} and $\mathcal{F}$ such that

$$\left| \int t(x,y)\bar{f}(y \mid x)dy - h(x) \right| \leq \eta \| t \|_\infty, \quad J \geq J_0 \text{ and } x \in \mathcal{I}.$$

After replacing $t^*(x,y)$ by $t^*(x,y) - h(x)$ and replacing $c^*(x)$ by $c^*(x) - h(x)$, we have that

$$\left| \int t(x,y)\bar{f}(y \mid x)dy \right| \leq \eta \| t \|_\infty, \quad J \geq J_0 \text{ and } x \in \mathcal{I}. \tag{45}$$

The argument used to prove (42) from (44) and (45) is similar to that used to prove (36), except that Lemma 3 is used instead of Lemma 1 and Theorem 12.8 of Schumaker (1981) is used instead of Page 167 of de Boor (1978).

Next it will be shown that, for each positive integer K, there is a positive integer J_0 and there is a positive constant M_{13}, both depending on $\mathcal{F}$, $M_1, \ldots, M_4$ and the order of $\mathcal{H}$ and $\mathcal{S}$, such that

$$\| \log(f) - \log(Q_\mathcal{T} f) \|_\infty \leq M_{13} \delta_\mathcal{F}(f), \quad J \geq J_0 \text{ and } f \in \mathcal{F}. \tag{46}$$

To this end, write

$$Q_\mathcal{S} f(y \mid x) = \exp\left(\sum_k \theta_k(x) B_k(y) - c(x) \right), \quad x, y \in \mathcal{I}.$$

From (21) we conclude that (as f varies over $\mathcal{F}$, etc.) the resulting functions $\theta_k(\cdot)$, $1 \le k \le K-1$, are uniformly bounded and equicontinuous, and there is a positive constant M_{14} such that

$$\max_{1\le k\le K-1} \delta_{\mathcal{H}}(\theta_k(\cdot)) \le M_{14}\delta_{\mathcal{T}}(f). \tag{47}$$

Observe that

$$\max_{1\le k\le K-1} \delta_{\mathcal{H}}(\theta_k(\cdot))$$

can be made arbitrary small by making J sufficiently large (see Page 167 of de Boor, 1978). According to (1), there is a positive constant M_{15} such that

$$\left|\log(f(y \mid x)) - \left(\sum_k \theta_k(x)B_k(y) - c(x)\right)\right| \le M_{15}\delta_{\mathcal{T}}(f), \qquad x, y \in \mathcal{I}. \tag{48}$$

It follows from (19) that

$$\int B_k(y)\left[\exp\left(\sum_m \theta_m(x)B_m(y) - c(x)\right) - f(y \mid x)\right] dy = 0$$

for $x \in \mathcal{I}$ and $1 \le k \le K$ and hence that

$$\sum_i H_j(x_i) \int B_k(y)\left[\exp\left(\sum_m \theta_m(x_i)B_m(y) - c(x_i)\right) - f(y \mid x_i)\right] dy = 0$$

for $1 \le j \le J$ and $1 \le k \le K$. Thus we conclude from (38) that

$$\sum_k H_j(x_i) \int B_k(y)\left[\exp\left(\sum_m \theta_m(x_i)B_m(y) - c(x_i)\right) - Q_{\mathcal{T}}f(y \mid x_i)\right] dy = 0$$

for $1 \le j \le J$ and $1 \le k \le K$.

For $1 \le k \le K-1$, choose $\bar{h}_k \in \mathcal{H}$ such that

$$| \theta_k(x) - \bar{h}_k(x) | = \delta_{\mathcal{H}}(\theta_k(\cdot)), \qquad x \in \mathcal{I}.$$

Set

$$\bar{c}(x) = \log\left(\int \exp\left(\sum_k \bar{h}_k(x)B_k(y)\right) dy\right), \qquad x \in \mathcal{I},$$

and define $\bar{f}$ on $\mathcal{I} \times \mathcal{I}$ by

$$\bar{f}(y \mid x) = \exp\left(\sum_k \bar{h}_k(x)B_k(y) - \bar{c}(x)\right).$$

Write

$$Q_{\mathcal{T}}f(y \mid x) = \exp\left(\sum_k h^*(x)B_k(y) - c^*(x)\right), \quad x, y \in \mathcal{I},$$

where $h^* \in \mathcal{H}$ for $1 \leq k \leq K-1$. It now follows by arguing as in the proofs of (36) and (42) that there is a positive constant M_{16} such that

$$| \theta_k(x) - h^*(x) | \leq M_{16} \max_{1 \leq k \leq K-1} \delta_{\mathcal{H}}(\theta_k(\cdot)), \quad 1 \leq k \leq K-1 \text{ and } x \in \mathcal{I}.$$

Thus there is a positive constant M_{17} such that

$$\left|\log(Q_{\mathcal{T}}f(y \mid x)) - \left(\sum_k \theta_k(x)B_k(y) - c(x)\right)\right| \leq M_{17} \max_{1 \leq k \leq K-1} \delta_{\mathcal{H}}(\theta_k(\cdot)). \tag{49}$$

The desired result (46) follows from (47)-(49).

Finally it will be shown that, for each positive integer J, there is a positive integer K_0 and there is a positive constant M_{18}, both depending on $\mathcal{F}$, $M_1, \ldots, M_4$ and the order of $\mathcal{H}$ and $\mathcal{S}$, such that

$$\| \log(f) - \log(Q_{\mathcal{T}}f) \|_\infty \leq M_{18}\delta_{\mathcal{T}}(f), \quad K \geq K_0 \text{ and } f \in \mathcal{F}. \tag{50}$$

To this end, let $\beta_1(\cdot), \ldots, \beta_J(\cdot)$ be the real-valued functions on $\mathcal{I}$ such that

$$\sum_i \left[\log(f(y \mid x_i)) - \sum_j \beta_j(y)H_j(x_i)\right]^2$$

minimizes

$$\sum_i \left[\log(f(y \mid x_i)) - \sum_j \beta_j H_j(x_i)\right]^2$$

for $y \in \mathcal{I}$. It follows from the appropriate analog of Lemma 2 that, as f varies over $\mathcal{F}$, etc., the resulting functions $\beta_1(\cdot), \cdots, \beta_J(\cdot)$ are uniformly bounded and equicontinuous, that there is a positive constant M_{19} such that

$$\max_{1\le j\le J} \delta_{\mathcal{S}}(\beta_j(\cdot)) \le M_{19}\delta_{\mathcal{T}}(f), \tag{51}$$

and that there is a positive constant M_{20} such that

$$\left|\log(f(y \mid x)) - \sum_j \beta_j(y)H_j(x)\right| \le M_{20}\delta_{\mathcal{T}}(f), \quad x, y \in \mathcal{I}. \tag{52}$$

Observe that

$$\max_{1\le j\le J} \delta_{\mathcal{S}}\ (\beta_j(\cdot))$$

can be made arbitrarily small by making K sufficiently large. For $1 \le j \le J$ choose $\bar{s}_j \in \mathcal{S}$ such that

$$\mid \beta_j(y) - \bar{s}_j(y) \mid = \delta_{\mathcal{S}}(\beta_j(\cdot)), \quad y \in \mathcal{I}. \tag{53}$$

Set

$$\bar{c}(x) = \log\left(\int \exp\left(\sum_j H_j(x)\bar{s}_j(y)dy\right)\right), \quad x \in \mathcal{I}.$$

There is a constant M_{21} such that

$$\mid \bar{c}(x) \mid \le M_{21}\delta_{\mathcal{T}}(f), \quad x \in \mathcal{I}. \tag{54}$$

Define $\bar{f}$ on $\mathcal{I} \times \mathcal{I}$ by $\bar{f}(y \mid x) = \exp(\sum_j H_j(x)\bar{s}_j(y) - \bar{c}(x))$. Write

$$Q_{\mathcal{T}}f(y \mid x) = \exp\left(\sum_j H_j(x)s_j^*(y) - c^*(x)\right), \quad x, y \in \mathcal{I},$$

where $s^* \in \mathcal{S}$ for $1 \le j \le J$. It follows as in the proofs of (36), (42) and (49) that there is a positive constant M_{22} such that

$$\mid \log(Q_{\mathcal{T}}f(y \mid x)) - \log(\bar{f}(y \mid x)) \mid \le M_{22} \max_{1\le j\le J} \delta_{\mathcal{S}}(\beta_j(\cdot)). \tag{55}$$

The desired result (50) follows from (51)-(55).

Inequality (2) follows from (41), (42), (46), and (50). ∎

REFERENCES

de Boor, C. (1976), A bound on the L_∞-norm of the L_2-approximation by splines in terms of a global mesh ratio, *Mathematics of Computation*, **30**, 765-771.

de Boor, C. (1978), *A Practical Guide to Splines*, Springer–Verlag, New York.

Descloux, J. (1972), On finite element matrices, *SIAM Journal on Numerical Analysis*, **9**, 260-265.

Schumaker, L. L. (1981), *Spline Functions: Basic Theory*, Wiley, New York.

Stone, C.J. (1988), Large-sample inference for logspline models, Technical Report No. 171, Department of Statistics, University of California, Berkeley.

Stone C.J. and Koo, C.-Y. (1986), Logspline density estimation, *Contemporary Mathematics*, **59**, 1-15, American Mathematical Society, Providence.

An Alternative to C_p Model Selection that Emphasizes the Quality of Coefficient Estimation

Roy E. Welsch*
Massachusetts Institute of Technology

1. INTRODUCTION

Often in the process of analyzing regression data and selecting variables, we have produced the C_p plots discussed by Mallows (1973). While frequently quite useful, these plots are most appropriate when the emphasis is on prediction (fit) rather than on the quality of the estimated coefficients since C_p is based on a sum of squares of errors in the fitted values. It seems more reasonable to look for a selection statistic based on the sum of squares of errors in the coefficient estimates when we are particularly interested in the coefficients.

AMS 1980 Subject Classifications: Primary 62J05, Secondary 62J07.

Key words and phrases: Selection of variables, regression, ridge regression, C_p plot.

*We would like to thank Karen Martel for her technical typing. This work was supported, in part, by National Science Foundation Grants No. 8420614-IST and No. 8706393-DMS and U.S. Army Research Office Grant No. DAAG29-84-K-0207.

Copyright © 1989 by Academic Press, Inc.
All rights of reproduction in any form reserved.
ISBN-0-12-058470-0

Although every textbook emphasizes that (in the presence of collinearity) we cannot hold everything else constant and argue that the impact of a small change in an explanatory variable on the response is proportional to the estimated coefficient for that variable, this seems to be what happens very often in actual regression analysis, especially in process control. This is not forecasting, but a type of model sensitivity analysis that depends on good coefficient estimates. In other situations, the coefficients are being used to check some theory and their values are important in confirming or rejecting the theory.

Section 2 of this paper discusses the R_p statistic, an alternative to C_p, based on the sum of squares of errors in the coefficient estimates. Section 3 examines how these selection criteria relate to principal components and ridge (or Bayesian) regression. Section 4 contains an example.

2. THE R_p STATISTIC

Consider the linear model

$$y = X\beta + \varepsilon$$

where y is $n \times 1$, X is $n \times (d+1)$, β is $(d+1) \times 1$ and ε is $n \times 1$ with $E(\varepsilon\varepsilon^T) = \sigma^2 I$. We will also assume that X is of full rank, includes as its first column a dummy intercept variable, and has been scaled so that the columns of X, including the intercept, have unit lengths. Let b denote the usual least-squares estimate of β. Our presumption is that the data analyst is interested in choosing a good estimate of β and may be willing to admit bias by forcing some components of the estimate to be zero (or perhaps some other specified value) in order to improve estimates of the remaining elements of β.

Let P denote a subset of the indices $\{1, 2, ..., d\}$ and let Q be the complementary subset. Suppose the number of elements in P and Q are p and q respectively, so that $p + q = d + 1$. Let Z_p denote the matrix consisting of the p columns of X with subscripts in P, and let T_p be the $(d+1) \times (d+1)$ matrix formed by placing the rows of $(Z_p^T Z_p)^{-1} Z_p^T X$ in the row locations corresponding to indices in P and rows of zeros elsewhere. Finally, denote by b_p the $(d+1) \times 1$ vector formed in a similar manner from the vector $(Z_p^T Z_p)^{-1} Z_p^T y$ and let I denote the $(d+1) \times (d+1)$ identity matrix.

Clearly, b_p uses the least-squares estimate for the coefficients denoted by P and zero for those denoted by Q. For any such estimate, a measure of estimated coefficient quality is the scaled sum of coefficient errors

$$W_p = \frac{1}{\sigma^2}(b_p - \beta)^T (b_p - \beta). \tag{1}$$

Then

$$E(W_p) = V_p + B_p \,/\sigma^2 \tag{2}$$

where V_p and B_p are the variance and bias contributions given by

$$V_p = \text{trace}\left(Z_p^T Z_p\right)^{-1} \tag{3}$$

and

$$B_p = \beta^T (T_p - I)^T (T_p - I)\beta. \tag{4}$$

Our goal is to find a way to estimate W_p by a statistic which we will call R_p. We shall require, at least, that

$$E(R_p) = E(W_p) \tag{5}$$

when σ^2 is known. A major problem occurs because we do not know β. Our best estimate of β in the sense of low bias is likely to be b, assuming that the full model of size $d+1$ was chosen with reasonable care. Hence we will replace β by b in (4). This does not satisfy (5) so we use instead

$$\begin{aligned} R_p = V_p &+ \left(b^T (T_p - I)^T (T_p - I)b\right)/\hat{\sigma}^2 \\ &-\text{trace}\left[(T_p - I)^T (T_p - I)(X^T X)^{-1}\right] \end{aligned} \tag{6}$$

where $\hat{\sigma}^2$ is an estimate of σ^2 (usually based on the full model). R_p as defined in (6) will satisfy (5) when σ^2 is known and replaces $\hat{\sigma}^2$ in (6). Our hope is that by finding a model with a relatively small R_p, we will have a model with relatively small W_p as well. For comparison, the C_p statistic is based on the quality of fit, namely

$$J_p = \frac{1}{\sigma^2}\left(b_p - \beta\right)^T X^T X \left(b_p - \beta\right) \tag{7}$$

with $E(J_p) = E(C_p)$ if

$$C_p = \frac{1}{\sigma^2} \sum_{i=1}^{n} \left(y_i - x_i^T b_p \right)^2 - n + 2p. \tag{8}$$

When computing C_p, σ^2 is replaced by $\hat{\sigma}^2$.

Since R_p is sensitive to the scale of the explanatory variables, we have chosen to make it invariant to simple column rescaling by always scaling the columns of X, including the intercept, to have unit length. This is, of course, not necessary and R_p could always be computed using the original units for the explanatory variables. It is troublesome, however, to have the "best" model depend on what units are used to measure the explanatory variables.

R_p can also be affected by centering. Since we may wish to explore collinear relationships with the intercept, it is useful to have both a centered and uncentered version of R_p. The centered version is obtained by transforming the X-matrix so that the non-constant columns have first been centered using column means and then scaled to have length one. The intercept is now treated separately so that

$$W_p = \frac{1}{\sigma^2} \left\{ \sum_{j \neq 0} \left[(b_p)_j - \beta_j \right]^2 + \left(\sqrt{n}\bar{y} - \beta_0 \right)^2 \right\}, \tag{9}$$

where the $\sqrt{n}$ arises because the constant column has length one. Let $\tilde{Z}_p$ and $\tilde{X}$ be the centered and scaled (length one) versions of Z_p and X, each with d columns. Then (3) becomes

$$\tilde{V}_p = 1 + \text{trace}\left(\tilde{Z}_p^T \tilde{Z}_p \right)^{-1} \tag{10}$$

and the formula for B_p is unchanged if Z_p is replaced by $\tilde{Z}_p$. The resulting formula for centered R_p is

$$\begin{aligned} \tilde{R}_p &= \tilde{V}_p + b^T (\tilde{T}_p - I)^T (\tilde{T}_p - I) b / \hat{\sigma}^2 \\ &- \text{trace}\left[(\tilde{T}_p - I)^T (\tilde{T}_p - I) (\tilde{X}^T \tilde{X})^{-1} \right] \end{aligned} \tag{11}$$

where b no longer contains the least-squares estimate of the intercept and I is now a $d \times d$ identity matrix.

R_p is somewhat more complicated than C_p and $V_{d+1} \neq d + 1$. It is still logical to plot R_p against p in order to determine which values of p give relatively small values of R_p. An example is given in Section 4.

3. RIDGE REGRESSION

Ridge regression is often used to "improve" coefficient estimation by allowing some bias in order to reduce the variance. We prefer to think of ridge regression as Bayesian regression with a particular (and often unrealistic) prior specification. However, the dispersion ratio, k, is usually left unspecified and many procedures, including the ridge trace, have been suggested for estimating this parameter. R_p provides another natural approach.

Mallows (1973) introduced C_k plots in order to combine the C_p statistic with ridge regression. This seems a bit unnatural since ridge regression is designed to improve coefficient estimation in the presence of severe collinearity and C_p is aimed at improving prediction or fit. Since R_p is aimed at coefficient estimation, it is better suited for linkage to ridge regression.

In ridge regression, the data is usually both centered and scaled so that only collinearity among the non-constant regressors is considered. In some instances it is important to consider possible collinearity with the constant term. However, ridge regression in its standardized form is not a good tool for addressing this problem. We will first discuss the centered ridge estimate and then consider an uncentered form.

Again in this section, let $\tilde{X}$ and $\tilde{y}$ denote the centered and scaled form of X and y. Let V be the matrix of eigenvectors of $\tilde{X}^T\tilde{X}$ and $\lambda_j\,(j = 1, ..., d)$ the eigenvalues. Define the principal component parameter set and least-squares estimates as

$$\alpha = V^T\beta \tag{12}$$

and

$$a = V^T b. \tag{13}$$

The ridge estimator is usually written as

$$b(k) = \left(\tilde{X}^T\tilde{X} + kI\right)^{-1}\tilde{X}^T\tilde{y} \tag{14}$$

when a zero prior mean is assumed for the non-intercept coefficients. Clearly $b = b(0)$. An alternative form is

$$\tilde{b}(k) = \left(\tilde{X}^T\tilde{X} + kI\right)^{-1}\left(\tilde{X}^T\tilde{X}\right)b. \tag{15}$$

There are many generalizations possible for the matrix kI.

If we replace $\tilde{V}_p$ by $1 + (\tilde{X}^T\tilde{X} + kI)^{-1}\tilde{X}^T\tilde{X}(\tilde{X}^T\tilde{X} + kI)^{-1}$ and $\tilde{T}_p$ by $(\tilde{X}^T\tilde{X} + kI)^{-1}\tilde{X}^T\tilde{X}$ in (11), we obtain in principal components notation

$$\text{(16)}\quad \tilde{R}_k = 1 + \sum_{j=1}^{d} \frac{\lambda_j}{(\lambda_j + k)^2} + \frac{1}{\hat{\sigma}^2} \sum_{j=1}^{d} \frac{k^2 a_j^2}{(\lambda_j + k)^2} - \sum_{j=1}^{d} \frac{k^2}{\lambda_j (\lambda_j + k)^2}.$$

Let

$$\text{(17)}\qquad \tilde{F} = (\tilde{X}^T\tilde{X} + kI)^{-1}\tilde{X}^T,$$

then

$$\text{(18)}\qquad \tilde{C}_k = \frac{1}{\hat{\sigma}^2}\tilde{y}^T(I - \tilde{X}\tilde{F})^T(I - \tilde{X}\tilde{F})\tilde{y} - n + 2 + 2\,\text{trace}\tilde{\text{X}}\tilde{\text{F}}$$

which in principal components form is

$$\text{(19)}\qquad \tilde{C}_k = 1 - d + \frac{1}{\hat{\sigma}^2}\sum_{j=1}^{d} \lambda_j \frac{k^2 a_j^2}{(\lambda_j + k)^2} + 2\sum_{j=1}^{d} \frac{\lambda_j}{\lambda_j + k}.$$

It is often useful to be able to graphically compare R_p with R_k (or C_p with C_k). To do this we must be able to relate k to p. Since k is related to the prior variance of β, it does not necessarily have any direct relationship to p. However, one reason we are considering ridge (or Bayesian) regression in this context is in order to reduce collinearity and improve coefficient estimation. When the regression is viewed in the principal components coordinate system, variables are being given different weights that are a function of the eigenvalues and the parameter k. In traditional principal components regression, coefficients are forced to be zero if their corresponding variance, λ_j^{-1}, is too large or equivalently, the eigenvalue is sufficiently small. After transforming to principal components (15) becomes

$$\text{(20)}\qquad a_j(k) = \left(\frac{\lambda_j}{\lambda_j + k}\right) a_j$$

and we see that ridge regression is doing something very similar to principal components regression by giving a low weight to a_j when λ_j is small and a higher weight when λ_j is large. The relative weighting is controlled by the value of k. It is therefore natural to add up the weights $\lambda_j/(\lambda_j + k)$ to see how many principal component parameters are left. This gives

$$\text{(21)}\qquad \tilde{p}(k) = 1 + \sum_{j=1}^{d} \frac{\lambda_j}{\lambda_j + k}$$

where one has been added to account for the intercept. When k is infinity, the prior value of zero is being used for all (principal component) coefficients except the intercept and $p(\infty) = 1$. When $k = 0$, all coefficients are being estimated (none set equal to zero) and $p(0) = d + 1$.

Mallows noted that the trace of the variance of the fitted values (except for σ^2) is just p, the number of variables in the model. Generalizing this to the ridge case, he argued that the number of principal component parameters in the model would be the trace of the variance of the ridge fitted values or

$$\tilde{p}_m(k) = 1 + \sum_{j=1}^{d} \left(\frac{\lambda_j}{\lambda_j + k}\right)^2. \tag{22}$$

Myers (1986, p. 252) noted that for least-squares the variance (except for σ^2) of the fitted values is just the hat matrix, $\tilde{X}(\tilde{X}^T\tilde{X})^{-1}\tilde{X}^T$, which transforms $\tilde{y}$ into $\hat{\tilde{y}}$. However, instead of looking at the variance in the ridge case, he considered the matrix that transforms $\tilde{y}$ into $\hat{\tilde{y}}$, $\tilde{X}(\tilde{X}^T\tilde{X} + kI)^{-1}\tilde{X}^T$. Its trace gives the same result as (21) when one is added for the intercept.

If we choose not to center the data, the "standard" ridge estimator does not apply. Recall that X has columns of unit length but is not centered, and let D be the $(d+1) \times (d+1)$ diagonal matrix:

$$\begin{bmatrix} 0 & & & & \\ & k & & & \\ & & k & & \\ & & & \ddots & \\ & & & & k \end{bmatrix}. \tag{23}$$

A ridge (Bayes) estimate would be

$$b(k) = (X^T X + D)^{-1} X^T y. \tag{24}$$

In effect, this says that our prior variance for the intercept is infinity. Unfortunately there is no simple formula for R_k, so we must use

$$R_k = \text{trace}\left[(X^T X + D)^{-1} X^T X (X^T X + D)^{-1}\right] \tag{25}$$

$$+ \frac{1}{\hat{\sigma}^2} b^T \left[(X^T X + D)^{-1} X^T X - I\right]^T \left[(X^T X + D)^{-1} X^T X - I\right] b$$

$$-\text{trace}\left\{\left[(X^TX+D)^{-1}X^TX-I\right]^T\left[(X^TX+D)^{-1}X^TX-I\right](X^TX)^{-1}\right\}.$$

The corresponding formula for uncentered C_k is

$$C_k = \frac{1}{\hat{\sigma}^2}y^T(I-XF)^T(I-XF)y - n + 2\text{trace}XF. \tag{26}$$

It is possible to relate k to p by utilizing the approach taken by Myers and setting p equal to the trace of the ridge "projection" matrix

$$p(k) = \text{trace}\mathrm{X}(\mathrm{X^TX+D})^{-1}\mathrm{X^T} = \text{trace}(\mathrm{X^TX+D})^{-1}\mathrm{X^TX}. \tag{27}$$

The Mallows idea (22) becomes

$$p_m(k) = \text{trace}\left[(\mathrm{X^TX+D})^{-1}\mathrm{X^TX}\right]^2. \tag{28}$$

4. EXAMPLES

As an example consider the tobacco data from Myers (1986, p. 260) where a detailed analysis can be found. Four important components of 30 tobacco blends were measured to develop a linear model for the response that measures the amount of heat given off by the tobacco during the smoking process. Thus, $n = 30$ and $d + 1 = 5$, including the intercept. In the following analyses, the intercept is always forced to be in the model.

Myers (1986, p. 261) also gives for the tobacco data the variance proportion collinearity diagnostics developed by Belsley, Kuh, and Welsch (1980). There is considerable evidence that a collinear relation exists between components one, three, and four and, to a lesser extent, between the intercept and variable two.

The "best" model (lowest C_p) for uncentered C_p involves all of the explanatory variables (Table 1). For R_p, the best model includes the intercept and variables one and two only. It is interesting to note that the second best model includes only components one and the intercept. Hence it appears that R_p is attempting to adjust for collinearity problems by first eliminating components three and four, and then two. C_p, after initially using all of the components, uses the intercept and components two and four in the second best model, and the intercept and component four in the third best

C_p model. Choosing from among these models would be difficult, but having R_p provides some additional insight. Of course, the centered results for C_p are the same as in the uncentered case. For R_p, the best two models in the uncentered case are the same as in the centered case, but changes occur after that.

The four sets of plots show the uncentered and centered versions of R_p and C_p with the associated R_k and C_k overlayed. Note that while centered and uncentered C_p are identical, centered and uncentered C_k are not. The top panel in each plot uses formula (21) or (27) to relate p to k for overlaying R_k or C_k and the bottom panel uses the Mallows approach (22) or (28).

The top plots of C_k and R_k in the uncentered case indicate low values of C_k and R_k near $p = 4$. The actual minimum for C_p is at $p = 5$. The Mallows approach for relating k to p (in the lower plots) actually does a better job for R_k, indicating a minimum between $p = 3$ and 4 when the actual minimum for R_p occurs at $p = 3$.

In the centered case, our approach (21) does somewhat better than the Mallows approach for C_k, since the minimum C_k is between $p = 4$ and 5 in the former case and $p = 3$ and 4 in the latter. For centered R_k, the reverse is again true. Other examples also indicate that choosing the number of parameters based on equations (21) and (22) is not perfect, but not particularly bad, either. Based on this limited evidence we would argue that the Mallows idea for $p(k)$ is better for R_p and (21) is better for C_p. These formulas could be used to reduce the computational burden by finding a $p(k^*)$ where k^* minimizes R_k or C_k. Then we would only look at models with $p(k^*)$ parameters to minimize R_p or C_p. Perhaps the best use of the values obtained at mimimum R_k or C_k is in the ridge estimate itself as an alternative to the k values produced by the ridge trace.

Mallows noted that when $\tilde{X}^T\tilde{X}$ is an identity matrix, the value chosen for k by minimizing C_k is close to the James-Stein estimator. The definition of R_k leads to exactly the same result as C_k when $\tilde{X}^T\tilde{X}$ is the identity matrix and therefore the same comment is true for R_k.

Where possible, we think that both R_p and C_p should be considered for the selection of variables and R_p should be given extra weight when the estimated coefficients are getting the most attention. We also feel that choosing k by minimizing $\tilde{R}_k$ or R_k has considerable merit and should compare favorably with other methods of selecting k.

REFERENCES

Belsley, D. A., Kuh, E., and Welsch, R. E. (1980), *Regression Diagnostics*, John Wiley and Sons, Inc., New York.

Mallows, C. L. (1973), Some Comments on C_p, *Technometrics*, **15**, 661-675.

Myers, R. H. (1986), *Classical and Modern Regression Analysis with Applications*, Duxbury Press, Boston.

R_p and C_p for the Tobacco Data

Uncentered

R_p	C_p	Rank of C_p	Number of Variables	Variables In Model
4233.90	6.07	6	3	1 2 0
6329.17	6.50	8	2	1 0
7364.22	5.00	1	5	1 2 3 4 0
7612.30	6.44	7	3	1 3 0
8631.07	5.39	4	4	1 2 4 0
9420.23	6.51	9	4	1 3 4 0
9545.11	5.85	5	4	1 2 3 0
13149.61	5.15	3	2	4 0
13815.31	6.68	10	3	1 4 0
13906.07	5.11	2	3	2 4 0
15504.47	6.92	11	3	3 4 0
16809.93	7.09	12	4	2 3 4 0
19594.12	556.12	16	1	0
31710.39	14.79	13	2	3 0
33853.90	16.78	14	3	2 3 0
43030.63	29.97	15	2	2 0

Centered

R_p	C_p	Rank of C_p	Number of Variables	Variables In Model
316.33	6.07	6	3	1 2 0
456.56	6.50	8	2	1 0
601.40	6.44	7	3	1 3 0
681.75	5.00	1	5	1 2 3 4 0
796.48	6.51	9	4	1 3 4 0
814.48	5.39	4	4	1 2 4 0
888.68	5.85	5	4	1 2 3 0
1235.21	5.15	3	2	4 0
1258.19	6.68	10	3	1 4 0
1385.31	5.11	2	3	2 4 0
1438.80	6.92	11	3	3 4 0
1613.25	556.12	16	1	0
1648.77	7.09	12	4	2 3 4 0
2723.23	29.97	15	2	2 0
3107.83	14.79	13	2	3 0
3210.96	16.78	14	3	2 3

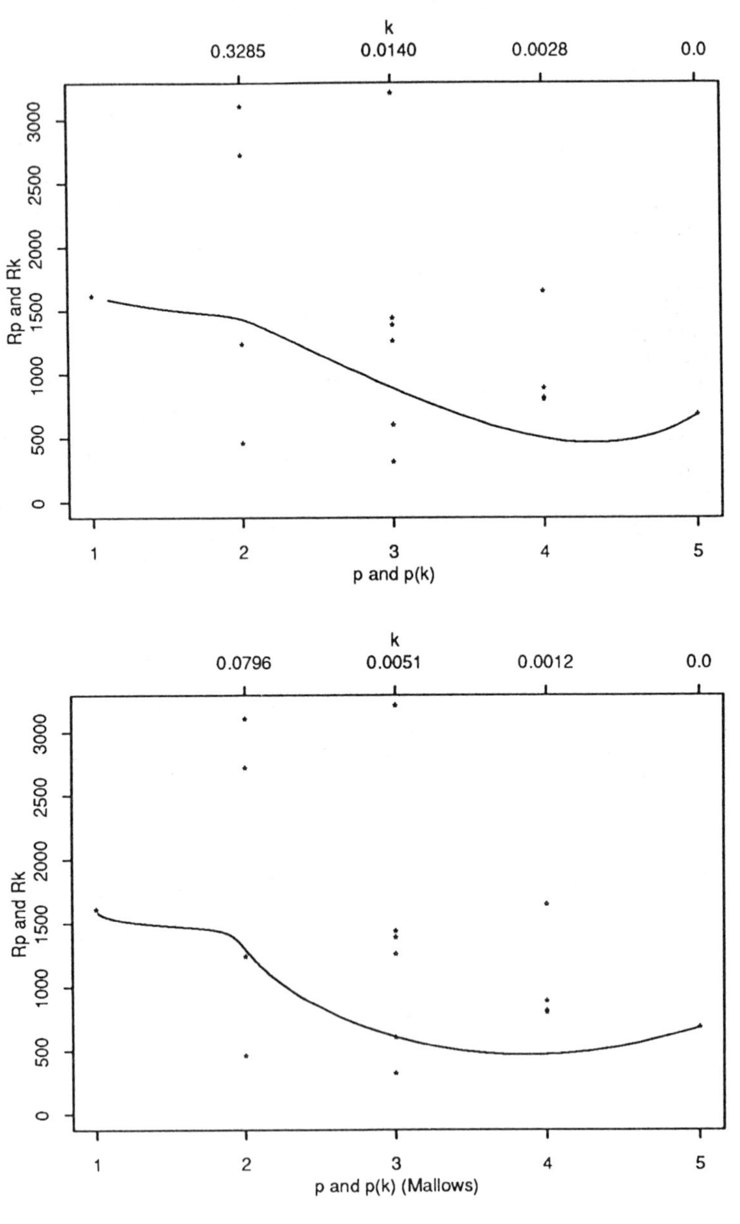
Rp and Rk (centered)
k
0.3285
0.0140
0.0028
0.0
Rp and Rk
0
500
1000
1500
2000
2500
3000
1
2
3
4
5
p and p(k)
k
0.0796
0.0051
0.0012
0.0
Rp and Rk
0
500
1000
1500
2000
2500
3000
1
2
3
4
5
p and p(k) (Mallows)

Figure 1 .

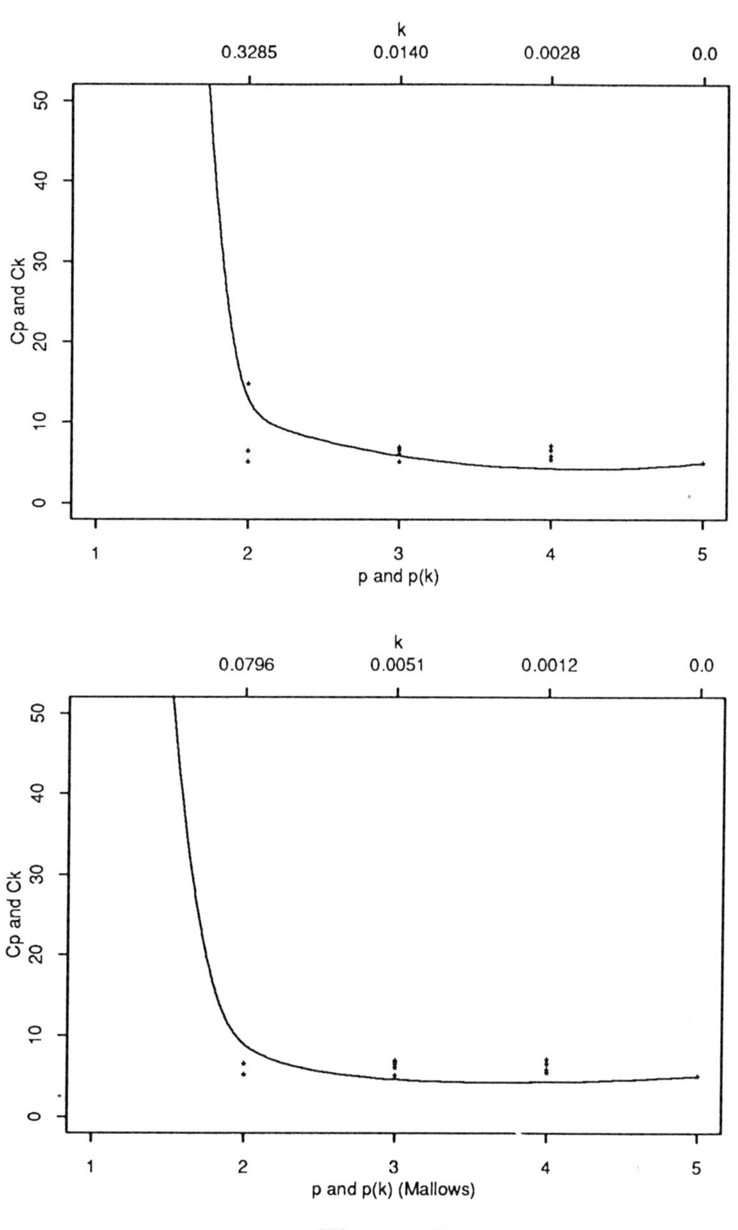
Cp and Ck (centered)
k
0.3285
0.0140
0.0028
0.0
Cp and Ck
0
10
20
30
40
50
1
2
3
4
5
p and p(k)
k
0.0796
0.0051
0.0012
0.0
Cp and Ck
0
10
20
30
40
50
1
2
3
4
5
p and p(k) (Mallows)

Figure 2 .

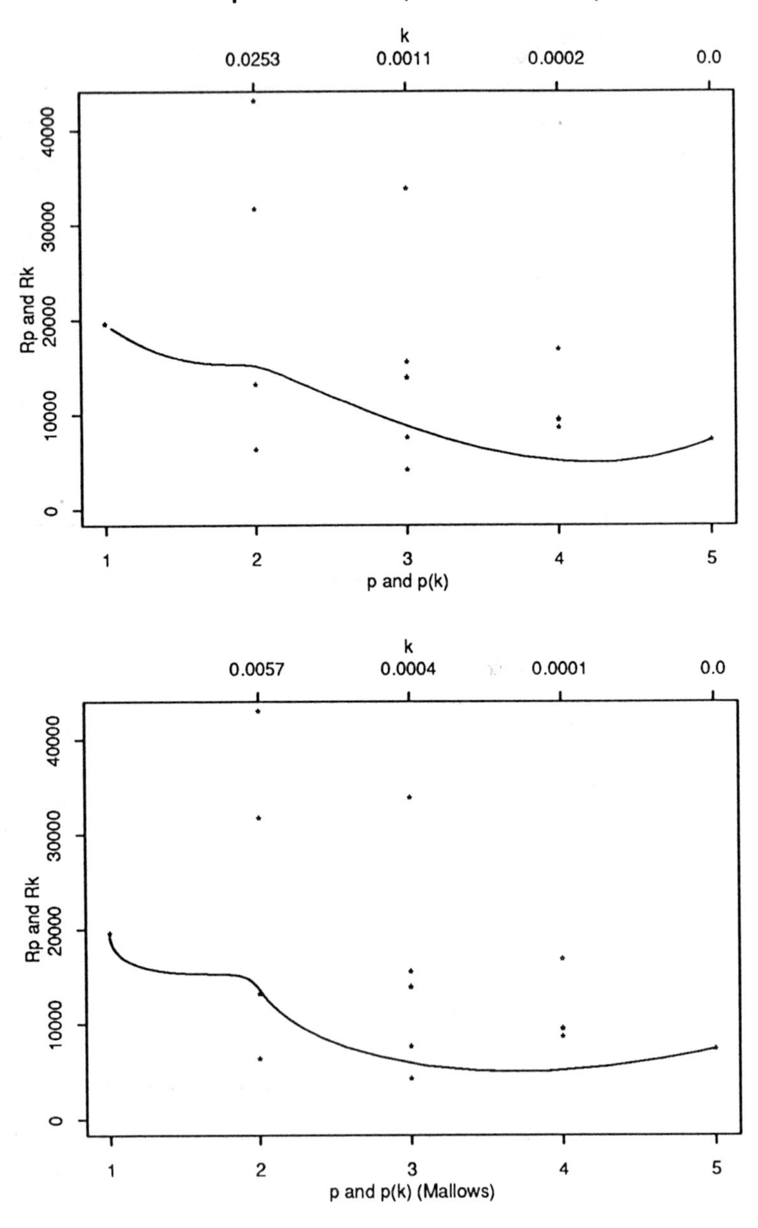
Rp and Rk (uncentered)
k
0.0253
0.0011
0.0002
0.0
40000
30000
20000
10000
0
Rp and Rk
1
2
3
4
5
p and p(k)
k
0.0057
0.0004
0.0001
0.0
40000
30000
20000
10000
0
Rp and Rk
1
2
3
4
5
p and p(k) (Mallows)

Figure 3 .

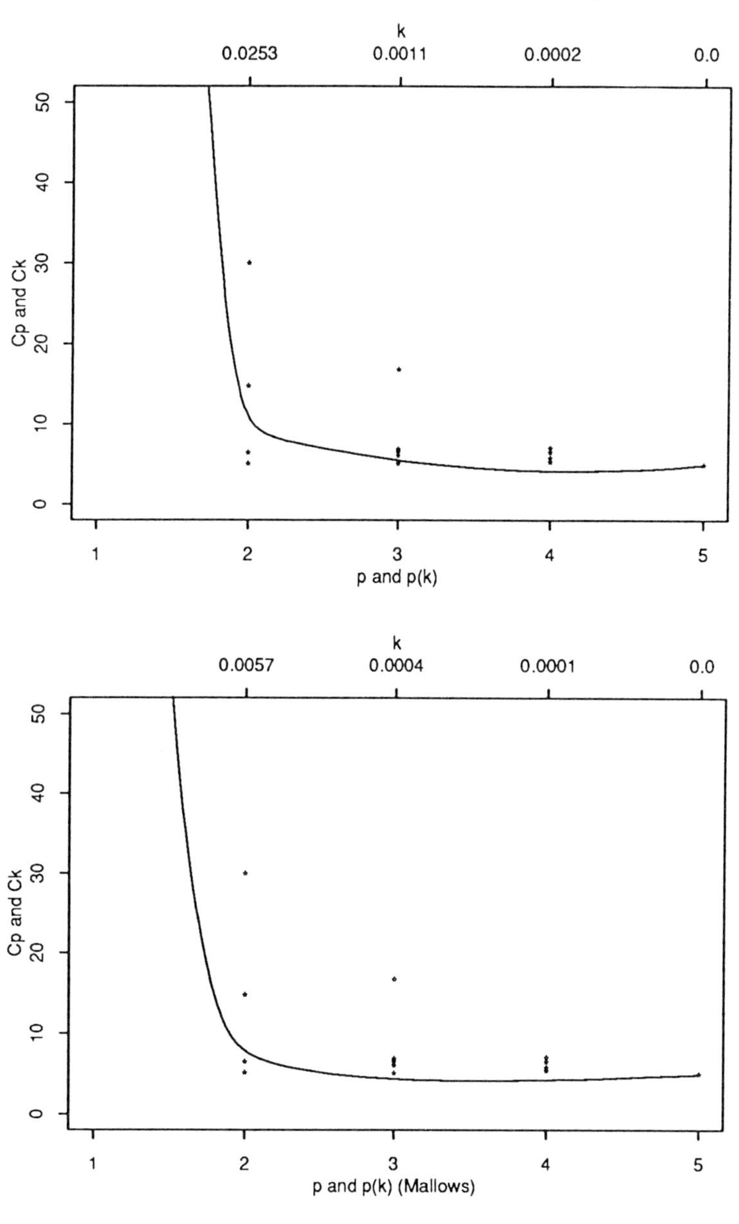
Cp and Ck (uncentered)
k
0.0253
0.0011
0.0002
0.0
Cp and Ck
0
10
20
30
40
50
1
2
3
4
5
p and p(k)
k
0.0057
0.0004
0.0001
0.0
Cp and Ck
p and p(k) (Mallows)

Figure 4 .